T0073641

Cancer Treatment and Research

Series Editor
Steven T. Rosen
Robert H. Lurie Comprehensive
    Cancer Center
Northwestern University
Chicago, IL
USA

For further volumes, go to
http://www.springer.com/series/5808

Lalitha Nagarajan

Editor

Acute Myelogenous Leukemia

Genetics, Biology and Therapy

 Springer

Editor

Lalitha Nagarajan
Department of Genetics
University of Texas MD
Anderson Cancer Center
Unit 1006
1515 Holcombe Boulevard
Houston TX 77030
USA
lnagaraj@mdanderson.org

ISSN 0927-3042
ISBN 978-0-387-69257-9   e-ISBN 978-0-387-69259-3
DOI 10.1007/978-0-387-69259-3
Springer New York Dordrecht Heidelberg London

Library of Congress Control Number: 2006939515

Springer is part of Springer Science+Business Media (www.springer.com)

Contents

Contributors

Gheath Alatrash, DO, PhD Division of Cancer Medicine, University of Texas MD Anderson Cancer Center, Houston, TX, USA

Michael Andreeff, MD, PhD Department of Blood and Marrow Transplantation, University of Texas MD Anderson Cancer Center, Houston, TX, USA, mandreef@mdanderson.org

Gautam Borthakur, MD Department of Leukemia, The University of Texas MD Anderson Cancer Center, Houston, TX, USA, gborthak@mdanderson.org

George A. Calin, MD, PhD Departments of Experimental Therapeutics and Cancer Genetics, The University of Texas MD Anderson Cancer Center, Houston, TX, USA, gcalin@mdanderson.org

Bing Carter, PhD Section of Molecular Hematology and Therapy, Department of Stem Cell Transplantation and Cellular Therapy, The University of Texas MD Anderson Cancer Center, Houston, TX, USA

Michael E. Engel, MD, PhD Division of Pediatric Hematology/Oncology, Department of Pediatrics, Monroe Carell Jr. Children's Hospital at Vanderbilt, Vanderbilt-Ingram Cancer Center, Nashville, TN, USA, mike.engel@vanderbilt.edu

Elihu E. Estey, MD Department of Leukemia, The University of Texas MD Anderson Cancer Center, Houston, TX, USA

Zeev Estrov, MD Department of Leukemia, University of Texas MD Anderson Cancer Center, Houston, TX, USA, zertrov@mdanderson.org

Brunangelo Falini, MD Institute of Hematology, University of Perugia, Perugia, Italy, faliniem@unipg.it

Julie M. Fortier, MD Division of Oncology, Stem Cell Biology Section, Washington University School of Medicine, St. Louis, MO, USA

Timothy A. Graubert, MD Division of Oncology, Stem Cell Biology Section, Washington University School of Medicine, St. Louis, MO, USA, tgrauber@dom.wustl.edu

David Grimwade, MD, PhD Department of Medical and Molecular Genetics, King 's College London School of Medicine, London, UK, david.grimwade@genetics.kcl.ac.uk

Fabien Guidez, PhD Department of Medical and Molecular Genetics, King's College London School of Medicine, London, UK

Scott W. Hiebert, PhD Division of Pediatric Hematology/Oncology, Department of Pediatrics, Monroe Carell Jr. Children's Hospital at Vanderbilt, Vanderbilt-Ingram Cancer Center, Nashville, TN, USA

Jean-Pierre J. Issa, MD Department of Leukemia, University of Texas MD Anderson Cancer Center, Houston, TX, USA, jissa@mdanderson.org

Bharti Jasra, MD Departments of Experimental Therapeutics and Cancer Genetics, The University of Texas MD Anderson Cancer Center, Houston, TX, USA

Marina Konopleva, MD, PhD Section of Molecular Hematology and Therapy, Department of Leukemia, The University of Texas MD Anderson Cancer Center, Houston, TX, USA

Soheil Meshinchi, MD, PhD Clinical Research Division, Fred Hutchinson Cancer Research Center, Seattle, WA, USA

Anita R. Mistry, PhD Department of Medical and Molecular Genetics, King 's College London School of Medicine, London, UK

Jeffrey J. Molldrem, MD Department of Stem Cell Transplantation and Cellular Therapy, University of Texas MD Anderson Cancer Center, Houston, TX, USA, jmolldre@mdanderson.org

Beatrice U. Mueller, MD Department of Internal Medicine, University Hospital, Bern, Switzerland, beatrice.mueller@insel.ch

Lalitha Nagarajan, PhD Department of Molecular Genetics, University of Texas MD Anderson Cancer Center, Houston, TX, USA, lnagaraj@mdanderson.org

Milena S. Nicoloso, MD Departments of Experimental Therapeutics and Cancer Genetics, The University of Texas MD Anderson Cancer Center, Houston, TX, USA

Yasuhiro Oki, MD Department of Leukemia, University of Texas MD Anderson Cancer Center, Houston, TX, USA

Thomas Pabst, MD Department of Oncology, University Hospital, Bern, Switzerland

Nallasivam Palanisamy, PhD Michigan Center for Translational Pathology, University of Michigan Health System, Ann Arbor, MI, USA, nallasiv@med.umich.edu

Ismael Samudio, PhD Section of Molecular Hematology and Therapy, Department of Stem Cell Transplantation and Cellular Therapy, The University of Texas MD Anderson Cancer Center, Houston, TX, USA

Ellen Solomon, PhD Department of Medical and Molecular Genetics, King's College London School of Medicine, London, UK

Derek L. Stirewalt, MD Clinical Research Division, Fred Hutchinson Cancer Research Center, Seattle, WA, USA, dstirewa@fhcrc.org

Peter J.M. Valk, PhD Department of Hematology, Erasmus University Medical Center Rotterdam, Rotterdam, The Netherlands, p.valk@erasmusmc.nl

Roel G.W. Verhaak, PhD Department of Hematology, Erasmus University Medical Center Rotterdam, Rotterdam, The Netherlands

The Leukemia Stem Cell

Zeev Estrov

Introduction and Historical Perspective

In a meeting held in the Charité Hospital in Berlin in 1909, Alexander Maximow postulated that all circulating blood cells arise from a single lymphocyte-like cell [68]. An almost identical hypothesis was proposed by Artur Pappenheim in 1917 [78]. This hypothesis was tested years later by numerous investigators who demonstrated that all hematopoietic cells arise from a single hematopoietic stem cell (HSC) [54, 75, 90, 116]. Our knowledge of the leukemogenic process has benefited from hematopoiesis research. Identification and characterization of the HSC led to the theory that leukemia is a stem cell disease, i.e., that leukemia arises from a neoplastic HSC.

As far as we know, leukemia has always existed. The first patient reported to exhibit symptoms most likely attributable to chronic lymphocytic leukemia was a 63-year-old Parisian lemonade salesman named Monsieur Vernis [108]. At first there was not much interest in this disease, and Armand Velpeau's report drew little attention. However, within a few years, other reports followed. The first cases of chronic myelogenous leukemia (CML) were reported by Donne and Craigie [30, 25], and the clinical characteristics of several patients with different forms of leukemia were published by John Hughes Bennett [5–9], who first used the term "leucocythemia." The word 'leukämie' was coined by Rudolph Virchow who by then had described several patients with this disease [110–113]. Several decades passed before the morphological features of leukemia cells were defined and the pathophysiology of leukemia deciphered.

The stem cell concept has been applied to non-hematopoietic tissues. Embryonic stem cells and non-hematopoietic, tissue-specific stem cells have been isolated, characterized, and extensively studied [60]. Cells isolated from embryonic tissue, gametes, and fertilized eggs have been found to have

Z. Estrov (✉)
The Department of Leukemia, The University of Texas MD, Anderson Cancer Center, Houston, TX, USA
e-mail: zestrov@mdanderson.org

L. Nagarajan (ed.), *Acute Myelogenous Leukemia*,
Cancer Treatment and Research 145, DOI 10.1007/978-0-387-69259-3_1,
© Springer Science+Business Media, LLC 2010

totipotent and pluripotent capacities [47]. Although these two capacities are dissimilar, the cells received identical names from different investigators, and to avoid confusion a common terminology was agreed upon. According to the recently established nomenclature (Table 1), adult tissue-specific stem cells (such as HSCs) that are capable of generating and replacingterminally differentiated cells within their own tissue boundaries are termed multipotent (adult) stem cells.

Table 1 Stem cell nomenclature

Totipotent stem cells: The zygote (fertilized egg) (*totus:* entire). These cells have the capacity to give rise to every tissue associated with the embryo and, eventually, the adult

Pluripotent stem cells: The blastocyst (starting 4 days after fertilization) (*plures:* several). These cells can give rise to all tissue cells except the embryo's outer layer (the trophoblast) and the embryo-supporting placenta

Multipotent stem cells: Somatic (adult, tissue-specific) stem cells

The Hematopoietic Stem Cell

Concept and Definition

The hematopoietic system is thought to originate from multipotent HSCs capable of reproducing themselves (through a process termed "self-renewal") and producing a hierarchy of downstream multilineage and unilineage progenitor cells that differentiate fully into mature cells [28, 58, 115]. The concept arose from studies exploring the formation of hematopoietic colonies in spleens of irradiated mice. It has been known since the late 1940s that exposing mice to ionizing radiation results in the generation of macroscopic hematopoietic spleen colonies [51]. In the early 1960s, Till and McCulloch [104] and their colleagues [3, 105] demonstrated that a spleen colony is generated from a single cell called a spleen colony-forming unit (CFU-S). It was demonstrated later that cells arising from a single CFU-S can rescue a lethally irradiated mouse, and subsequent limiting-dilution studies established that a single cell is capable of repopulating the entire hematopoietic system [54, 90, 72]. These studies defined the HSC as a cell that possesses self-renewal and clonogenic abilities and the capacity to differentiate into multiple lineages. HSCs with the capacity for both long-term and short-term repopulation of the mouse hematopoietic system have been identified. Similar studies with human bone marrow cells, performed in the severe combined immunodeficiency (SCID) mouse model, yielded similar results. HSCs are thought to be rare quiescent cells that, upon demand, give rise to progenitor cells (characterized in vitro and termed CFUs), which are destined to generate fully differentiated cells. The division of HSCs is thought to be either symmetric, producing two identical daughter cells that are either both stem cells or both progenitor cells, or asymmetric, producing an HSC and a progenitor with diminished self-renewal capacity but with the ability to enact clonal expansion and maintain the circulating blood cell population [28, 74, 86, 110].

Isolation of HSCs

Because it was thought that a single cell is capable of repopulating the hematopoietic system, several investigators have attempted to isolate HSCs by using phenotypic cell-surface markers associated with defined lineages and developmental stages of hematopoietic cells. Using high-speed multi-parameter flow cytometry, distinct cell populations were purified and collected for functional analysis. Several cell-surface proteins thought to be specific to primitive cells, such as CD34, CD133, and CD150, and combinations of cell-surface markers such as $Lin^-c-kit^{high}Thy^{low}Sca^+$ [76, 93, 103], $Lin^-Thy^+CD34^+CD38^{-/low}$), or the SLAM family receptors [57] have been used to identify HSCs. Other techniques to isolate HSCs utilize cellular physical characteristics or enzymatic activity, such as the extrusion of Hoechst dye 33342 (like the activity of P-glycoprotein that is encoded by the *MDR* gene) identifying "side population cells" and the use of BODIPY aminoacetaldehyde to assess aldehyde dehydrogenase activity [39, 46, 56, 57, 96].

HSC Characterization

To test the stem cell characteristics of a cellular population, it is necessary to purify the cells to functional homogeneity and to demonstrate that every single purified cell is capable of reconstituting the entire hematopoietic system. In addition, it is necessary to show that every single cell of a phenotypically identical cellular population isolated from one animal is capable of reconstituting the hematopoietic system in another animal, i.e., that the stem cells that repopulated the hematopoietic system self-renew (Fig. 1). These goals cannot be reproducibly reached using any of the currently available stem cell fractionation techniques because not every cell of the phenotypically identical cell population is a functional stem cell. Rather, the end products of these fractionation assays are cells that are enriched with HSCs. Remarkably, some cells of the "stem cell-depleted" cellular fraction are also functional stem cells, as demonstrated in studies with CD34 fractionation. CD34 was the first HSC "marker," and CD34 fractionation has been used clinically for HSC enrichment. Remarkably, hematopoietic stem cell transplantation has been successfully performed with a population of CD34-negative cells [120, 121], suggesting that functional HSCs are present within the CD34-negative cell fraction. Thus, a stem cell should be defined by its functional capacity, not by phenotype, surface marker expression, or other cellular characteristics.

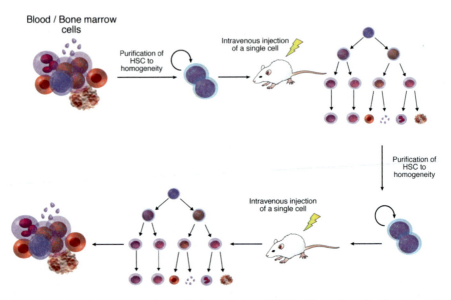

Fig. 1 Functional assessment of a purified population of HSCs. To confirm that fractionated cells are HSCs, every single cell of that homogenous population should have the capacity to repopulate the hematopoietic system in a lethally irradiated animal and, in addition, every single blood or bone marrow cell isolated from the irradiated animal by the same method should be capable of reconstituting all blood and bone marrow elements of an identical lethally irradiated animal (second-generation reconstitution)

The Hematopoietic Stem Cell Niche

Boneand bone marrow are intrinsically linked. Morphological and functional studies suggest that HSCs are located proximal to the endosteal surface of trabecularbone (reviewed in [71]). Several studies have demonstrated that osteoblast cells are required for this localization. Furthermore, HSCs express a calcium-sensing receptor. Stem cells lacking this receptor failed to localize to the endostealniche and did not function normally after transplantation, highlighting the importance of the ionic mineral contentof the bone itself and of the bone-derived matrix in the lodgmentand retention of HSCs within the endosteal niche. However, HSCs function in the absence of osteoblasts. Extramedullary hematopoiesis is frequently found in patients with myeloproliferative disorders and in transgenic mice where osteoblast cells have been ablated, the marrowis aplastic, and extensive extramedullary hematopoiesis occurs. Thus, HSCs can survive and function in tissues thathave no osteoblasts. The contributions of other cellular elements, suchas stromal cells, osteoclasts, or perivascular cells, have been characterized. For example, it has been shown that HSCs can be recruited to a "vascularniche" that servesas an "extramedullary niche." Experiments with parabiotic mice have demonstrated that HSCs circulate between the blood and the bone marrow. Therefore, the existence of multiple types of HSC niches is not surprising. Whether different niches affect HSC function differently has yet to be determined.

The Leukemia Stem Cell

Cancer Stem Cell: Concept and Definition

The cancer stem cell hypothesis was proposed about 150 years ago by Rudolf Virchow and Julius Cohnheim, who argued that cancer results from the activation of dormant embryonic tissue remnants [23, 109].

The notion that leukemia stem cells (LSCs) might exist emerged from experiments performed in the early 1970s showing that only a subset of leukemia cells was capable of in vitro proliferation [70, 73, 98]. These in vitro studies, together with the in vivo studies described below, suggested that among the entire population of leukemia cells there are rare cells with the potential for self-renewal that drive the expansion of the leukemic clone. These LSCs are thought to exhibit characteristics similar to those of normal HSCs, with the exception that LSCs do not necessarily differentiate into different lineages. The current dogma is that all leukemia cells are derived from LCSs and that the descendents of these cells are clones.

It is thought that LSCs, like normal HSCs, give rise to differentiated daughter progenitor cells (termed CFU-Ls) that differentiate into leukemic blasts that lose their self-renewal capacity (Fig. 2). However, defects in the cellular machinery of CFU-Ls usually eliminate their ability to differentiate fully into morphologically and phenotypically mature cells. As a result, the leukemic cell population consists of undifferentiated and variably differentiated leukemia cells. The

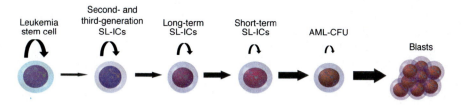

Fig. 2 Proposed hierarchal differentiation of LSCs: Rare LSCs with self-renewal capacity (*curved arrows*) give rise to SCID leukemia-initiating cells (SL-ICs) capable of initiating leukemia in second- and third-generation mice. These SL-ICs give rise to long-term SL-ICs capable of repopulating the marrow of identical second-generation mice. The short-term SL-ICs are capable of repopulating the marrow of the injected animal and give rise to AML colony-forming units (AML-CFU) that have a low rate of self-renewal. During this hierarchal differentiation process, the proliferation capacity (*straight arrows*) of the leukemic cells increases (adopted from [49] and [83]). This model is based on the assumption that human leukemia cells that engraft in SCID and NOD/SCID mice carrying LSC characteristics

degree of differentiation of leukemia cells has been used to identify their lineage (for example, myeloid versus lymphoid). Indeed, the current clinical classification of leukemias is based on the presence of "differentiation markers" in the leukemia blasts [10, 107].

Models of Leukemogenesis

The currently accepted paradigm of leukemogenesis rests on the theory that leukemia arises from a single cell and is maintained by a small population of LSCs [37, 49, 69]. It is thought that normal HSCs themselves can undergo leukemic transformation. First, in normal HSCs, the machinery for self-renewal is already activated, and second, stem cells persist for a long time; therefore, the opportunity to accumulate mutations in these cells is greater than that in mature, short-lived cell types. Nonetheless, restricted progenitors could potentially be transformed, either by acquiring mutations that cause them to self-renew or by inheriting existing mutations from the preleukemic stem cells and then acquiring additional mutations themselves. Indeed, Jamieson et al. [53] reported that in chronic myelogenous leukemia, the LSC is the differentiated progenitor CFU-granulocyte-macrophage (CFU-GM).

Two models of leukemogenesis have long been proposed. According to the stochastic model, leukemia consists of a homogenous population of immature cells and a few cells that can either self-renew or proliferate in a stochastic manner [61, 86, 104]. In contrast, in the hierarchy model, leukemia consists of a heterogeneous population, within which only a small percentage of LSCs generate leukemia clones and sustain the disease. Recently, a third model was proposed [83]. According to this model, mature leukemia cells can de-differentiate and regain functional LSC capacity [53]. Whereas the first two models hold that only a few immature cells sustain leukemia, the third model allows that mature cells can regain self-renewal capacity. One might envision leukemia as a disease in which the phenotypically mature and immature neoplastic cells, characterized by genomic instability, multiply and form new leukemic clones selected to expand based on their proliferation and survival capacity. Such cells might at times either be quiescent or be proliferate; and during this process, they may differentiate or self-renew without adhering to the "rules" of the normal hematopoietic system.

LSC Phenotype and Function

The phenotypic and functional properties of normal HSCs have been used to study LSCs [15, 29, 42] Wissman 2000). Classic studies of human acute myeloid leukemia (AML) have shown that AML consists of a heterogeneous population of cells with a small percentage of quiescent (non-cycling) cells [22, 114], of which only 0.1–1% harbor the capacity to initiate leukemia when injected into SCID or non-obese diabetic (NOD)/SCID mice [15, 62]. In most AML subtypes, except for promyelocytic leukemia (AML-M3), the cells capable of transplanting leukemia in NOD/SCID mice have been shown to have a $CD34+CD38^-$ or $CD34^+CD38^{+low}$ phenotype similar to that of normal HSC, whereas more mature $CD34+CD38+$ leukemic blasts failed to engraft [12, 15]. Using clonal tracking of retroviral-transduced normal and leukemic cells in NOD/SCID mice, it was demonstrated that both normal and LSC compartments were composed of individual stem cell classes that differed in their repopulating and self-renewal capacities [44, 49]. These elegant studies demonstrated that a fractionated population of phenotypically identical human leukemia cells engrafting in NOD/SCID mice consists of cells with different capacities. Some cells, termed NOD/SCID leukemia-initiating cells (SL-ICs), could be harvested and, when injected into a second and third generation of identical mice, engrafted and induced leukemia. A "marker" to determine whether a certain cell is a short-term or long-term SL-IC has not been identified.

Other combinations of cell-surface markers [75] and other techniques, including the extrusion of Hoechst dye 33342, have been used to identify LSCs [16, 39, 56] Several studies suggest that HSCs and LSCs share some cell-surface markers but not others. For example, HSCs and LSCs both express CD34, CD71, and HLA-DR. However, Thy-1 (CD90) and c-Kit (CD117) are expressed on HSCs but not on LSCs, and CD123 the interleukin-3 receptor-α are expressed on LSCs but not on HSCs [12, 13, 55]. Remarkably, cytogenetically abnormal LSCs have been found in the $CD34^+CD90^+$ populations in several patients with AML, and in rare cases, $CD34^-$ cells as well as $CD34^+$ cells have successfully engrafted and initiated human leukemia in mice [17, 85, 102]. In acute promyelocytic leukemia, unlike in other myeloid leukemias, the characteristic translocation has been observed in $CD34^-CD38^+$ but not in $CD34^+CD38^-$ cells [15, 41].

The NOD/SCID mouse studies described above suggest that AML, like normal hematopoiesis, is indeed organized as a hierarchy, which is initiated and maintained by an LSC that gives rise to SL-ICs with short-term and long-term repopulating capacities that, in turn, give rise to cells with abnormal differentiation programs, leading ultimately to the production of blasts and abnormally differentiated leukemia cells [28] (Fig. 2). However, Taussig et al. [101] showed recently that cells expressing well-established mature myeloid markers (CD33 or CD13) could function as SL-ICs, thereby questioning the validity of surface marker expression as a predictor of stem cell function. Taussig's data agree with previous studies that have demonstrated that cells devoid of self-renewal capacity, such as committed progenitors and mature cells, could be targets for leukemic transformation. For example, activation of promoter elements of several mature myeloid-specific human genes (like MRP8, CD11b, or cathepsin G) induces human leukemias in transgenic mouse models [18, 52, 119]. Recently, Cozzio et al. [23] showed that the leukemic fusion gene MLL-ENL, which results from the t(11;19) translocation, induced the same type of leukemia whether transduced into HSCs, common myeloid progenitors (CMPs), or granulocyte-macrophage progenitors (GMPs). Furthermore, the fusion gene MOZ-TIF2 has also recently been shown to contribute to the transformation of both HSCs and committed myeloid progenitors. Thus, the activation of an appropriate oncogene in a mature hematopoietic cell could transform the cell into an LSC with self-renewal capacity. In contrast, So et al. [92], using a different fusion partner of MLL, GAS7, showed that only the transduction of murine HSC, and not CMP or GMP, resulted in the production of mixed-lineage leukemias in mice. Therefore, a universal phenotype for LSCs may not exist, and patient-to-patient variation in cell-surface protein expression may be the rule.

Of note, a similar heterogeneity exists in HSC gene expression: genes thought to form a stem cell gene signature have been identified; however, dissimilar data suggest that the existence of such a signature may be premature [50, 80, 82]. Because stem cell-specific properties, such as self-renewal, quiescence, and proliferation, are not governed by genes that are specific to stem cells, it has been proposed that there is a "stem cell state" rather than a "stem cell portrait" in the hematopoietic system [125]. The "stem cell state" may represent a transient and potentially reversible state that cells can assume in response to the correct trigger. Using a similar concept, one can argue that there is an "LSC state" rather than an "LSC phenotype," a concept that is supported by the well-described lineage "infidelity" in human leukemias [91]. Whereas defining the sets of conditions that pertain to the LSC state may allow the development of strategies to eliminate LSCs or render them inconsequential, the identification of specific genes and surface markers expressed solely by LSCs may be difficult or impossible.

Taken together, these data suggest that LSCs should be characterized by function rather than by phenotype and be referred to as "leukemia-initiating cells." It is unclear whether studies performed using SCID or NOD/SCID mouse models reflect normal human physiology and/or pathophysiology. Thus, the overall conclusion that can be drawn from those studies is that the capability of a leukemia cell to initiate and sustain leukemia in humans, regardless of its phenotype, physical characteristics, or function in immune-deficient mice, is the most important biological feature of the cells that we refer to as LSCs.

Molecular Pathways Regulating HSCs and LSCs

The current dogma is that self-renewal is the hallmark property of stem cells in both normal and neoplastic tissues. Therefore, the molecular pathways that regulate this self-renewal capacity have been studied extensively. Several genes, transcription factors, and cell-cycle regulators modulate the self-renewal, proliferation, and differentiation of HSCs [95, 123]. The genes SCL, GATA-2,

LMO-2, and AML-1 (also known as CBFA2 or RUNX1) govern the transcriptional regulation of early hematopoiesis, and the deregulation of these genes through chromosomal aberrations plays a key role in leukemogenesis. For example, the gene encoding the transcription factor SCL is the most frequent target of chromosomal rearrangements in children with T-cell acute lymphoblastic leukemia [4]. SCL is normally expressed in HSCs and immature progenitors and is downregulated during differentiation, and its abnormal activation might initiate malignant transformation [64]. Similarly, abnormal activation of AML-1, a gene that is required for normal hematopoiesis, as a result of translocation t(8;21) and the generation of the fusion protein AML-ETO is thought to be leukemogenic [67, 77]. The fusion protein AML-ETO has been shown to induce stem cell self-renewal [26]. However, increased self-renewal is of no apparent pathogenic consequence, presumably because secondary mutations are necessary for the expression of the leukemic phenotype [26].

Other transcription factors such as the Homeobox (Hox) genes, which include HoxB4, and the Wnt-signaling pathway have well-described roles in regulating the self-renewal and differentiation of HSCs [85, 124]. HoxB4 is abundantly expressed in HSCs, plays a role in HSC expansion, and declines as terminal differentiation proceeds [88]. Of note, deregulated expression of Hox family members such as HoxA9 is commonly observed in AML [38, 63]. The Wnt-signaling pathway, whose activity is critical to the development of several organs, plays an important role in the regulation of hematopoietic stem and progenitor cell function [85, 94]. Overexpression of ?-catenin, a downstream activator of the Wnt signaling pathway, expands the HSC pool [86], and activation of the Wnt pathway increases the expression of transcription factors and cell-cycle regulators important in HSC renewal, such as HoxB4 and Notch-1 [31, 85].

The Notch/Jagged pathway modulates extracellular regulatory signals controlling HSC fate [2]. Members of the Notch family have critical roles in keeping HSCs in an undifferentiated state and may act as gatekeepers for factors governing self-renewal and lineage commitment [81]. Of interest, the gene encoding the Notch receptor is rearranged by recurrent chromosomal translocations in some patients with T-cell acute lymphoblastic leukemia [32]. Transcription factors and cell-cycle regulators associated with oncogenesis, such as Bmi-1 and Sonic hedgehog (Shh), regulate the proliferation HSCs and LSCs [99, 106]. Bmi-1, a member of the Polycomb family [65], is expressed in normal and leukemic HSCs and regulates self-renewal by modulating the activity of genes governing proliferation, survival, and lineage commitment [66, 79]. Although direct evidence for the role of Shh in the regulation of self-renewal is lacking, in vitro studies demonstrated HSC self-renewal activity of Shh in combination with various growth factors [11]. Recent data suggest that PTEN (phosphatase and tensin homologue deleted on chromosome ten), a negative regulator of the phosphatidylinositol 3 kinase (PI3K) pathway, has essential roles in restricting the activation of HSCs, in lineage fate determination, and in the prevention of leukemogenesis [122]. Furthermore, it has been shown recently that PTEN dependence could distinguish HSCs from LSCs [118].

Taken together, these data suggest that differential expression of several transcription factors controls the fate of HSCs and plays a critical role in the determination of self-renewal, differentiation, and lineage commitment. These pathways are under the control of various intracellular stimuli as well as cytokines and stromal factors from adjacent cells in the bone marrow. Further studies of these transcription factors in HSCs and the mechanisms causing their deregulation are likely to provide us with better targets for therapy.

Cellular HSC and LSC Regulators

Within the hematopoietic microenvironment, early progenitors are thought to be maintained and compartmentalized in a specific location in the endosteal lining of the bone cavities. However, extramedullary niches have also been described, as discussed above. Extramedullary disease is occasionally found in several leukemias. Although it is well established that the leukemogenic process is bone marrow derived, it is not clear whether, as in normal HSCs, an LSC niche exists.

Normal hematopoietic and leukemia cells proliferate in response to several cytokines and hematopoietic growth factors [34, 84]. Thus, the regulation of normal and leukemic hematopoiesis is the result of multiple processes involving cell–cell and cell–extracellular matrix interactions and the actions of specific growth factors and other cytokines as well as intrinsic modulators of hematopoietic development. Despite some important progress, the genetic and cellular factors that influence stem cells to differentiate into developmentally restricted progenitor cells or to self-renew to replace cells that have become committed to differentiation are still poorly understood.

Clinical Significance

LSCs, Drug Resistance, and Relapse

Despite the development of several new agents that effectively reduce the tumor burden in patients with leukemia, relapse continues to be the most common cause of death, particularly in patients with AML. Therefore, most, if not all, patients who attain complete remission still harbor disease-sustaining cells. Because these residual leukemia cells cannot be detected by conventional techniques (such as light microscopy), the term minimal residual disease (MRD) is used to describe them [33]. Current efforts directed at detecting and quantifying MRD are based on the assumption that eradication of disease-sustaining cells will improve treatment outcomes [35, 100]. The question of why neoplastic cells with disease-sustaining capacity survive chemotherapy has prompted extensive research in recent decades.

Normal HSCs are mostly quiescent and, because of that, protected from cell-cycle-dependent insults. Similarly, LSCs are thought to be quiescent [42] and, therefore, resistant to endogenous or exogenous apoptotic stimuli [27, 59]. Several studies have provided data that support this theory, including one by Guzman et al. [45], which demonstrated that as much as 96% of the LSC population, as defined by the phenotype $CD34^+/CD38^-/CD123^+$, was in the G0 phase of the cell cycle. Thus, the current dogma that the resting status of the putative LSCs protects them from the commonly used cell-cycle-specific chemotherapeutic agents ignores the fact that non-cell-cycle-specific agents are routinely used to treat all forms of leukemia.

Resistance to drugs and toxins by cells in the quiescent phase is thought to be mediated through the expression of ATP-associated transporters [27]. High levels of ATP-binding cassette transporters have been found in both normal and cancer stem cells but not in lineage-committed progenitor cells [20, 40, 89]. Because of this property, the efflux of fluorescent dyes such as Hoechst 33342 was thought to isolate HSCs [36, 39]. Thus, quiescent LSCs inherently possessing drug resistance mechanisms or acquiring them through mutations [27] are thought to survive chemotherapy and sustain the disease [1, 42, 45, 48, 86, 103]. Remarkably, this theory has been generally accepted, despite data showing that HSC may be non-quiescent [117], that ATP-associated transporters are present in mature non-quiescent cells [87], and that drug and dye efflux can be found in cells that do not exhibit stem cell capacity [19, 97].

It is likely that secondary events, such as the development of mutations, further contribute to the intrinsic resistance properties of LSCs. Quiescent LSCs and non-quiescent leukemia-sustaining cells may carry the initial mutations leading to genomic instability resulting in secondary mutagenic events that contribute to the development of a more resistant phenotype. Alternatively, random secondary mutations or mutations occurring as a result of selective pressure caused by therapy may contribute to disease progression and drug resistance. This has been seen in patients with CML treated with imatinib mesylate, in whom mutations of the ATP-binding site of BCR-ABL are well documented [21].

LSCs and Strategies to Cure Leukemia

As discussed above, relapse continues to be the most frequent cause of treatment failure, particularly in AML. The clinical benefits of early detection of MRD remain uncertain and are under investigation. The most commonly used assay to identify residual cells expressing a leukemia-specific molecular abnormality is polymerase chain reaction (PCR). This technique has several limitations, however, including false-positive and false-negative results, lack of standardization among laboratories, and insufficient sensitivity (i.e., the inability to detect clinically significant MDR that results in clinical relapse). More importantly, PCR cannot distinguish LSCs from their terminally differentiated progeny, i.e., it cannot distinguish between leukemia-sustaining cells and clinically insignificant MRD [35, 83, 100].

Rather than MRD "detection and monitoring," characterization of the molecular and biologic features of the cells that initiate and maintain leukemia is the essential step in the development of novel strategies to cure this disease. These cells have dissimilar morphologic features, no specific phenotype, no specific cell-surface or molecular markers, and they may have various degrees of abnormal maturation and may or may not be quiescent. They also may or may not fit the current definition of an LSC. Function is the only common denominator of these cells. Leukemia-sustaining cells may be different in different subtypes of leukemia and may vary from patient to patient because of the large variety of mutagenic events that initiate and drive this disease. Whether or not the cells that initiate and sustain leukemia fit the current agreed-upon features of "stem cells," these are the cells on which we should concentrate our efforts.

Acknowledgment I thank Dawn Chalaire for editing this manuscript.

References

1. Ailles LE, Humphries RK, Thomas TE, Hogge DE. Retroviral marking of acute myelogenous leukemia progenitors that initiate long-term culture and growth in immunodeficient mice. *Exp Hematol.* 1999;27:1609–1620.
2. Artavanis-Tsakonas S, Rand MD, Lake RJ. Notch signaling: cell fate control and signal integration in development. *Science.* 1999;284:770–776.
3. Becker A, McCulloch EA, Till JE. Cytological demonstration of the clonal nature of spleen colonies derived from transplanted mouse bone marrow cells. *Nature.* 1963;197:452–454.
4. Begley CG, Green AR. The SCL gene: from case report to critical hematopoietic regulator. *Blood.* 1999;93:2760–2770.
5. Bennett JH. On leucocythemia, or blood containing an unusual number of colourless corpuscles. *Monthly J Med Sci.* 1851a;12:17–38.
6. Bennett JH. On leucocythemia, or white cell blood. *Monthly J Med Sci.* 1851b;12:312–326.
7. Bennett JH. On leucocythemia, or white cell blood. *Monthly J Med Sci.* 1851c;13:97–111.
8. Bennett JH. On leucocythemia, or white cell blood. *Monthly J Med Sci.* 1851d;13:317–326.
9. Bennett JH. Case of hypertrophy of the spleen and liver, in which death took place from suppuration of the blood. *Edinb Med Surg J.* 1845;64:413–423.
10. Bennett JM, Catovsky D, Daniel MT, et al. Proposed revised criteria for the classification of acute myeloid leukemia. A report of the French-American-British Cooperative Group. *Ann Intern Med.* 1985;103:620–625.
11. Bhardwaj G, Murdoch B, Wu D, et al. Sonic hedgehog induces the proliferation of primitive human hematopoietic cells via BMP regulation. *Nat Immunol.* 2001;2:172–180.
12. Blair A, Hogge DE, Ailles LE, Lansdorp PM, Sutherland HJ. Lack of expression of Thy-1 (CD90) on acute myeloid leukemia cells with long-term proliferative ability in vitro and in vivo. *Blood.* 1997;89:3104–3112.

13. Blair A, Sutherland HJ. Primitive acute myeloid leukemia cells with long-term proliferative ability in vitro and in vivo lack surface expression of c-kit (CD117). *Exp Hematol.* 2000;28:660–671.
14. Bonnet D. Normal and leukaemic stem cells. *Br J Haematol.* 2005;130:469–479.
15. Bonnet D, Dick JE. Human acute myeloid leukemia is organized as a hierarchy that originates from a primitive hematopoietic cell. *Nat Med.* 1997;3:730–737.
16. Bornhauser M, Eger L, Oelschlaegel U, et al. Rapid reconstitution of dendritic cells after allogeneic transplantation of CD133+ selected hematopoietic stem cells. *Leukemia.* 2005;19:161–165.
17. Brendel C, Mohr B, Schimmelpfennig C, et al. Detection of cytogenetic aberrations both in CD90 (Thy-1)-positive and (Thy-1)-negative stem cell (CD34) subfractions of patients with acute and chronic myeloid leukemias. *Leukemia.* 1999;13:1770–1775.
18. Brown D, Kogan S, Lagasse E, et al. A PML/RAR alpha transgene initiates murine acute promyelocytic leukemia. *Proc Natl Acad Sci USA.* 1997;94:2551–2556.
19. Camargo FD, Chambers SM, Drew E, McNagny KM, Goodell MA. Hematopoietic stem cells do not engraft with absolute efficiencies. *Blood.* 2006;107:501–507.
20. Chaudhary PM, Roninson IB. Expression and activity of P-glycoprotein, a multidrug efflux pump, in human hematopoietic stem cells. *Cell.* 1991;66:85–94.
21. Chu S, Xu H, Shah NP, et al. Detection of BCR-ABL kinase mutations in CD34+ cells from chronic myelogenous leukemia patients in complete cytogenetic remission on imatinib mesylate treatment. *Blood.* 2005;105:2093–2098.
22. Clarkson B, Strife A, Fried J, et al. Studies of cellular proliferation in human leukemia. IV. Behavior of normal hemotopoietic cells in 3 adults with acute leukemia given continuous infusions of 3H-thymidine for 8 or 10 days. *Cancer.* 1970;26:1–19.
23. Cohnheim J. Ueber entzundung und eiterung. *Path Anat Physiol Klin Med.* 1867;40:1–79.
24. Cozzio A, Passegue E, Ayton PM, Karsunky H, Cleary ML, Weissman IL. Similar MLL-associated leukemias arising from self-renewing stem cells and short-lived myeloid progenitors. *Genes Dev.* 2003;17:3029–3035.
25. Craigie D. Case of disease of spleen, in which death took place in consequence of the presence of purulent matter in the blood. *Edinb Med Surg J.* 1845;64:400–412.
26. de Guzman CG, Warren AJ, Zhang Z, et al. Hematopoietic stem cell expansion and distinct myeloid developmental abnormalities in a murine model of the AML1-ETO translocation. *Mol Cell Biol.* 2002;22:5506–5517.
27. Dean M, Fojo T, Bates S. Tumour stem cells and drug resistance. *Nat Rev Cancer.* 2005;5:275–284.
28. Dick JE. Stem cells: self-renewal writ in blood. *Nature.* 2003;423:231–233.
29. Domen J, Weissman IL. Self-renewal, differentiation or death: regulation and manipulation of hematopoietic stem cell fate. *Mol Med Today.* 1999;5:201–208.
30. Donne A. De l'origine des globules du sang, de leur mode de formation, dee leur fin. *C R Acad Sci.* 1842;14:366.
31. Duncan AW, Rattis FM, DiMascio LN, et al. Integration of notch and Wnt signaling in hematopoietic stem cell maintenance. *Nat Immunol.* 2005;6:314–322.
32. Ellisen LW, Bird J, West DC, et al. TAN-1, the human homolog of the Drosophila notch gene, is broken by chromosomal translocations in T lymphoblastic neoplasms. *Cell.* 1991;66:649–661.
33. Estrov Z, Grunberger T, Dube ID, Wang YP, Freedman MH. Detection of residual acute lymphoblastic leukemia cells in cultures of bone marrow obtained during remission. *N Engl J Med.* 1986;315:538–542.
34. Estrov Z, Kurzrock R, Talpaz M. Interruption of endogenous growth regulatory networks: a novel approach to inhibition of leukemia cell proliferation. *FORUM Trends Exp Clin Med.* 1993;3:306–318.
35. Faderl S, Talpaz M, Kantarjian HM, Estrov Z. Should polymerase chain reaction analysis to detect minimal residual disease in patients with chronic myelogenous leukemia be used in clinical decision making? *Blood.* 1999;93:2755–2759.

36. Feuring-Buske M, Hogge DE. Hoechst 33342 efflux identifies a subpopulation of cyto-genetically normal CD34(+)CD38(-) progenitor cells from patients with acute myeloid leukemia. *Blood.* 2001;97:3882–3889.
37. Fialkow PJ, Singer JW, Raskind WH, et al. Clonal development, stem-cell differentiation, and clinical remissions in acute nonlymphocytic leukemia. *N Engl J Med.* 1987;317:468–473.
38. Golub TR, Slonim DK, Tamayo P, et al. Molecular classification of cancer: class discovery and class prediction by gene expression monitoring. *Science.* 1999;286:531–537.
39. Goodell MA, Brose K, Paradis G, Conner AS, Mulligan RC. Isolation and functional properties of murine hematopoietic stem cells that are replicating in vivo. *J Exp Med.* 1996;183:1797–1806.
40. Gottesman MM, Fojo T, Bates SE. Multidrug resistance in cancer: role of ATP-dependent transporters. *Nat Rev Cancer.* 2002;2:48–58.
41. Grimwade D, Enver T. Acute promyelocytic leukemia: where does it stem from? *Leukemia.* 2004;18:375–384.
42. Guan Y, Gerhard B, Hogge DE. Detection, isolation, and stimulation of quiescent primitive leukemic progenitor cells from patients with acute myeloid leukemia (AML). *Blood.* 2003;101:3142–3149.
43. Guan Y, Hogge DE. Proliferative status of primitive hematopoietic progenitors from patients with acute myelogenous leukemia (AML). *Leukemia.* 2000;14:2135–2141.
44. Guenechea G, Gan OI, Dorrell C, Dick JE. Distinct classes of human stem cells that differ in proliferative and self-renewal potential. *Nat Immunol.* 2001;41:75–82.
45. Guzman ML, Neering SJ, Upchurch D, et al. Nuclear factor-kappaB is constitutively activated in primitive human acute myelogenous leukemia cells. *Blood.* 2001;98:2301–2307.
46. Hess DA, Meyerrose TE, Wirthlin L, et al. Functional characterization of highly purified human hematopoietic repopulating cells isolated according to aldehyde dehydrogenase activity. *Blood.* 2004;104:1648–1655.
47. Hochedlinger K, Jaenisch R. Nuclear transplantation, embryonic stem cells, and the potential for cell therapy. *N Engl J Med.* 2003;349:275–286.
48. Holyoake T, Jiang X, Eaves C, Eaves A. Isolation of a highly quiescent subpopulation of primitive leukemic cells in chronic myeloid leukemia. *Blood.* 1999;94:2056–2064.
49. Hope KJ, Jin L, Dick JE. Acute myeloid leukemia originates from a hierarchy of leukemic stem cell classes that differ in self-renewal capacity. *Nat Immunol.* 2004;5:738–743.
50. Ivanova NB, Dimos JT, Schaniel C, Hackney JA, Moore KA, Lemischka IR. A stem cell molecular signature. *Science.* 2002;298:601–604.
51. Jacobson LO, Marks EK, Gaston EO, Simmons EL, Robson MJ, Eldredge JH. Role of the spleen in radiation injury. *Proc Exp Biol Med.* 1949;70:7440.
52. Jaiswal S, Traver D, Miyamoto T, Akashi K, Lagasse E, Weissman IL. Expression of BCR/ABL and BCL-2 in myeloid progenitors leads to myeloid leukemias. *Proc Natl Acad Sci USA.* 2003;100:10002–10007.
53. Jamieson CH, Ailles LE, Dylla SJ, et al. Granulocyte-macrophage progenitors as candidate leukemic stem cells in blast-crisis CML. *N Engl J Med.* 2004;351:657–667.
54. Jones RJ, Wagner JE, Calando P, Zicha MS, Sharkis SJ. Separation of pluripotent haematopoietic stem cells from spleen colony-forming cells. *Nature.* 1990;347:188–189.
55. Jordan CT, Upchurch D, Szilvassy SJ, et al. The interleukin-3 receptor alpha chain is a unique marker for human acute myelogenous leukemia stem cells. *Leukemia.* 2000;14:1777–1784.
56. Kania G, Corbeil D, Fuchs J, et al. Somatic stem cell marker prominin-1/CD133 is expressed in embryonic stem cell-derived progenitors. *Stem Cells.* 2005;23:791–804.
57. Kiel MJ, Yilmaz OH, Iwashita T, Yilmaz OH, Terhorst C, Morrison SJ. SLAM family receptors distinguish hematopoietic stem and progenitor cells and reveal endothelial niches for stem cells. *Cell.* 2005;121:1109–1121.
58. Kondo M, Wagers AJ, Manz MG, et al. Biology of hematopoietic stem cells and progenitors: implications for clinical application. *Annu Rev Immunol.* 2003;21:759–806.

59. Konopleva M, Zhao S, Hu W, et al. The anti-apoptotic genes Bcl-X(L) and Bcl-2 are over-expressed and contribute to chemoresistance of non-proliferating leukaemic CD34+ cells. *Br J Haematol.* 2002;118:521–534.

60. Korbling M, Estrov Z. Adult stem cells for tissue repair – a new therapeutic concept? *N Engl J Med.* 2003;349:570–582.

61. Korn AP, Henkelman RM, Ottensmeyer FP, Till JE. Investigations of a stochastic model of haemopoiesis. *Exp Hematol.* 1973;1:362–375.

62. Lapidot T, Sirard C, Vormoor J, et al. A cell initiating human acute myeloid leukaemia after transplantation into SCID mice. *Nature.* 1994;367:645–648.

63. Lawrence HJ, Rozenfeld S, Cruz C, et al. Frequent co-expression of the HOXA9 and MEIS1 homeobox genes in human myeloid leukemias. *Leukemia.* 1999;13:1993–1999

64. Lecuyer E, Hoang T. SCL: from the origin of hematopoiesis to stem cells and leukemia. *Exp Hematol.* 2004;32:11–24.

65. Lessard J, Baban S, Sauvageau G. Stage-specific expression of polycomb group genes in human bone marrow cells. *Blood.* 1998;91:1216–1224.

66. Lessard J, Sauvageau G. Bmi-1 determines the proliferative capacity of normal and leukaemic stem cells. *Nature.* 2003;423:255–260.

67. Licht JD. AML1 and the AML1-ETO fusion protein in the pathogenesis of t(8;21) AML. *Oncogene.* 2001;20:5660–5679.

68. Maximow A. Der lymphozyte als gemeinsame stammzelle der verschiedenen blutele-mente in der embryonalen entwicklung und im poetfetalen leben der säugetiere. Demon-strationsvortrag, gehalten in der asserordentlichen sitzung der Berliner Hämatologoschen Gesellschaft am 1, Juni 1909.

69. McCulloch EA, Howatson AF, Buick RN, Minden MD, Izaguirre CA. Acute myelo-blastic leukemia considered as a clonal hemopathy. *Blood Cells.* 1979;5:261–282.

70. Minden MD, Buick RN, McCulloch EA. Separation of blast cell and T-lymphocyte pro-genitors in the blood of patients with acute myeloblastic leukemia. *Blood.* 1979;54:186–195.

71. Moore KA, Lemischka IR. Stem cells and their niches. *Science.* 2006;311:1880–1885.

72. Moore MA. Converging pathways in leukemogenesis and stem cell self-renewal. *Exp Hematol.* 2005;33:719–737.

73. Moore MA, Metcalf D. Cytogenetic analysis of human acute and chronic myeloid leukemic cells cloned in agar culture. *Int J Cancer.* 1973;11:143–152.

74. Moore MA, Shapiro F. Regulation and function of hematopoietic stem cells. *Curr Opin Hematol.* 1994;3:180–186.

75. Morrison SJ, Uchida N, Weissman IL. The biology of hematopoietic stem cells. *Annu Rev Cell Dev Biol.* 1995;11:35–71.

76. Muller-Sieburg CE, Townsend K, Weissman IL, Rennick D. Proliferation and differen-tiation of highly enriched mouse hematopoietic stem cells and progenitor cells in response to defined growth factors. *J Exp Med.* 1988;167:1825–1840.

77. Mulloy JC, Cammenga J, MacKenzie KL, Berguido FJ, Moore MA, Nimer SD. The AML1-ETO fusion protein promotes the expansion of human hematopoietic stem cells. *Blood.* 2002;99:15–23.

78. Pappenheim A. Prinzipen der neuren morphologichschen haematozytologie nach zyto-genetischer grundlage. *Folia Haematol.* 1917;21:91–101.

79. Park IK, Qian D, Kiel M, et al. Bmi-1 is required for maintenance of adult self-renewing haematopoietic stem cells. *Nature.* 2003;423:302–305.

80. Phillips RL, Ernst RE, Brunk B, et al. The genetic program of hematopoietic stem cells. *Science.* 2000;288:1635–1640.

81. Pui JC, Allman D, Xu L, et al. Notch1 expression in early lymphopoiesis influences B versus T lineage determination. *Immunity.* 1999;11:299–308.

82. Ramalho-Santos M, Yoon S, Matsuzaki Y, Mulligan RC, Melton DA. "Stemness": transcriptional profiling of embryonic and adult stem cells. *Science.* 2002;298:597–600.

83. Ravandi F, Estrov Z. Eradication of leukemia stem cells as a new goal of therapy in leukemia. *Clin Cancer Res.* 2006;12:340–344.

84. Ravandi F, Talpaz M, Kantarjian H, Estrov Z. Cellular signaling pathways: new targets in leukemia therapy. *Br J Haematol.* 2002;116:57–77.

85. Reya T, Duncan AW, Ailles L, et al. A role for Wnt signalling in self-renewal of haematopoietic stem cells. *Nature.* 2003;423:409–414.

86. Reya T, Morrison SJ, Clarke MF, Weissman IL. Stem cells, cancer, and cancer stem cells. *Nature.* 2001;414:105–111.

87. Robey RW, Zhan Z, Piekarz RL, Kayastha GL, Fojo T, Bates SE. Increased MDR1 expression in normal and malignant peripheral blood mononuclear cells obtained from patients receiving depsipeptide (FR901228, FK228, NSC630176). *Clin Cancer Res.* 2006;5:547–555.

88. Sauvageau G, Thorsteinsdottir U, Eaves CJ, et al. Overexpression of HOXB4 in hematopoietic cells causes the selective expansion of more primitive populations in vitro and in vivo. *Genes Dev.* 1995;9:1753–1765.

89. Scharenberg CW, Harkey MA, Torok-Storb B. The ABCG2 transporter is an efficient Hoechst 33342 efflux pump and is preferentially expressed by immature human hematopoietic progenitors. *Blood.* 2002;99(2):507–512.

90. Sharkis SJ, Collector MI, Barber JP, Vala MS, Jones RJ. Phenotypic and functional characterization of the hematopoietic stem cell. *Stem Cells.* 1997;(suppl. 1):41–44.

91. Smith LJ, Curtis JE, Messner HA, Senn JS, Furthmayr H, McCulloch EA. Lineage infidelity in acute leukemia. *Blood.* 1983;61:1138–1145.

92. So CW, Karsunky H, Passegue E, Cozzio A, Weissman IL, Cleary ML. MLL-GAS7 transforms multipotent hematopoietic progenitors and induces mixed lineage leukemias in mice. *Cancer Cell.* 2003;3:161–171.

93. Spangrude GJ, Heimfeld S, Weissman IL. Purification and characterization of mouse hematopoietic stem cells. *Science.* 1988;241:58–62.

94. Staal FJ, Clevers HC. WNT signalling and haematopoiesis: a WNT-WNT situation. *Nat Rev Immunol.* 2005;5:21–30.

95. Stein MI, Zhu J, Emerson SG. Molecular pathways regulating the self-renewal of hematopoietic stem cells. *Exp Hematol.* 2004;32:1129–1136.

96. Storms RW, Goodell MA, Fisher A, Mulligan RC, Smith C. Hoechst dye efflux reveals a novel CD7(+)CD34(−) lymphoid progenitor in human umbilical cord blood. *Blood.* 2000;96:2125–2133.

97. Storms RW, Trujillo AP, Springer JB, et al. Isolation of primitive human hematopoietic progenitors on the basis of aldehyde dehydrogenase activity. *Proc Natl Acad Sci USA.* 1999;96:9118–9123.

98. Sutherland HJ, Lansdorp PM, Henkelman DH, Eaves AC, Eaves CJ. Functional characterization of individual human hematopoietic stem cells cultured at limiting dilution on supportive marrow stromal layers. *Proc Natl Acad Sci USA.* 1990;87:3584–3588.

99. Taipale J, Beachy PA. The hedgehog and Wnt signalling pathways in cancer. *Nature.* 2001;411:349–354.

100. Talpaz M, Estrov Z, Kantarjian H, Ku S, Foteh A, Kurzrock R. Persistence of dormant leukemic progenitors during interferon-induced remission in chronic myelogenous leukemia. Analysis by polymerase chain reaction of individual colonies. *J Clin Invest.* 1994;94:1383–1389.

101. Taussig DC, Pearce DJ, Simpson C, et al. Hematopoietic stem cells express multiple myeloid markers: implications for the origin and targeted therapy of acute myeloid leukemia. *Blood.* 2005;106:4086–4092.

102. Terpstra W, Ploemacher RE, Prins A, et al. Fluorouracil selectively spares acute myeloid leukemia cells with long-term growth abilities in immunodeficient mice and in culture. *Blood.* 1997;88:1944–1950.

103. Terpstra W, Prins A, Ploemacher RE, et al. Long-term leukemia-initiating capacity of a CD34-subpopulation of acute myeloid leukemia. *Blood.* 1996;87:2187–2194.
104. Till JE, McCulloch EA. A direct measurement of the radiation sensitivity of normal mouse bone marrow cells. *Radiat Res.* 1961;14:213–222.
105. Till JE, McCulloch EA, Siminovitch LA. Stochastic model of stem cell proliferation, based on the growth of spleen colony-forming cells. *Proc Natl Acad Sci USA.* 1964;51:29–36.
106. van der Lugt NM, Alkema M, Berns A, Deschamps J. The polycomb-group homolog Bmi-1 is a regulator of murine Hox gene expression. *Mech Dev.* 1996;58:153–164.
107. Vardiman JW, Harris NL, Brunning RD. The World Health Organization (WHO) classification of the myeloid neoplasms. *Blood.* 2002;100:2292–2302.
108. Velpeau A. Sur la resorption du pusuaet sur l'alternation du sang dans les maladies clinique de persection nenemant. Premier observation. Revue Medical Francaise et Étrangère. *Rev Med.* 1827;2:216–240.
109. Virchow R. Editorial. *Physiol Klin Med.* 1855;3:23.
110. Virchow R. Weisses Blut und Milztumoren. *Medicinische Zeitung.* 1847a;16:9–15.
111. Virchow R. Weisses Blut (leukämie). *Virchow Arch Pathol Anat.* 1847b;1:563–569.
112. Virchow R. Weisses Blut und Milztumoren. *Medicinische Zeitung.* 1846;15:157–163.
113. Virchow R. Weisses Blut. In: Froriep LFv, Froriep R, eds. *Neue Notizen aus dem Gebiete der Natur- und Heilkunde.* Vol. 36. Berlin; 1845:151–156.
114. Wantzin GL. Nuclear labelling of leukaemic blast cells with tritiated thymidine triphosphate in 35 patients with acute leukaemia. *Br J Haematol.* 1977;37:475–482.
115. Warner JK, Wang JC, Hope KJ, Jin L, Dick JE. Concepts of human leukemic development. *Oncogene.* 2004;23:7164–7177.
116. Weissman IL. Translating stem and progenitor cell biology to the clinic: barriers and opportunities. *Science.* 2000;287:1442–1446.
117. Wilpshaar J, Bhatia M, Kanhai HH, et al. Engraftment potential of human fetal hematopoietic cells in NOD/SCID mice is not restricted to mitotically quiescent cells. *Blood.* 2002;100:20–27.
118. Yilmaz OH, Valdez R, Theisen BK, et al. Pten dependence distinguishes haematopoietic stem cells from leukaemia-initiating cells. *Nature.* 2006;441(7092):475–482.
119. Yuan Y, Shen H, Franklin DS, Scadden DT, Cheng T. In vivo self-renewing divisions of haematopoietic stem cells are increased in the absence of the early G1-phase inhibitor, p18INK4C. *Nat Cell Biol.* 2004;5:436–442.
120. Zanjani ED, Almeida-Porada G, Livingston AG, Flake AW, Ogawa M. Human bone marrow CD34- cells engraft in vivo and undergo multilineage expression that includes giving rise to CD34+ cells. *Exp Hematol.* 1998;26:1022–1023.
121. Zanjani ED, Almeida-Porada G, Livingston AG, Zeng H, Ogawa M. Reversible expression of CD34 by adult human bone marrow long-term engrafting hematopoietic stem cells. *Exp Hematol.* 2003;31(5):406–412.
122. Zhang J, Grindley JC, Yin T, et al. PTEN maintains haematopoietic stem cells and acts in lineage choice and leukaemia prevention. *Nature.* 2006;441(7092):518–522.
123. Zhu J, Emerson SG. Hematopoietic cytokines, transcription factors and lineage commitment. *Oncogene.* 2002;21:3295–3313.
124. Zhu J, Giannola DM, Zhang Y, Rivera AJ, Emerson SG. NF-Y cooperates with USF1/2 to induce the hematopoietic expression of HOXB4. *Blood.* 2003;102:2420–2427.
125. Zipori D. The nature of stem cells: state rather than entity. *Nat Rev Genet.* 2004;5:873–878.

Epigenetic Mechanisms in AML – A Target for Therapy

Yasuhiro Oki and Jean-Pierre J. Issa

Abstract Epigenetics refers to a stable, mitotically perpetuated regulatory mechanism of gene expression without an alteration of the coding sequence. Epigenetic mechanisms include DNA methylation and histone tail modifications. Epigenetic regulation is part of physiologic development and becomes abnormal in neoplasia, where silencing of critical genes by DNA methylation or histone deacetylation can contribute to leukemogenesis as an alternative to deletion or loss-of-function mutation. In acute myelogenous leukemia (AML), aberrant DNA methylation can be observed in multiple functionally relevant genes such as *p15, p73, E-cadherin, ID4, RARβ2*. Abnormal activities of histone tail-modifying enzymes have also been seen in AML, frequently as a direct result of chromosomal translocations. It is now clear that these epigenetic changes play a significant role in development and progression of AML, and thus constitute important targets of therapy. The aim of targeting epigenetic effector protein or "epigenetic therapy" is to reverse epigenetic silencing and reactivate various genes to induce a therapeutic effect such as differentiation, growth arrest, or apoptosis. Recent clinical studies have shown the relative safety and efficacy of such epigenetic therapies.

Introduction

Carcinogenesis is a multistep process at the molecular level [64], driven by genetic alterations such as gene mutation and deletion, resulting in activation of oncogenes or inactivation of tumor suppressor genes [64]. Epigenetic changes have also been shown to play a significant role in the malignant transformation of cells [84, 12]. Epigenetics refers to a stable, mitotically perpetuated regulatory mechanism of gene expression without an alteration of the gene coding

J.-P.J. Issa (✉)
Department of Leukemia, University of Texas MD Anderson Cancer Center,
Houston, TX, USA
e-mail: jissa@mdanderson.org

L. Nagarajan (ed.), *Acute Myelogenous Leukemia*,
Cancer Treatment and Research 145, DOI 10.1007/978-0-387-69259-3_2,
© Springer Science+Business Media, LLC 2010

sequence. Epigenetic mechanisms include DNA methylation and histone tail modifications such as acetylation and methylation [84, 12]. Epigenetic changes can lead to carcinogenesis by silencing critical genes [84, 12]. DNA methylation is very stable and maintained once established, except in special states such as embryogenesis [146, 81]. Histone modifications are more dynamic biochemical changes in the context of expression regulation [146, 81] but are also involved in stable gene silencing. Another, more flexible mechanism of epigenetic regulation is through small regulatory non-coding antisense RNAs, which can achieve transcriptional or posttranscriptional gene silencing [9], although the role of these processes in carcinogenesis is unknown. Over the past decades, alterations in DNA methylation and histone modifications in leukemogenesis have been well described and are now recognized as targets of therapy for AML and other hematological malignancies.

DNA Methylation and DNA Methyltransferase (DNMT)

The addition of a methyl-group to cytosine forming 5-methylcytosine in DNA has genetic and epigenetic effects on cellular development, differentiation, and carcinogenesis [84, 12]. In mammalian DNA, cytosine methylation is restricted to cytosine followed by guanosine (the CpG dinucleotide) [84, 12]. DNA methylation is accomplished by DNA methyltransferases (DNMTs), which catalyze the covalent addition of a methyl group to the $5'$ position of cytosine from a donor S-adenosylmethionine [67]. Three different proteins, DNMT1, DNMT3A, and DNMT3B, have been shown to have DNA methyltransferase catalytic activity in mammalian cells [123, 122, 9]. In general, DNMT1 serves as a maintenance DNMT, while DNMT3A and 3B serve as de novo DNMTs introducing methyl groups to previously unmethylated CpG sites [123, 122]. DNMT3L, another type of DNMT, does not have catalytic activity but has been identified as a stimulator of the catalytic activity of DNMT3A and DNMT3B [21, 147, 11]. Once DNA methylation is established in a CpG nucleotide, it is maintained after cell division through the activity of DNMTs, which localize to replication foci to work on newly synthesized hemi-methylated DNA [132, 102]. Recent studies suggest that DNA methylation status is determined via complex mechanisms where DNMTs interact with each other and with other proteins to induce DNA methylation [134, 53]. The major target of DNA methylation in normal mammalian cells is repeated transposable sequences, but it also plays a key role in imprinting and X chromosome inactivation in women [84, 12].

CpG sites are rare in the human genome relative to their predicted frequency, presumably because they were eliminated during evolution through C to T mutations of methylcytosine [13]. On the other hand, the human genome contains small regions with clusters of CpG sites, called "CpG islands," where the frequency of CpG is higher than expected [13]. About half of all human genes have CpG islands in their promoter regions, and these are not usually methylated in normal tissues, regardless of the transcriptional status of the gene. Methylation in a CpG island is associated with changes in chromatin organization and consequent repression of gene transcription (Fig. 1). One mechanism by which gene silencing is achieved is that of methylation of cytosine residues in CpG dinucleotides triggers the binding of methyl-binding proteins to DNA, which attracts histone deacetylases and histone methylases that eventually modify the structure of histones into a condensed chromatin state [119]. Condensation of the chromatin prevents specific transcription factors or DNA-dependent RNA polymerase from having access to the promoter region to cause gene silencing [84, 12]. Histone H3 lysine 9 (H3K9) methylation appears to trigger further DNA methylation through a feedback loop, thus reinforcing gene silencing [7, 149]. Since CpG methylation is maintained after cell division, gene silencing by DNA methylation is also maintained and is essentially stable once it is established. It can be only reversed physiologically and reset in early embryogenesis [116]. About half of human genes do not have CpG islands in their promoters. In these cases, DNA methylation can mark the silenced state but can be reversed physiologically by activation of gene expression [16]. The extent to which non-CpG island methylation plays a role in carcinogenesis remains unclear.

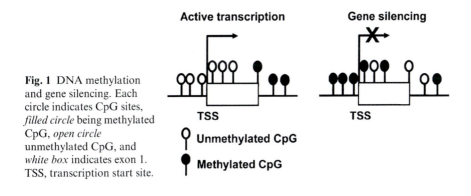

Fig. 1 DNA methylation and gene silencing. Each circle indicates CpG sites, *filled circle* being methylated CpG, *open circle* unmethylated CpG, and *white box* indicates exon 1. TSS, transcription start site.

DNA Methylation in AML

Cancers have altered patterns of DNA methylation. The global DNA methylation level is often decreased in malignant cells. Simultaneously, hypermethylation also occurs in specific regions of the genome [84, 12]. Hypomethylation was initially postulated as a mechanism of carcinogenesis through activation of oncogenes [50]. It is also known that hypomethylation is associated with chromosomal instability in vitro, and this may play a role in carcinogenesis [24]. On the other hand, aberrant DNA hypermethylation can clearly contribute to carcinogenesis by silencing tumor suppressor genes, and thus play an important role as an alternative mechanism to deletion or loss-of-function mutation for eliminating expression of functional proteins [84, 12].

Hypermethylation in some genes can be observed in normal tissues during the process of aging [75]. Other genes are methylated exclusively in cancer cells and they may arise from a combination of gene-specific predisposition to methylation and rare chance events that lead to gene silencing and a selective advantage for affected cells [75]. Different malignant diseases are associated with unique DNA methylation patterns. For example, the *RB1* gene can be found methylated primarily in retinoblastoma [135], and *VHL* gene methylation mainly occurs in renal cell carcinoma [69]. In AML, promoter methylation of $p15^{INK4B}$ has long been known [68], and the methylation of this gene in MDS appears to be associated with poor prognosis and a higher chance of developing AML [154]. Other frequently methylated genes in AML are summarized in Table 1. It should be noted that only a few of those genes have known tumor suppressor function and that some are not expressed in normal hematopoietic cells [151]. Promoter hypermethylation in cancer therefore is not necessarily limited to silencing of critical genes. Also, in some cases, methylation can be detected experimentally, but does not lead to significant gene silencing because of low density of methylation or because the involved CpG island is not in the promoter region of the genes. Therefore, only some of the methylated CpG islands are pathophysiologically significant to the neoplastic process. Nevertheless, frequent gene-specific methylation is of potential value for disease classification and/or prognostication [151]. Many cancers are characterized by intense methylation of multiple genes simultaneously, a phenomenon termed CpG island methylator phenotype [76, 91, 133], and this characterizes AML as well. Moreover, frequent aberrant gene methylation is generally associated with poor prognosis [76, 91, 133], including in AML [151].

Table 1 Genes methylated in AML

Gene name	Methylation prevalence	Function	References
p15	31‑71%	Cyclin-dependent kinase inhibitor, TGF-beta-induced growth arrest. Putative tumor suppressor	[151, 77, 125, 54, 80, 45, 27, 28, 111]
p16	0–38%	Cyclin-dependent kinase inhibitor. Tumor suppressor	[151, 62] [111]
p73	10–13%	Participates in the apoptotic response to DNA damage. Putative tumor suppressor	[54, 45]
HIC1	83% at intron 2. 0% at promoter	Transcription factor. Putative tumor suppressor	[110]
ID4	87%	DNA–binding inhibitor. Putative tumor suppressor	[169]

Table 1 (continued)

Gene name	Methylation prevalence	Function	References
RARβ2	18–20%	Nuclear transcriptional regulator, which limits growth of many cell types by regulating gene expression	[54, 45]
DAPK	3–61%	Serine/threonine kinase which acts as a positive regulator of apoptosis.	[54, 45]
CDH1 (E-cadherin)	13–69%	Calcium-dependent cell adhesion protein and a potent invasion suppressor role	[125, 54, 45, 111]
SHP1	52%	Negative regulator of Jak/STAT signaling pathway	[26]
MGMT	5%	Repair alkylated guanine in DNA	[54]
FHIT	14%	Involved in purine metabolism. Putative tumor suppressor	[80]
CRBP1	28%	Carrier protein involved in the transport of retinol. Putative tumor suppressor	[48]
ER	40–54%	Estrogen receptor	[151, 125, 45, 111]
SOCS1	39%	Negative regulator of cytokine signaling	[45]
WIT-1	49%	Unknown	[128]
MYOD1	61%	Transcription factor which regulates muscle cell differentiation by inducing cell cycle arrest	[151]
PITX2	64%	Transcription factor that regulates procollagen lysyl hydroxylase gene expression	[151]
GPR37	47%	G-protein-coupled receptor 37 precursor	[151]
SDC4	56%	Cell surface proteoglycan that bears heparan sulfate	[151]
MEIS1	64% of AML1-ETO AML	Homeodomain genes which cooperates with HoxA9	[95]
THBS1	25%	Adhesive glycoprotein that mediates cell-to-cell and cell-to-matrix interactions	[151]
Calcitonin	71%	Peptide hormone that reduces serum calcium	[111]

The causes of abnormal methylation in cancer remain poorly defined. High levels of DNMT activity in primary cancer cells have been reported in many studies [46, 79, 120, 38, 57, 82] but not in all [99, 43]. In one study, high expression of DNMT3a and DNMT3b genes in AML analyzed by gene expression profiling was associated with poor prognosis [17]. When interpreting these results, it is important to be aware that DNMT expression is regulated with the growth state of cells and that rapidly dividing cells have high levels of expression of DNMT [148]. Furthermore, it is not yet clear whether overexpression of DNMT is responsible for increased promoter methylation leading to transcriptional repression. In one study in AML, overexpression of DNMT1 and 3b was associated with *p15^{INK4B}* gene hypermethylation [114]. In another study, however, high levels of expression of DNMT1 in MDS were not associated with *p15^{INK4B}* hypermethylation [3]. Thus, it remains unclear whether DNMT overexpression contributes functionally to malignant transformation.

Histone Modifications

The DNA of eukaryotic cells is packaged into chromatin. The primary subunit of chromatin, the nucleosome, consists of an octamer of core histone proteins, i.e., H3/H4 tetramer and two H2A/H2B dimers, surrounded by 146 bp of DNA [51]. A change in local chromatin architecture alters the accessibility of transcription factors to DNA (Fig. 2) [51]. The chromatin structure is largely determined by posttranslational modifications of core histone proteins, which in turn affect gene expression [51]. In general, increased acetylation is associated

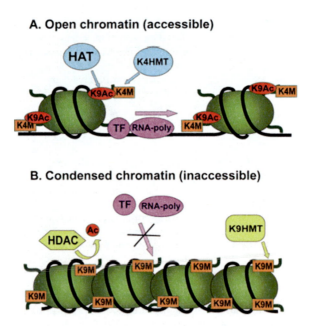

Fig. 2 Histone modifications and chromatin condensation (color). (**A**) Open chromatin. Histone H3 lysine 9 is acetylated and histone H3 lysine 4 is methylated. Nucleosomes (*green ovals*) are loosely spaced, and DNA (*black strings*) is accessible to transcription factor and RNA polymerase. The gene therefore is transcriptionally active.(**B**) Condensed chromatin. Lysine 9 is deacetylated, and methylated. Nucleosomes are tightly packed, and the gene becomes transcriptionally inactive. Ac indicates acetylation; K4M, lysine 4 methylation; K9Ac, lysine 9 acetylation; K9M, lysine 9 methylation; HAT, histone acetyltransferase; HDAC, histone deacetylase; K4HMT, lysine 4 histone methyltransferase; K9HMT, lysine 9 histone methyltransferase; RNA-poly, RNA polymerase; and TF, transcription factor

with increased transcriptional activity, whereas decreased acetylation is associated with repression of gene expression [51]. Recent studies have revealed that histone tails undergo many other modifications such as methylation, phosphorylation, ubiquitination, and sumoylation [51, 90]. Such histone tail modifications are referred to as "histone code," which collectively can characterize the transcriptional status of the gene [81].

Histone deacetylases (HDACs) and the family of histone acetyl transferases (HATs) are involved in determining the state of acetylation of histones [138, 137, 152]. In addition to deacetylation of histones, HDACs regulate physiology by deacetylating transcription factors such as p53, E2F1, and others, [106, 83, 109] or other proteins such as α-tubulin, importin-a7, and others [63, 164, 8, 14, 34, 52, 72]. Histone methylases and demethylases regulate methylation of specific residues resulting in activation of gene expression (histone H3 lysine 4 [H3K4]) or repression of gene expression (histone H3 lysine 9 and 27) [173, 131, 100]. The polycomb group family of proteins, which are essential to the stem cell phenotype, are also involved in determining chromatin structure through interactions with H3K27 [100].

Histone Alterations in AML

Histones can be altered in AML through three general mechanisms: (i) direct genetic alterations in histone modifiers, (ii) recruitment of HDAC by mutated proteins or fusion proteins, and (iii) recruitment by DNA methylation (Table 2).

(i) Direct genetic alterations in histone modifiers

A histone methyltransferase, MLL (also known as HRX or ALL1), on chromosome 11q23 is often fused with other genes leading to leukemogenesis

Table 2 Genetic alterations associated with aberrant histone modifications

Type of gene alteration	Partner proteins	References
Direct genetic alterations in histone modifiers		
MLL (histone methyltransferase)	ENL, ELL, AF10, AFX, FKHRL1, CBP, AF1p, GAS7, gephyrin. Tandem duplication of MLL gene itself is also frequently observed	[98, 141, 40, 97, 96, 107, 39, 142, 143, 44, 19, 18]
CBP (histone acetyltransferase)	MOZ	[15]
P300 (histone acetyltransferase)	MLL	[74]
Genetic alterations leading to abnormal recruitment of histone modifiers		
AML1	MTG8, TEL, MTG16	[136, 42, 70, 73]
CBFβ	MYH11	[136]
RARα	PML, PZLF, NPM1, NuMA, STAT5b	[61, 105, 65, 25, 130, 159, 5]

through chromatin modulation [32, 174, 150, 41]. The N-terminal fragment of MLL contains a DNA-binding site [168] and a transcriptional repression domain [172]. The repression domain consists of two different subdomains, one containing the DNA methyltransferase homology domain, which recruits repressor complexes including polycomb group proteins such as HPC2, Bmi1, and the corepressor CtBP [53, 36]. The other repression domain recruits HDAC1 and HDAC2 [165]. The N-terminal fragment of MLL also contains plant homeodomain (PHD) zinc fingers, which are involved in transcriptional regulation and chromatin-remodeling activity through interaction with Cyp33, a known suppressor of HoxA9 and HoxC8 transcription [49]. One of the critical functions of MLL is the maintenance of expression of HoxA9 and HoxC8 by binding to their promoter regions and keeping the chromatin open for transcription [113, 118]. The C-terminus of MLL contains a transcriptional-activation domain that binds to a HAT, CBP [47]. Furthermore, the C-terminal SET domain has histone H3 lysine 4 (H3K4)-specific histone methyltransferase (HMT) activity, and trimethylation of H3K4 is associated with transcriptional activation [113, 118]. In general, the different MLL chimeric proteins include the DNA-binding site and MT domains at the N-terminus but lack the SET domain [172]. It is thought that these result in inappropriately maintained expression of HOX genes involved in self-renewal, leading to leukemogenesis.

A histone acetyltransferase, CBP can be directly disrupted by chromosomal translocations in AML, resulting in the formation of fusion oncoprotein transcription factors. A rare chromosomal translocation observed in AML, t(8;16)(p11;p13) produces a fusion protein of CBP with the MOZ (*monocytic leukemia zinc finger*) gene [15]. This fusion results in a small deletion of the N-terminal 266 amino acids of CBP leaving the rest of the molecule intact [15]. Interestingly, the MOZ gene also has a putative acetyltransferase domain that is retained in the MOZ–CBP fusion [15]. It is believed that the leukemogenic effect is not merely from the disruption of the normal function of CBP but that the fusion protein results in aberrant recruitment of CBP to MOZ-regulated genes' promoters leading to abnormal expression of those genes.

Another histone acetyltransferase, p300, was found to be fused to the MLL gene in an AML patient carrying a t(11;22)(q23; q13) translocation [74]. This results in a fusion product that preserves most of the p300 molecules. The authors suggested that the basis for the leukemogenesis of t(11; 22)-AML is the inability of p300 to regulate cell cycle and cell differentiation after fusion with MLL.

(ii) Genetic alterations leading to abnormal recruitment of histone modifiers

Common translocations in AML result in fusion proteins that affect gene transcription by HAT/HDAC recruitment. For example, translocations involving the AML1 gene (also known as RUNX1 or CBFA2), such as t(8;21)(q22;q22), result in a fusion with other genes (ETO in the case of t(8;21)) [136, 42]. The N-terminus of AML1 binds to the promoter region of its target genes such as IL3 [153], GM-CSF [144], and MPO [6, 157, 112]. The C-terminus of AML1 normally interacts with a co-activator complex containing p300, which has HAT activity, and will result in histone acetylation and transcriptional activation [136, 42]. In AML with an AML1/ETO translocation, the N-terminus of AML1 is maintained and can interact with DNA, but the p300 binding site of the C-terminus is replaced by ETO, which attracts a co-repressor complex containing N-CoR/Sin3/HDACs instead of HAT, resulting in transcriptional repression and a block of myeloid differentiation [136, 42]. Other AML1 partner genes such as TEL and MTG16 also mediate transcription repression [70, 73]. Similarly in AML with inv(16)(p13q22), the beta subunit of core-binding factor (CBFβ), which partners with AML1, is fused with smooth muscle myosin heavy chain gene (MYH11), leading to transcriptional repression of AML1-transactivated genes and suppression of myeloid differentiation [136].

Another translocation that results in altered histone modifications is t(15;17)(q21;q21), characteristic of acute promyelocytic leukemia. This translocation fuses PML with RARα (retinoic acid receptor alpha). RARα is a ligand-dependent transcriptional activator that binds as a heterodimer with members of the RXR (retinoid X receptor) family of nuclear receptors [170]. In the absence of a ligand, the heterodimer binds either N-CoR [93, 71] or SMRT [23, 22], which then form a co-repressor complex containing Sin3 and HDAC [2, 94, 66]. When bound to a ligand (retinoic acid), RARα functions as a transcription activator. The fusion PML–RARα is constitutively complexed with co-repressor molecules at physiologic levels of retinoic acid resulting in repression of target gene expression through histone deacetylases [61, 105, 65].

(iii) Histone modification by DNA methylation

Histone alterations can also be recruited by DNA methylation. Methylated CpG sites trigger the binding of methyl-binding proteins to DNA, which attracts histone deacetylases and histone methylases. This leads to H3K9 deacetylation and methylation, as well as demethylation of H3K4 [119].

Epigenetic Therapy

The aim of epigenetic therapy is to reverse epigenetic silencing and reactivate various genes hoping for a therapeutic effect such as differentiation, growth arrest, or apoptosis. Indeed, pharmacologic inhibition of DNMT or HDAC in vitro has been shown to result in reactivation of gene expression for genes silenced either physiologically or pathologically [11, 108]. The concentrations of these agents required for this effect can readily be achieved in vivo [11, 108]. There have been extensive clinical studies in this field over the past decade, focusing on inhibitors of DNMT and HDAC.

DNMT Inhibitors

5-Azacytidine (azacitidine) and its deoxy analogue 5-aza-2'-deoxycytidine (decitabine) are the two nucleoside analogues that have been studied most extensively as DNMT inhibitors (Table 3). Both have been recently approved by the Food and Drug Administration (FDA) for the treatment of MDS. Decitabine is initially phosphorylated by deoxycytidine kinase, and eventually becomes decitabine triphosphate, which is incorporated into DNA. Azacitidine is phosphorylated and activated by uridine–cytidine kinase and is mainly incorporated into RNA and markedly inhibits protein synthesis [29]. Azacitidine diphosphate is also reduced by ribonucleotide reductase to the corresponding deoxynucleotide diphosphate, decitabine diphosphate, which is further phosphorylated by nucleoside diphosphate kinases to decitabine triphosphate [29]. Incorporation of a high concentration of decitabine triphosphate into DNA can inhibit DNA synthesis [104, 103, 31]. At lower concentrations, decitabine triphosphate covalently binds to DNMT after it has been incorporated into DNA, which eventually causes degradation of DNMT without DNA synthesis arrest [1, 35]. DNA replication in the absence of DNMT leads to hypomethylation induction and gene reactivation [29].

Table 3 Summary of selected clinical trials of azacitidine and decitabine in AML and MDS

Diseases	Dose	Disease (N)	CR (%)	PR (%)	References
Azacitidine	1000–1500 mg/m^2/ course	Relapsed AML (>200)	15–30	2–10	Reviewed in [58]
	Combination with other chemotherapies	Relapsed AML (66)	40	12	Reviewed in [156]
	250 mg/m^2 × 2 days with etoposide and amsacrine	Relapsed AML (17)	39	–	[145]
	75 mg/m^2/d SQ (summary of phase II and III studies)	MDS (270)	6	9	[86]
	350–700 mg/m^2/course followed by butyrate	MDS (16)	19	6	[59]
Decitabine	37–67 mg/kg at 1 mg/kg/h	Pediatric-relapsed AML (6)	33	17	[115]
	90–120 mg/m^2 Q8H for 3 days	Untreated high-risk AML (12)	25	8	[126]
	90 mg/m^2 Q8H for 3 days	Relapsed AML (8)	13	–	[89]

Table 3 (continued)

Diseases	Dose	Disease (N)	CR (%)	PR (%)	References
	125 mg/m^2 Q12H × 6 days with amsacrine or idarubicin	Relapsed AML(30 and 33)	26 and 45	–	[163]
	5–20 mg/m^2 IV × 5–10 days	Relapsed AML (35) and MDS (7)	14 and 29	–	[77]
	45 mg/m^2 daily × 3 days	MDS (66)	20	29	[161]
	15 mg/m^2 Q8H for 3 days versus supportive care	MDS (89 versus 81)	9 versus 0	8 versus 0	[87]
	100 mg/ m^2/course over 5 or 10 days	MDS (96)	34	38 (including hematologic improvement)	[88]

5-Azacytidine (Azacitidine)

During the 1970s and 1980s, azacitidine was investigated as a treatment for various solid tumors [156, 158] and hematologic malignancies [139, 101, 155, 10]. In these early clinical trials, azacitidine was mostly used as a cytotoxic agent. Single-agent azacitidine used at relatively high dose (600–1500 mg/m^2/ course) in refractory AML resulted in about 45% overall responses [58]. Combination therapy with other chemotherapeutic agents resulted in an overall response rate of 30–60% [156, 145]. High-dose azacitidine was toxic, however, and this approach was abandoned.

In parallel, in vitro studies have shown that a lower dose of azacitidine induces cell differentiation and apoptosis by promoting the expression of genes that are silenced by hypermethylation [85]. Low-dose azacitidine has been applied to clinical settings, mostly in MDS. A phase III study where patients with MDS were randomly assigned to receive 75 mg/m^2 a day of azacitidine subcutaneously for 7 days, repeated on a 28-day cycle, or supportive care only [140] yielded a response rate of 16% (CR 6%, PR 10%). This led to approval of azacitidine by the FDA for treatment of MDS.

5-Aza-2′-deoxycytidine (Decitabine)

Decitabine was initially evaluated two decades ago in cancer treatment and has shown significant antitumor activity in hematologic malignancies [4]. In an early study by Momparler et al. [115], the overall response rate of single-agent decitabine (37–81 mg/kg over 40–60 h; the length of infusion was increased stepwise) in relapsed pediatric leukemia was 37%, including a CR of 22%. Petti et al. [126] reported a 25% CR rate in a pilot study of single-agent decitabine (90–120 mg/m^2 every 8 h for 3 days) in the treatment of 12 untreated patients with AML with poor prognosis. In relapsed and refractory adult AML, 90 mg/m^2 of single-agent decitabine every 8 h for 3 days resulted in only one CR among eight patients treated [89]. Combination therapy studies of decitabine against refractory or relapsed AML were performed by the EORTC, with a regimen of 125 mg/m^2 of decitabine every 12 h for 6 days (total 1500 mg/m^2) with amsacrine (120 mg/m^2 on days 6 and 7; in 30 patients), or idarubicin (12 mg/m^2 on days 5–7; in 33 patients) [163]. The CR rates were 26 and 45%, respectively. The median disease-free survival was 8 months.

On the basis of in vitro studies showing that lower-dose decitabine produced more hypomethylation than high doses [85], low doses of decitabine have been evaluated in hematological malignancies [171, 162, 161]. Following promising results in early phase trials, a randomized phase III study was performed in patients with MDS [87], yielding an overall response rate in the decitabine group of 17% including 9% CR. This trial led to the recent approval of decitabine in MDS.

Although low-dose decitabine is clinically active, its optimal dosing schedule is not known. In a phase I trial in relapsed or refractory hematologic malignancies, decitabine activity was found to be significant at low doses, but response were lost with dose escalation, consistent with its mechanism of action [77].

Mechanisms of Response to Hypomethylating Agents

Given the dual nature of this class of agents (cytotoxicity and hypomethylation), the in vivo mechanisms of responses need to be elucidated to guide further trials. Yang et al. examined global DNA methylation changes surrogated by LINE and Alu methylation [167] in the peripheral blood of patients with leukemia treated with decitabine [166]. There was a dose-dependent linear decrease in methylation on day 5 at low doses of 5–20 mg/m^2/day with no significant increase in hypomethylation beyond 20 mg/m^2/day suggesting a plateau effect. In another study where global methylation changes were analyzed during decitabine treatment in patients with MDS [124], LINE hypomethylation was a good pharmacodynamic surrogate of decitabine's hypomethylating activity but was not connected with clinical activity. Thus, while global methylation decreases after decitabine therapy, it is not a good predictor of responses.

Daskalakis et al. showed in patients with MDS that $p15^{INK4B}$ hypomethylation induction several weeks after therapy was associated with a clinical response [37]. However, in other studies [166, 124, 78], immediate induction of $p15^{INK4B}$ hypomethylation (i.e., day 5–10 after treatment) was not correlated with subsequent response.

To determine whether hypomethylation leads to gene induction, $p15^{INK4B}$ expression levels were analyzed in patients with MDS enrolled on a study of decitabine in three dosing schedules [124]. The expression levels of $p15^{INK4B}$ were induced significantly after treatment, and the induction was higher in responders than non-responders. It is thus possible that the key to decitabine responses is not hypomethylation per se, but sustained tumor suppressor gene hypomethylation and activation. This deserves confirmation in larger trials.

HDAC Inhibitors

HDAC inhibitors reverse the deacetylation of histone tails and activate the expression of selected genes [108]. They were originally discovered based on screens for agent that induce cellular differentiation in vitro. Several structural classes of HDAC inhibitors have now been identified, and some have been evaluated in clinical trials.

Vorinostat (suberoylanilide hydroxamic acid, SAHA) is a hydroxamic acid that is highly potent in vitro. In a recent phase I trial of oral vorinostat for hematologic malignancies [56] several responses were observed in AML (9 of 31 patients, 29%), including 1 CR (duration, >6 months), 2 CRs without platelet recovery (duration, 4–6 weeks), 1 PR, and 5 complete marrow responses (blasts less than 5%).

Depsipeptide is a cyclic tetrapeptide that is also a potent HDAC inhibitor in vitro [117]. It has significant clinical activity in cutaneous T-cell lymphoma [127]. In a multicenter phase II trial [121], depsipeptide was administered to 18 patients with refractory or relapsed AML. Two patients had disappearance of bone marrow blasts in the setting of a normocellular marrow, with concomitant recovery of near-normal hematopoiesis following 1–2 cycles of therapy. The responses, however, were short lived.

Valproic acid (VPA) is an antiepileptic agent that has been shown to inhibit HDAC activity at low levels [60, 160, 33]. In a phase I study, high doses of VPA were administered to result in serum concentrations of 50–100 μg/mL in patients with MDS, with, or without ATRA [92]. Responses were observed in 8 (44%) of 18 patients given VPA monotherapy. More limited activity was seen in AML [129]. Thus, HDAC inhibitors form a class of epigenetic acting agents that have promising activity in AML. However, though histone acetylation has been demonstrated in vivo [83], it has been shown no correlation with response, and there remains a distinct possibility that responses to HDAC inhibitors are related to non-histone acetylation or to other mechanisms [83].

Combination Epigenetic Therapy

Elucidation of multiple interacting mechanisms of gene silencing has led to an interest in combining drugs that affect multiple epigenetic pathways. For example, DNA methylation inhibitors and HDAC inhibitors are synergistic in activating gene expression [20] and clinical trials of this approach have started. A combination of decitabine with VPA has shown promising activity in AML, with a response rate of 22% in patients with AML and MDS [55]. There is also interest in developing drugs that affect other epigenetic pathways such as methyl binding proteins, histone methylation, and other histone deacetylases such as SIRT1. It is likely that such inhibitors will enter the clinic in next few years, and it will be interesting to combine them with existing drugs.

Conclusions

It is now clear that epigenetic changes play a significant role in the development and progression of AML. These epigenetic changes can be important targets of treatment, and recent clinical studies have shown the relative safety and efficacy of such epigenetic therapies. Although epigenetic modulation is effective in the treatment of AML, precise in vivo effects of these drugs have not been well described. Also, events downstream of gene expression induction such as apoptosis, senescence, and immunomodulation remain to be clarified. Further basic, translational, and clinical studies are essential to move the field forward.

References

1. Adams RL, Burdon RH. DNA methylation in eukaryotes. *CRC Crit Rev Biochem.* 1982;13(4):349–354.
2. Alland L, Muhle R, Hou H, Jr, et al. Role for N-CoR and histone deacetylase in Sin3-mediated transcriptional repression. *Nature.* 1997;387(6628):49–55.
3. Aoki E, Ohashi H, Uchida T, Murate T, Saito H, Kinoshita T. Expression levels of DNA methyltransferase genes do not correlate with p15INK4B gene methylation in myelodysplastic syndromes. *Leukemia.* 2003;17(9):1903–1904.
4. Aparicio A, Weber JS. Review of the clinical experience with 5-azacytidine and 5-aza-2'-deoxycytidine in solid tumors. *Curr Opin Investig Drugs.* 2002;3(4):627–633.
5. Arnould C, Philippe C, Bourdon V, Gr goire MJ, Berger R, Jonveaux P. The signal transducer and activator of transcription STAT5b gene is a new partner of retinoic acid receptor alpha in acute promyelocytic-like leukaemia. *Hum Mol Genet.* 1999;8(9):1741–1749.
6. Austin GE, Zhao WG, Regmi A, Lu JP, Braun J. Identification of an upstream enhancer containing an AML1 site in the human myeloperoxidase (MPO) gene. *Leuk Res.* 1998;22(11):1037–1048.
7. Bachman KE, Park BH, Rhee I, et al. Histone modifications and silencing prior to DNA methylation of a tumor suppressor gene. *Cancer Cell.* 2003;3(1):89–95.
8. Bannister AJ, Miska EA, Gorlich D, Kouzarides T. Acetylation of importin-alpha nuclear import factors by CBP/p300. *Curr Biol.* 2000;10(8):467–470.
9. Bartel DP. MicroRNAs: genomics, biogenesis, mechanism, and function. *Cell.* 2004;116 (2):281–297.
10. Bellet RE, Mastrangelo MJ, Engstrom PF, Strawitz JG, Weiss AJ, Yarbro JW. Clinical trial with subcutaneously administered 5-azacytidine (NSC-102816). *Cancer Chemother Rep.* 1974;58(2):217–222.
11. Bender CM, Zingg JM, Jones PA. DNA methylation as a target for drug design. *Pharm Res.* 1998;15(2):175–187.
12. Bird A. DNA methylation patterns and epigenetic memory. *Genes Dev.* 2002;16(1):6–21.

13. Bird AP. CpG-rich islands and the function of DNA methylation. *Nature*. 1986;321 (6067):209–213.
14. Blander G, Zalle N, Daniely Y, Taplick J, Gray MD, Oren M. DNA damage-induced translocation of the Werner helicase is regulated by acetylation. *J Biol Chem*. 2002;277 (52):50934–50940.
15. Borrow J, Stanton VP, Jr., Andresen JM, et al. The translocation t(8;16)(p11;p13) of acute myeloid leukaemia fuses a putative acetyltransferase to the CREB-binding protein. *Nat Genet*. 1996;14(1):33–41.
16. Bruniquel D, Schwartz RH. Selective, stable demethylation of the interleukin-2 gene enhances transcription by an active process. *Nat Immunol*. 2003;4(3):235–240.
17. Bullinger L, Dohner K, Bair E, et al. Use of gene-expression profiling to identify prognostic subclasses in adult acute myeloid leukemia. *N Engl J Med*. 2004;350(16): 1605–1616.
18. Caligiuri MA, Strout MP, Lawrence D, et al. Rearrangement of ALL1 (MLL) in acute myeloid leukemia with normal cytogenetics. *Cancer Res*. 1998;58(1):55–59.
19. Caligiuri MA, Strout MP, Oberkircher AR, Yu F, de la Chapelle A, Bloomfield CD. The partial tandem duplication of ALL1 in acute myeloid leukemia with normal cytogenetics or trisomy 11 is restricted to one chromosome. *Proc Natl Acad Sci USA*. 1997;94(8):3899–3902.
20. Cameron EE, Bachman KE, Myohanen S, Herman JG, Baylin SB. Synergy of demethylation and histone deacetylase inhibition in the re-expression of genes silenced in cancer. *Nat Genet*. 1999;21(1):103–107.
21. Chedin F, Lieber MR, Hsieh CL. The DNA methyltransferase-like protein DNMT3L stimulates de novo methylation by Dnmt3a. *Proc Natl Acad Sci USA*. 2002;99(26): 16916–16921.
22. Chen JD, Evans RM. A transcriptional co-repressor that interacts with nuclear hormone receptors. *Nature*. 1995;377(6548):454–457.
23. Chen JD, Umesono K, Evans RM. SMRT isoforms mediate repression and anti-repression of nuclear receptor heterodimers. *Proc Natl Acad Sci USA*. 1996;93(15): 7567–7571.
24. Chen RZ, Pettersson U, Beard C, Jackson-Grusby L, Jaenisch R. DNA hypomethylation leads to elevated mutation rates. *Nature*. 1998;395(6697):89–93.
25. Chen Z, Brand NJ, Chen A, et al. Fusion between a novel Kruppel-like zinc finger gene and the retinoic acid receptor-alpha locus due to a variant t(11;17) translocation associated with acute promyelocytic leukaemia. *EMBO J*. 1993;12(3):1161–1167.
26. Chim CS, Wong AS, Kwong YL. Epigenetic dysregulation of the Jak/STAT pathway by frequent aberrant methylation of SHP1 but not SOCS1 in acute leukaemias. *Ann Hematol*. 2004;83(8):527–532.
27. Chim CS, Wong AS, Kwong YL. Epigenetic inactivation of INK4/CDK/RB cell cycle pathway in acute leukemias. *Ann Hematol*. 2003;82(12):738–742.
28. Christiansen DH, Andersen MK, Pedersen-Bjergaard J. Methylation of p15INK4B is common, is associated with deletion of genes on chromosome arm 7q and predicts a poor prognosis in therapy-related myelodysplasia and acute myeloid leukemia. *Leukemia*. 2003;17(9):1813–1819.
29. Christman JK. 5-Azacytidine and 5-aza-2′-deoxycytidine as inhibitors of DNA methylation: mechanistic studies and their implications for cancer therapy. *Oncogene*. 2002;21 (35):5483–5495.
30. Cihak A, Vesely J. Prolongation of the lag period preceding the enhancement of thymidine and thymidylate kinase activity in regenerating rat liver by 5-azacytidine. *Biochem Pharmacol*. 1972;21(24):3257–3265.
31. Cihak A, Vesely J, Skoda J. Azapyrimidine nucleosides: metabolism and inhibitory mechanisms. *Adv Enzyme Regul*. 1985;24:335–354.
32. Cimino G, Moir DT, Canaani O, et al. Cloning of ALL-1, the locus involved in leukemias with the t(4;11)(q21;q23), t(9;11)(p22;q23), and t(11;19)(q23;p13) chromosome translocations. *Cancer Res*. 1991;51(24):6712–6714.

33. Cinatl J, Jr, Cinatl J, Driever PH, et al. Sodium valproate inhibits in vivo growth of human neuroblastoma cells. *Anticancer Drugs.* 1997;8(10):958–963.
34. Cohen HY, Lavu S, Bitterman KJ, et al. Acetylation of the C terminus of Ku70 by CBP and PCAF controls Bax-mediated apoptosis. *Mol Cell.* 2004;13(5):627–638.
35. Creusot F, Acs G, Christman JK. Inhibition of DNA methyltransferase and induction of Friend erythroleukemia cell differentiation by 5-azacytidine and 5-aza-2'-deoxycytidine. *J Biol Chem.* 1982;257(4):2041–2048.
36. Cross SH, Meehan RR, Nan X, Bird A. A component of the transcriptional repressor MeCP1 shares a motif with DNA methyltransferase and HRX proteins. *Nat Genet.* 1997;16(3):256–259.
37. Daskalakis M, Nguyen TT, Nguyen C, et al. Demethylation of a hypermethylated P15/INK4B gene in patients with myelodysplastic syndrome by 5-Aza-2'-deoxycytidine (decitabine) treatment. *Blood.* 2002;100(8):2957–2964.
38. De Marzo AM, Marchi VL, Yang ES, Veeraswamy R, Lin X, Nelson WG. Abnormal regulation of DNA methyltransferase expression during colorectal carcinogenesis. *Cancer Res.* 1999;59(16):3855–3860.
39. DiMartino JF, Ayton PM, Chen EH, Naftzger CC, Young BD, Cleary ML. The AF10 leucine zipper is required for leukemic transformation of myeloid progenitors by MLL-AF10. *Blood.* 2002;99(10):3780–3785.
40. DiMartino JF, Miller T, Ayton PM, et al. A carboxy-terminal domain of ELL is required and sufficient for immortalization of myeloid progenitors by MLL-ELL. *Blood.* 2000;96(12):3887–3893.
41. Djabali M, Selleri L, Parry P, Bower M, Young BD, Evans GA. A trithorax-like gene is interrupted by chromosome 11q23 translocations in acute leukaemias. *Nat Genet.* 1992;2(2):113–118.
42. Durst KL, Hiebert SW. Role of RUNX family members in transcriptional repression and gene silencing. *Oncogene* 2004;23(24):4220–4224.
43. Eads CA, Danenberg KD, Kawakami K, Saltz LB, Danenberg PV, Laird PW. CpG island hypermethylation in human colorectal tumors is not associated with DNA methyltransferase overexpression. *Cancer Res.* 1999;59(10):2302–2306.
44. Eguchi M, Eguchi-Ishimae M, Greaves M. The small oligomerization domain of gephyrin converts MLL to an oncogene. *Blood.* 2004;103(10):3876–3882.
45. Ekmekci CG, Gutierrez MI, Siraj AK, Ozbek U, Bhatia K. Aberrant methylation of multiple tumor suppressor genes in acute myeloid leukemia. *Am J Hematol.* 2004;77(3):233–240.
46. el-Deiry WS, Nelkin BD, Celano P, et al. High expression of the DNA methyltransferase gene characterizes human neoplastic cells and progression stages of colon cancer. *Proc Natl Acad Sci USA.* 1991;88(8):3470–3474.
47. Ernst P, Wang J, Huang M, Goodman RH, Korsmeyer SJ. MLL and CREB bind cooperatively to the nuclear coactivator CREB-binding protein. *Mol Cell Biol.* 2001;21(7):2249–2258.
48. Esteller M, Guo M, Moreno V, et al. Hypermethylation-associated inactivation of the cellular retinol-binding-protein 1 gene in human cancer. *Cancer Res.* 2002;62(20):5902–5905.
49. Fair K, Anderson M, Bulanova E, Mi H, Tropschug M, Diaz MO. Protein interactions of the MLL PHD fingers modulate MLL target gene regulation in human cells. *Mol Cell Biol.* 2001;21(10):3589–3597.
50. Feinberg AP, Vogelstein B. Hypomethylation of ras oncogenes in primary human cancers. *Biochem Biophys Res Commun.* 1983;111(1):47–54.
51. Felsenfeld G, Groudine M. Controlling the double helix. *Nature.* 2003;421(6921):448–453.
52. Fuino L, Bali P, Wittmann S, et al. Histone deacetylase inhibitor LAQ824 down-regulates Her-2 and sensitizes human breast cancer cells to trastuzumab, taxotere, gemcitabine, and epothilone B. *Mol Cancer Ther.* 2003;2(10):971–984.

53. Fuks F, Burgers WA, Brehm A, Hughes-Davies L, Kouzarides T. DNA methyltransferase Dnmt1 associates with histone deacetylase activity. *Nat Genet*. 2000;24(1):88–91.
54. Galm O, Wilop S, Luders C, et al. Clinical implications of aberrant DNA methylation patterns in acute myelogenous leukemia. *Ann Hematol*. 2005;84 Suppl 13:39–46.
55. Garcia-Manero G, Kantarjian H, Sanchez-Gonzalez B, et al. Final results of a phase I/II study of the combination of the hypomethylating agent 5-aza-2'-deoxycytidine (DAC) and the histone deacetylase inhibitor valproic acid (VPA) in patients with leukemia. *Blood*. 2005;106(11):(abstr 408).
56. Garcia-Manero G, Yang H, Sanchez-Gonzalez B, et al. Final results of a phase I study of the histone deacetylase inhibitor vorinostat (suberoylanilide hydroxamic acid, SAHA), in patients with leukemia and myelodysplastic syndrome. *Blood*. 2005;106(11):(abstr 2801).
57. Girault I, Tozlu S, Lidereau R, Bieche I. Expression analysis of DNA methyltransferases 1, 3A, and 3B in sporadic breast carcinomas. *Clin Cancer Res*. 2003;9(12):4415–4422.
58. Glover AB, Leyland-Jones BR, Chun HG, Davies B, Hoth DF. Azacitidine: 10 years later. *Cancer Treat Rep*. 1987;71(7–8):737–746.
59. Gore S, Baylin SB, Dauses T, et al. Changes in promoter methylation and gene expression in patients with MDS and MDS-AML treated with 5-azacitidine and sodium phenylbutyrate. *Blood*. 2004;104(11):(abstr 469).
60. Gottlicher M, Minucci S, Zhu P, et al. Valproic acid defines a novel class of HDAC inhibitors inducing differentiation of transformed cells. *EMBO J*. 2001;20(24):6969–6978.
61. Grignani F, De Matteis S, Nervi C, et al. Fusion proteins of the retinoic acid receptor-alpha recruit histone deacetylase in promyelocytic leukaemia. *Nature*. 1998;391(6669):815–818.
62. Guo SX, Taki T, Ohnishi H, et al. Hypermethylation of p16 and p15 genes and RB protein expression in acute leukemia. *Leuk Res*. 2000;24(1):39–46.
63. Haggarty SJ, Koeller KM, Wong JC, Grozinger CM, Schreiber SL. Domain-selective small-molecule inhibitor of histone deacetylase 6 (HDAC6)-mediated tubulin deacetylation. *Proc Natl Acad Sci USA*. 2003;100(8):4389–4394.
64. Hanahan D, Weinberg RA. The hallmarks of cancer. *Cell*. 2000;100(1):57–70.
65. He LZ, Guidez F, Triboli C, et al. Distinct interactions of PML-RARalpha and PLZF-RARalpha with co-repressors determine differential responses to RA in APL. *Nat Genet*. 1998;18(2):126–135.
66. Heinzel T, Lavinsky RM, Mullen TM, et al. A complex containing N-CoR, mSin3 and histone deacetylase mediates transcriptional repression. *Nature*. 1997;387(6628):43–48.
67. Herman JG, Baylin SB. Gene silencing in cancer in association with promoter hypermethylation. *N Engl J Med*. 2003;349(21):2042–2054.
68. Herman JG, Civin CI, Issa JP, Collector MI, Sharkis SJ, Baylin SB. Distinct patterns of inactivation of p15INK4B and p16INK4A characterize the major types of hematological malignancies. *Cancer Res*. 1997;57(5):837–841.
69. Herman JG, Latif F, Weng Y, et al. Silencing of the VHL tumor-suppressor gene by DNA methylation in renal carcinoma. *Proc Natl Acad Sci USA*. 1994;91(21): 9700–9704.
70. Hiebert SW, Sun W, Davis JN, et al. The t(12;21) translocation converts AML-1B from an activator to a repressor of transcription. *Mol Cell Biol*. 1996;16(4):1349–1355.
71. Horlein AJ, Naar AM, Heinzel T, et al. Ligand-independent repression by the thyroid hormone receptor mediated by a nuclear receptor co-repressor. *Nature*. 1995;377(6548):397–404.
72. Hubbert C, Guardiola A, Shao R, et al. HDAC6 is a microtubule-associated deacetylase. *Nature*. 2002;417(6887):455–458.
73. Ibanez V, Sharma A, Buonamici S, et al. AML1-ETO decreases ETO-2 (MTG16) interactions with nuclear receptor corepressor, an effect that impairs granulocyte differentiation. *Cancer Res*. 2004;64(13):4547–4554.
74. Ida K, Kitabayashi I, Taki T, et al. Adenoviral E1A-associated protein p300 is involved in acute myeloid leukemia with t(11;22)(q23;q13). *Blood*. 1997;90(12):4699–4704.

75. Issa JP. Aging, DNA methylation and cancer. *Crit Rev Oncol Hematol.* 1999;32(1):31–43.
76. Issa JP. CpG island methylator phenotype in cancer. *Nat Rev Cancer.* 2004;4(12): 988–993.
77. Issa JP, Garcia-Manero G, Giles FJ, et al. Phase 1 study of low-dose prolonged exposure schedules of the hypomethylating agent 5-aza-2′-deoxycytidine (decitabine) in hematopoietic malignancies. *Blood.* 2004;103(5):1635–1640.
78. Issa JP, Gharibyan V, Cortes J, et al. Phase II study of low-dose decitabine in patients with chronic myelogenous leukemia resistant to imatinib mesylate. *J Clin Oncol.* 2005;23(17):3948–3956.
79. Issa JP, Vertino PM, Wu J, et al. Increased cytosine DNA-methyltransferase activity during colon cancer progression. *J Natl Cancer Inst.* 1993;85(15):1235–1240.
80. Iwai M, Kiyoi H, Ozeki K, et al. Expression and methylation status of the FHIT gene in acute myeloid leukemia and myelodysplastic syndrome. *Leukemia.* 2005;19(8):1367–1375.
81. Jenuwein T, Allis CD. Translating the histone code. *Science.* 2001;293(5532):1074–1080.
82. Jin F, Dowdy SC, Xiong Y, Eberhardt NL, Podratz KC, Jiang SW. Up-regulation of DNA methyltransferase 3B expression in endometrial cancers. *Gynecol Oncol.* 2005;96(2): 531–538.
83. Johnstone RW, Licht JD. Histone deacetylase inhibitors in cancer therapy: is transcription the primary target? *Cancer Cell.* 2003;4(1):13–8.
84. Jones PA, Baylin SB. The fundamental role of epigenetic events in cancer. *Nat Rev Genet.* 2002;3(6):415–428.
85. Jones PA, Taylor SM. Cellular differentiation, cytidine analogs and DNA methylation. *Cell.* 1980;20(1):85–93.
86. Kaminskas E, Farrell A, Abraham S, et al. Approval summary: azacitidine for treatment of myelodysplastic syndrome subtypes. *Clin Cancer Res.* 2005;11(10):3604–3608.
87. Kantarjian H, Issa JP, Rosenfeld CS, et al. Decitabine improves patient outcomes in myelodysplastic syndromes: results of a phase III randomized study. *Cancer.* 2006;106(8):1794–1803.
88. Kantarjian H, Oki Y, Garcia-Manero G, et al. Results of a randomized study of three schedules of low-dose decitabine in higher risk myelodysplastic syndrome and chronic myelomonocytic leukemia. *Blood.* 2007;109(1):52–57.
89. Kantarjian HM, O'Brien SM, Estey E, et al. Decitabine studies in chronic and acute myelogenous leukemia. *Leukemia.* 1997;11(suppl. 1):S35–36.
90. Khorasanizadeh S. The nucleosome: from genomic organization to genomic regulation. *Cell.* 2004;116(2):259–272.
91. Kondo Y, Issa JP. Epigenetic changes in colorectal cancer. *Cancer Metastasis Rev.* 2004; 23(1–2):29–39.
92. Kuendgen A, Strupp C, Aivado M, et al. Treatment of myelodysplastic syndromes with valproic acid alone or in combination with all-trans retinoic acid. *Blood.* 2004;104(5): 1266–1269.
93. Kurokawa R, Soderstrom M, Horlein A, et al. Polarity-specific activities of retinoic acid receptors determined by a co-repressor. *Nature.* 1995;377(6548):451–454.
94. Laherty CD, Yang WM, Sun JM, Davie JR, Seto E, Eisenman RN. Histone deacetylases associated with the mSin3 corepressor mediate mad transcriptional repression. *Cell.* 1997;89(3):349–356.
95. Lasa A, Carnicer MJ, Aventin A, et al. MEIS 1 expression is downregulated through promoter hypermethylation in AML1-ETO acute myeloid leukemias. *Leukemia.* 2004;18(7):1231–1237.
96. Lavau C, Du C, Thirman M, Zeleznik-Le N. Chromatin-related properties of CBP fused to MLL generate a myelodysplastic-like syndrome that evolves into myeloid leukemia. *EMBO J.* 2000;19(17):4655–4664.
97. Lavau C, Luo RT, Du C, Thirman MJ. Retrovirus-mediated gene transfer of MLL-ELL transforms primary myeloid progenitors and causes acute myeloid leukemias in mice. *Proc Natl Acad Sci USA.* 2000;97(20):10984–10989.

98. Lavau C, Szilvassy SJ, Slany R, Cleary ML. Immortalization and leukemic transformation of a myelomonocytic precursor by retrovirally transduced HRX-ENL. *EMBO J.* 1997;16(14):4226–4237.

99. Lee PJ, Washer LL, Law DJ, Boland CR, Horon IL, Feinberg AP. Limited up-regulation of DNA methyltransferase in human colon cancer reflecting increased cell proliferation. *Proc Natl Acad Sci USA.* 1996;93(19):10366–10370.

100. Lee TI, Jenner RG, Boyer LA, et al. Control of developmental regulators by Polycomb in human embryonic stem cells. *Cell.* 2006;125(2):301–313.

101. Levi JA, Wiernik PH. A comparative clinical trial of 5-azacytidine and guanazole in previously treated adults with acute nonlymphocytic leukemia. *Cancer.* 1976;38 (1):36–41.

102. Li E. Chromatin modification and epigenetic reprogramming in mammalian development. *Nat Rev Genet.* 2002;3(9):662–673.

103. Li LH, Olin EJ, Buskirk HH, Reineke LM. Cytotoxicity and mode of action of 5-azacytidine on L1210 leukemia. *Cancer Res.* 1970;30(11):2760–2769.

104. Li LH, Olin EJ, Fraser TJ, Bhuyan BK. Phase specificity of 5-azacytidine against mammalian cells in tissue culture. *Cancer Res.* 1970;30(11):2770–2775.

105. Lin RJ, Nagy L, Inoue S, Shao W, Miller WH, Jr, Evans RM. Role of the histone deacetylase complex in acute promyelocytic leukaemia. *Nature.* 1998;391(6669):811–814.

106. Lindemann RK, Gabrielli B, Johnstone RW. Histone-deacetylase inhibitors for the treatment of cancer. *Cell Cycle.* 2004;3(6):779–788.

107. Luo RT, Lavau C, Du C, et al. The elongation domain of ELL is dispensable but its ELL-associated factor 1 interaction domain is essential for MLL-ELL-induced leukemogenesis. *Mol Cell Biol.* 2001;21(16):5678–5687.

108. Marks P, Rifkind RA, Richon VM, Breslow R, Miller T, Kelly WK. Histone deacetylases and cancer: causes and therapies. *Nat Rev Cancer.* 2001;1(3):194–202.

109. Marks PA, Rifkind RA, Richon VM, Breslow R. Inhibitors of histone deacetylase are potentially effective anticancer agents. *Clin Cancer Res.* 2001;7(4):759–60.

110. Melki JR, Vincent PC, Clark SJ. Cancer-specific region of hypermethylation identified within the HIC1 putative tumour suppressor gene in acute myeloid leukaemia. *Leukemia.* 1999;13(6):877–883.

111. Melki JR, Vincent PC, Clark SJ. Concurrent DNA hypermethylation of multiple genes in acute myeloid leukemia. *Cancer Res.* 1999;59(15):3730–3740.

112. Meyers S, Downing JR, Hiebert SW. Identification of AML-1 and the (8;21) translocation protein (AML-1/ETO) as sequence-specific DNA-binding proteins: the runt homology domain is required for DNA binding and protein-protein interactions. *Mol Cell Biol.* 1993;13(10):6336–6345.

113. Milne TA, Briggs SD, Brock HW, et al. MLL targets SET domain methyltransferase activity to Hox gene promoters. *Mol Cell.* 2002;10(5):1107–1117.

114. Mizuno S, Chijiwa T, Okamura T, et al. Expression of DNA methyltransferases DNMT1, 3A, and 3B in normal hematopoiesis and in acute and chronic myelogenous leukemia. *Blood.* 2001;97(5):1172–1179.

115. Momparler RL, Rivard GE, Gyger M. Clinical trial on 5-aza-2'-deoxycytidine in patients with acute leukemia. *Pharmacol Ther.* 1985;30(3):277–286.

116. Morgan HD, Santos F, Green K, Dean W, Reik W. Epigenetic reprogramming in mammals. *Hum Mol Genet.* 2005;14 Spec No 1:R47–58.

117. Nakajima H, Kim YB, Terano H, Yoshida M, Horinouchi S. FR901228, a potent antitumor antibiotic, is a novel histone deacetylase inhibitor. *Exp Cell Res.* 1998;241 (1):126–133.

118. Nakamura T, Mori T, Tada S, et al. ALL-1 is a histone methyltransferase that assembles a supercomplex of proteins involved in transcriptional regulation. *Mol Cell.* 2002;10(5):1119–1128.

119. Nan X, Ng HH, Johnson CA, et al. Transcriptional repression by the methyl-CpG-binding protein MeCP2 involves a histone deacetylase complex. *Nature.* 1998;393 (6683):386–389.

120. Nass SJ, Ferguson AT, El-Ashry D, Nelson WG, Davidson NE. Expression of DNA methyl-transferase (DMT) and the cell cycle in human breast cancer cells. *Oncogene.* 1999;18(52):7453–7461.
121. Odenike OM, Alkan S, Sher D, et al. The histone deacetylase inhibitor depsipeptide has differential activity in specific cytogenetic subsets of acute myeloid leukemia (AML). *Blood.* 2004;104(11):(abstr 264).
122. Okano M, Bell DW, Haber DA, Li E. DNA methyltransferases Dnmt3a and Dnmt3b are essential for de novo methylation and mammalian development. *Cell.* 1999;99(3): 247–257.
123. Okano M, Xie S, Li E. Cloning and characterization of a family of novel mammalian DNA (cytosine-5) methyltransferases. *Nat Genet.* 1998;19(3):219–220.
124. Oki Y, Kantarjian H, Davis J, et al. Hypomethylation induction in MDS after treatment with decitabine at three different doses. *J Clin Oncol.* 2005;23(16S):(abstr 6546).
125. Olesen LH, Aggerholm A, Andersen BL, et al. Molecular typing of adult acute myeloid leukaemia: significance of translocations, tandem duplications, methylation, and selective gene expression profiling. *Br J Haematol.* 2005;131(4):457–467.
126. Petti MC, Mandelli F, Zagonel V, et al. Pilot study of 5-aza-2'-deoxycytidine (Decitabine) in the treatment of poor prognosis acute myelogenous leukemia patients: preliminary results. *Leukemia.* 1993;(7 suppl. 1):36–41.
127. Piekarz RL, Robey R, Sandor V, et al. Inhibitor of histone deacetylation, depsipeptide (FR901228), in the treatment of peripheral and cutaneous T-cell lymphoma: a case report. *Blood.* 2001;98(9):2865–2868.
128. Plass C, Yu F, Yu L, et al. Restriction landmark genome scanning for aberrant methylation in primary refractory and relapsed acute myeloid leukemia; involvement of the WIT-1 gene. *Oncogene.* 1999;18(20):3159–3165.
129. Raffoux E, Chaibi P, Dombret H, Degos L. Valproic acid and all-trans retinoic acid for the treatment of elderly patients with acute myeloid leukemia. *Haematologica.* 2005;90(7):986–988.
130. Redner RL, Rush EA, Faas S, Rudert WA, Corey SJ. The t(5;17) variant of acute promyelocytic leukemia expresses a nucleophosmin-retinoic acid receptor fusion. *Blood.* 1996;87(3):882–886.
131. Richards EJ, Elgin SC. Epigenetic codes for heterochromatin formation and silencing: rounding up the usual suspects. *Cell* 2002;108(4):489–500.
132. Robertson KD. DNA methylation, methyltransferases, and cancer. *Oncogene.* 2001;20 (24):3139–3155.
133. Roman-Gomez J, Jimenez-Velasco A, et al. Promoter hypermethylation of cancer-related genes: a strong independent prognostic factor in acute lymphoblastic leukemia. *Blood.* 2004;104(8):2492–2498.
134. Rountree MR, Bachman KE, Baylin SB. DNMT1 binds HDAC2 and a new co-repressor, DMAP1, to form a complex at replication foci. *Nat Genet.* 2000;25(3): 269–277.
135. Sakai T, Toguchida J, Ohtani N, Yandell DW, Rapaport JM, Dryja TP. Allele-specific hypermethylation of the retinoblastoma tumor-suppressor gene. *Am J Hum Genet.* 1991;48(5):880–888.
136. Scandura JM, Boccuni P, Cammenga J, Nimer SD. Transcription factor fusions in acute leukemia: variations on a theme. *Oncogene.* 2002;21(21):3422–3444.
137. Schreiber SL, Bernstein BE. Signaling network model of chromatin. *Cell.* 2002;111(6): 771–778.
138. Schubeler D, MacAlpine DM, Scalzo D, et al. The histone modification pattern of active genes revealed through genome-wide chromatin analysis of a higher eukaryote. *Genes Dev.* 2004;18(11):1263–1271.
139. Shnider BI, Baig M, Colsky J. A phase I study of 5-azacytidine (NSC-102816). *J Clin Pharmacol.* 1976;16(4):205–212.

140. Silverman LR, Demakos EP, Peterson BL, et al. Randomized controlled trial of azacitidine in patients with the myelodysplastic syndrome: a study of the cancer and leukemia group B. *J Clin Oncol.* 2002;20(10):2429–2440.

141. Slany RK, Lavau C, Cleary ML. The oncogenic capacity of HRX-ENL requires the transcriptional transactivation activity of ENL and the DNA binding motifs of HRX. *Mol Cell Biol.* 1998;18(1):122–129.

142. So CW, Cleary ML. Common mechanism for oncogenic activation of MLL by forkhead family proteins. *Blood.* 2003;101(2):633–6739.

143. So CW, Cleary ML. MLL-AFX requires the transcriptional effector domains of AFX to transform myeloid progenitors and transdominantly interfere with forkhead protein function. *Mol Cell Biol.* 2002;22(18):6542–6552.

144. Speck NA. Core binding factor and its role in normal hematopoietic development. *Curr Opin Hematol.* 2001;8(4):192–196.

145. Steuber CP, Holbrook T, Camitta B, Land VJ, Sexauer C, Krischer J. Toxicity trials of amsacrine (AMSA) and etoposide +/- azacitidine (AZ) in childhood acute non-lymphocytic leukemia (ANLL): a pilot study. *Invest New Drugs.* 1991;9(2):181–184.

146. Strahl BD, Allis CD. The language of covalent histone modifications. *Nature.* 2000;403 (6765):41–45.

147. Suetake I, Shinozaki F, Miyagawa J, Takeshima H, Tajima S. DNMT3L stimulates the DNA methylation activity of Dnmt3a and Dnmt3b through a direct interaction. *J Biol Chem.* 2004;279(26):27816–27823.

148. Szyf M, Bozovic V, Tanigawa G. Growth regulation of mouse DNA methyltransferase gene expression. *J Biol Chem.* 1991;266(16):10027–10030.

149. Tamaru H, Selker EU. A histone H3 methyltransferase controls DNA methylation in Neurospora crassa. *Nature.* 2001;414(6861):277–283.

150. Tkachuk DC, Kohler S, Cleary ML. Involvement of a homolog of Drosophila trithorax by 11q23 chromosomal translocations in acute leukemias. *Cell.* 1992;71(4):691–700.

151. Toyota M, Kopecky KJ, Toyota MO, Jair KW, Willman CL, Issa JP. Methylation profiling in acute myeloid leukemia. *Blood.* 2001;97(9):2823–2829.

152. Turner BM. Memorable transcription. *Nat Cell Biol.* 2003;5(5):390–393.

153. Uchida H, Zhang J, Nimer SD. AML1A and AML1B can transactivate the human IL-3 promoter. *J Immunol.* 1997;158(5):2251–2558.

154. Uchida T, Kinoshita T, Nagai H, et al. Hypermethylation of the p15INK4B gene in myelodysplastic syndromes. *Blood.* 1997;90(4):1403–1409.

155. Vogler WR, Winton EF, Gordon DS, Raney MR, Go B, Meyer L. A randomized comparison of postremission therapy in acute myelogenous leukemia: a Southeastern Cancer Study Group trial. *Blood.* 1984;63(5):1039–1045.

156. Von Hoff DD, Slavik M, Muggia FM. 5-Azacytidine. A new anticancer drug with effectiveness in acute myelogenous leukemia. *Ann Intern Med.* 1976;85(2):237–245.

157. Wang SW, Speck NA. Purification of core-binding factor, a protein that binds the conserved core site in murine leukemia virus enhancers. *Mol Cell Biol.* 1992;12(1): 89–102.

158. Weiss AJ, Metter GE, Nealon TF, et al. Phase II study of 5-azacytidine in solid tumors. *Cancer Treat Rep.* 1977;61(1):55–58.

159. Wells RA, Catzavelos C, Kamel-Reid S. Fusion of retinoic acid receptor alpha to NuMA, the nuclear mitotic apparatus protein, by a variant translocation in acute promyelocytic leukaemia. *Nat Genet.* 1997;17(1):109–113.

160. Werling U, Siehler S, Litfin M, Nau H, Gottlicher M. Induction of differentiation in F9 cells and activation of peroxisome proliferator-activated receptor delta by valproic acid and its teratogenic derivatives. *Mol Pharmacol.* 2001;59(5):1269–1276.

161. Wijermans P, Lubbert M, Verhoef G, et al. Low-dose 5-aza-2'-deoxycytidine, a DNA hypomethylating agent, for the treatment of high-risk myelodysplastic syndrome: a multicenter phase II study in elderly patients. *J Clin Oncol.* 2000;18(5):956–962.

162. Wijermans PW, Krulder JW, Huijgens PC, Neve P. Continuous infusion of low-dose 5-Aza-2'-deoxycytidine in elderly patients with high-risk myelodysplastic syndrome. *Leukemia.* 1997;11(1):1–5.
163. Willemze R, Suciu S, Archimbaud E, et al. A randomized phase II study on the effects of 5-Aza-2'-deoxycytidine combined with either amsacrine or idarubicin in patients with relapsed acute leukemia: an EORTC Leukemia Cooperative Group phase II study (06893). *Leukemia.* 1997;11(Suppl 1):S24–27.
164. Wolf D, Rodova M, Miska EA, Calvet JP, Kouzarides T. Acetylation of beta-catenin by CREB-binding protein (CBP). *J Biol Chem.* 2002;277(28):25562–25567.
165. Xia ZB, Anderson M, Diaz MO, Zeleznik-Le NJ. MLL repression domain interacts with histone deacetylases, the polycomb group proteins HPC2 and BMI-1, and the corepressor C-terminal-binding protein. *Proc Natl Acad Sci USA.* 2003;100(14):8342–8347.
166. Yang AS, Doshi KD, Choi SW, et al. DNA methylation changes after 5-aza-2 deoxycytidine therapy in patients with leukemia. *Cancer Res.* 2006. In press.
167. Yang AS, Estecio MR, Doshi K, Kondo Y, Tajara EH, Issa JP. A simple method for estimating global DNA methylation using bisulfite PCR of repetitive DNA elements. *Nucleic Acids Res.* 2004;32(3):e38.
168. Yokoyama A, Kitabayashi I, Ayton PM, Cleary ML, Ohki M. Leukemia proto-oncoprotein MLL is proteolytically processed into 2 fragments with opposite transcriptional properties. *Blood.* 2002;100(10):3710–3718.
169. Yu L, Liu C, Vandeusen J, et al. Global assessment of promoter methylation in a mouse model of cancer identifies ID4 as a putative tumor-suppressor gene in human leukemia. *Nat Genet.* 2005;37(3):265–274.
170. Yu VC, Delsert C, Andersen B, et al. RXR beta: a coregulator that enhances binding of retinoic acid, thyroid hormone, and vitamin D receptors to their cognate response elements. *Cell.* 1991;67(6):1251–1266.
171. Zagonel V, Lo Re G, Marotta G, et al. 5-Aza-2'-deoxycytidine (Decitabine) induces trilineage response in unfavourable myelodysplastic syndromes. *Leukemia.* 1993;7 (suppl. 1):30–35.
172. Zeleznik-Le NJ, Harden AM, Rowley JD. 11q23 translocations split the "AT-hook" cruciform DNA-binding region and the transcriptional repression domain from the activation domain of the mixed-lineage leukemia (MLL) gene. *Proc Natl Acad Sci USA.* 1994;91(22):10610–10614.
173. Zhang Y, Reinberg D. Transcription regulation by histone methylation: interplay between different covalent modifications of the core histone tails. *Genes Dev.* 2001;15 (18):2343–2360.
174. Ziemin-van der Poel S, McCabe NR, et al. Identification of a gene, MLL, that spans the breakpoint in 11q23 translocations associated with human leukemias. *Proc Natl Acad Sci USA.* 1991;88(23):10735–10739.

Chromosomal Translocations in AML: Detection and Prognostic Significance

Nallasivam Palanisamy

Introduction

Clonal chromosome abnormalities are hallmarks of various cancer types. Non-random chromosome translocations have been identified in hematological malignancies over five decades due to their ability to yield informative metaphases. Among the various chromosome aberrations commonly found in different cancer types including deletions, duplications, and aneuploidy, balanced reciprocal translocations have been identified with remarkable specificity in hematological malignancies and soft tissue sarcomas. Recurrent chromosome aberrations are used as markers for diagnosis, prognosis, and treatment follow-up. The fusion and deregulated genes cloned from the site of translocation breakpoints are implicated in tumorigenesis. It has been well established that common molecular consequences of non-random reciprocal translocations result in the formation of a fusion gene from the breakpoints in the introns of two different genes on the same or different chromosome. Most of the fusion genes described in hematological malignancies are transcription factor genes and tyrosine kinases, conferring proliferative advantage to the leukemic clone.

Using conventional karyotypic analysis on hematological malignancies more than 275 genes involved in chromosome rearrangements have been identified [32]. The specificity of chromosome translocations within a histological type led to sub-classification based on chromosome aberrations. To date, about 500 such tumor-specific recurrent translocations have been identified. Cytogenetically, acute myelogenous leukemia remains the most extensively studied leukemia. While a complete analysis of all the known reciprocal translocation and the fusion genes is beyond the scope of the present chapter, a brief review of the most common chromosomal aberrations, the method for diagnosis, and future prospects are discussed here.

N. Palanisamy (✉)
Department of Pathology, Michigan Center for Translational Pathology,
University of Michigan, Ann Arbor, MI 48105, USA
e-mail: nallasiv@med.umich.edu

L. Nagarajan (ed.), *Acute Myelogenous Leukemia*,
Cancer Treatment and Research 145, DOI 10.1007/978-0-387-69259-3_3,
© Springer Science+Business Media, LLC 2010

Acute Myeloid Leukemia (AML)

As discussed extensively in other parts of this monograph, AML is characterized by the accumulation of immature bone marrow precursor in the marrow and peripheral blood. G-band karyotype analysis reveals clonal chromosome aberrations in more than 75% of patients [11,35]. Characteristic chromosome abnormalities include t (8; 21), inv (16) or t (16; 16), t (15; 17), and promiscuous translocations involving MLL gene at 11q23. The fusion genes identified at the translocation breakpoints are the contributing factors for the development of leukemia. Approximately 200 different karyotypic changes have been found to be recurring changes in AML [34]. In contrast to readily identifiable translocations, karyotypically normal patients may possess subtle aberrations not detected by conventional cytogenetics [11, 19, 26,]. For example, additional cryptic abnormalities have been reported in AML with isolated trisomy 8 using high-resolution array comparative genomic hybridization [39] emphasizing the limitations of conventional G-band karyotypes.

Classification of AML

Conventional classification of AML has been based on morphological characteristics and extent of cell maturation based on French-American-British (FAB) classification system. The current model of the FAB consists of eight major groups (M0–M7) classified based on predominant differentiation pathway and the degree of maturation with additional cytochemical criteria (Table 1). Subsequently, World Health Organization (WHO) classification of hematological disorders used clinical data and biologic characteristics, such as morphology, cytogenetics, molecular genetics, and immunological markers and in particular cytogenetics, and classified AML into four main groups which include AML with recurrent cytogenetic translocations; AML with multilineage dysplasia; AML with myelodysplastic syndrome, therapy related; and AML not otherwise categorized. The first group included patients with balanced recurrent translocations. The most frequent abnormalities in this group include AML with t(8;21) (q22;q22) AML1/CBFalpha/ETO, acute promyelocytic leukemia: AML with t(15;17)(q22;q12) and variant PML/RARalpha, AML with abnormal bone marrow eosinophils inv(16)(p13;q22)/t(16;16)(p13;q22) CBFbeta/MYH1, and AML with 11q23 MLL abnormalities. The second group defined by the presence of multilineage dysplasia syndrome (MDS) which typically affects adults. It is characterized by gain or loss of major segments of chromosomes: $-5/-5q$, $-7/$ del(7q), $+8$, $+9$, $+11$, del(11q), del(12p), del(17p), -18, $+19$, del(20q), $+21$. Rare translocations include t(1;7) and t(2;11) and aberrations of 3q21 and 3q26. The third group encompasses chromosomal aberrations associated with exposure to mutagens or chemo/radiotherapy. The fourth group consists of patients diagnosed as AML not otherwise categorized and did not satisfy the previous three categories.

Table 1 FAB classification of acute myeloid leukemia

FAB subgroup	Type of leukemia	Prognosis	Frequency in AML (5%)	Morphology
M0	Acute myeloblastic leukemia without maturation	Worse	<5	Immature myeloblasts, lacks definite myeloid differentiation by conventional morphologic or cytochemical analyses; myeloid differentiation evidenced by ultrastructural demonstration of peroxidase-positive granules and/or immunoreactivity
M1	Acute myeloblastic leukemia with minimal maturation	Average	20	Immature myeloblasts predominate; <10% promyelocytes, myelocytes, or monocytes; Auer rods may be present
M2	Acute myeloblastic leukemia with maturation	Better	30	Immature myeloblasts predominate, but more maturation than in M1 (>10% promyelocytes/ myelocytes); <20% monocytic cells; Auer rods may be present ; most cells peroxisdase positive
M3 and M3v	Acute promyelocytic leukemia	Best	10	Promyelocytes predominate; marked granulation in more than 30% cells; often bundles of Auer rods; granules not visible by light microscopy in M3v
M4 and M4EO	Acute myelomonocytic leukemia	Average/ better	25	Mixture of abnormal monocytoid cells (>20%) and myeloblasts/ promyelocytes (>20%); 1–30% eosinophilic cells in M4EO
M5 a and b	Acute monocytic leukemia	Average	10	Monocytoid cells predominate (>80%); in M5a, >80% nonerythroid cells are immature monoblasts; in M5b, >20% are more mature (monocytes)

Table 1 (continued)

FAB subgroup	Type of leukemia	Prognosis	Frequency in AML (5%)	Morphology
M6	Acute erythroleukemia	Worse	<5	Myeloblasts and erythroblasts (>50%) predominate; abnormal multinucleated erythroblasts containing PAS-positive blocks
M7	Acute megakaryoblastic leukemia	Worse	<5	Megakaryocytic cells as shown by platelet peroxidase activity on electron microscopy or by tests with platelet-specific antibodies; often myelofibrosis and increased bone marrow reticulin
Others	Undifferentiated acute leukemia, mixed-lineage leukemia, hypocellular AML	Not known	–	

In addition, two other systems, based on cytogenetic profiles, are created for accurate assessment of prognoses by typing favorable and unfavorable genetic alterations into prediction classes to facilitate optimal therapy selection. These were the UK MRC (United Kingdom Medical Research Council) and SWOG (Southwest Oncology Group). They define three and four categories, respectively. In MRC, favorable prognoses include presence of t(8;21), inv 16, and t(15;17), irrespective of additional abnormalities [54]. The prognoses are adverse if the karyotype include at least five unrelated abnormalities (complex karyotype) or one of the following abnormalities, −5, del(5q), −7, abnormalities 3q.

In SWOG, favorable abnormalities included t(8;21), inv 16, and t(15;17). However, the prognosis of t(8;21) is circumstantial as it is modulated by the presence of other aberrations. Intermediate prognostic markers involve detection of a normal karyotype or presence of +6, +8, −Y or 12p abnormalities. Adverse cytogenetic markers include detection of a complex karyotype, or possession of −5/del(5q), −7/del(7q), abnormalities of 3q, 9q, 11q, 20q, 21q, or 17p, t(6;9) or t(9;22). Despite the various classification methods, cytogenetic markers play an important role in determining the prognosis for clinical remission, overall survival, and disease-free survival. Among the patient groups with favorable and unfavorable cytogenetic markers, which determine the duration of remission before relapse, identification of additional markers at the genomic and expression level in both groups may help to better understand the nature of different course of the disease.

Cytogenetic Analysis in AML

In the 1970s reliable identification of normal and abnormal human chromosomes began with the invention of the banding method [10]. Further techniques like C-banding [1] or silver staining of the nucleolus organizing regions [18] and GTG banding [47] were introduced in the following years. At the same time, an alternative method utilizing in situ hybridization was introduced [16]. Further improvements, with the introduction of fluorescent probes in 1986 [23], permitted the analysis of several genomic loci simultaneously using differentially labeled probes. Conventional cytogenetic banding methods are laborious and time consuming for detecting translocations under routine clinical settings. Molecular methods such as Southern blotting, polymerase chain reaction (PCR), and real time-polymerase chain reaction (RT-PCR) can be used to detect gene rearrangements and fusion transcripts, but these methods have sensitivity and specificity issues. Southern blotting requires the presence of at least 10% of tumor cells. Among the various molecular methods, fluorescence in situ hybridization (FISH) emerged as an adjunct method for precision identification of translocations at interphase level. FISH method is relatively simple and large number of interphase cells can be scored in a single experiment. Highly sensitive probes using dual fusion approach have been described for translocations in hematological malignancies [15]. Identification of specific type of translocation at the time of diagnosis will be useful for the clinicians to select appropriate treatment options for better management of patients. Limitations in resolution of conventional cytogenetic methods such as G-band karyotype analysis prevent the visualization of many subtle chromosomal aberrations in many cancer types. To improve microscopic visualization, highly sensitive and specific two color FISH probes are developed for reliable detection of translocation and associated abnormalities at the interphase level. FISH method eliminates the need to collect intact fresh specimens, which is necessary to preserve RNA to perform RT-PCR. RT-PCR requires designing primers from the exact sequence region to get reliable results. Any deviation from the expected fragment size amplification need further sequencing and other molecular characterization, which is time consuming and laborious. Complex rearrangements other than the known breakpoints, and additional deletions, as described for t(9;22) in chronic myelogenous leukemia [14] lead to false-negative results by RT-PCR and genomic PCR-based assays.

FISH is instrumental in the identification of new chromosomal translocations and cryptic aberrations as well as identification of marker chromosomes. Recent improvements in genome analysis, such as completion of the Human Genome Project, expanded the possibilities to develop probes for any given genomic location in the human genome. Better labeling and imaging systems have also increased the efficacy of the method but maintained its cost-effectiveness. The utility of FISH further increased with the development of different types of probes suitable to detect specific type of aberrations (Table 2). With the

Table 2 Different types of FISH probes used to detect various structural and numerical chromosome aberrations

Type	Application
Centromere probes	Enumeration of chromosomal copy number
Chromosome paint probes	Characterization of marker chromosomes, chromosome translocation detection
Micro-deletion unique sequence probes	Identification of small submicroscopic deletions in interphase and metaphase
Amplification detection Probes	Detection of amplified segments of chromosomes and specific gene amplifications
Telomerere-specific probes	Detection of small terminal rearrangements near chromosome telomeres
Translocation junction unique sequence probes	For detecting the presence of chromosome translocations using metaphase and interphase cells

availability of prior knowledge for any aberrations, FISH probes can be designed to detect a specific chromosome aberrations associated with cryptic changes as found in AML M4E0 type and other leukemia subtypes where inversion, translocation, and cryptic deletions are known to occur at the translocation breakpoints (Fig. 1). These sub-microscopic deletions are detected only after the application of fluorescence in situ hybridization (FISH) probes for translocation detection. Among the various sub-microscopic deletions reported in leukemias, deletion of the 5′ region of the ABL gene and the 3′ region of BCR in chronic myeloid leukemia (CML) and acute lymphoblastic leukemia (ALL), as well as the 5′ region of ETO in acute myeloid leukemia (AML) French-American-British type M2 associated with t(8;21), 3′MLL in AML and ALL, and 3′ of CBFB in AML associated with inv(16) (Fig. 1). While it has been widely reported that submicroscopic deletions of the derivative 9 in CML have an adverse prognostic impact, the clinical significance, if any, of deletions associated with t(8;21), inv(16)/ t(16;16), or MLL rearrangement is yet to be determined. Analysis of 39 patients diagnosed with AML who had cytogenetically detectable inv(16)/t(16;16) by using a FISH probe for the CBFB region detected three patients with deletions in CBFB region on 16p13 (8%), all associated with inv(16), bringing the number of cases reported so far to seven. The prognostic significance of this finding remains unclear.

G-band karyotyping and FISH analysis unraveled the complex chromosome abnormalities in AML, which lead to stratification of patients into clinical subgroups. Cytogenetic aberrations serve as an important markers for diagnosis, follow-up, and assess minimal residual disease, nevertheless, search for additional recurrent submicroscopic changes including deletions, duplications, and small inversions need to be conducted using high-resolution methods. Due to the poor resolution of the G-band karyotype, SKY, and requirement of prior knowledge about the aberrations for FISH analysis many aberrations go undetected by G-band karyotyping, spectral karyotyping, and low-resolution comparative genomic hybridization (CGH) but are detectable using high-density

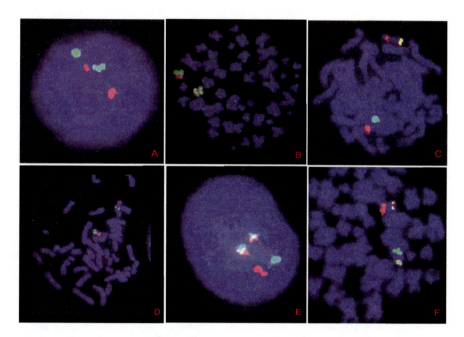

Fig. 1 Dual color and dual fusion FISH probe analysis of inversion, translocation, and deletion in inv(16)(p13q22) (Cancer Genetics, Inc. New Jersey, USA). (**A, D**) Normal interphase and metaphase cells showing two *green* (CBFB) and two *red* signals (MYH11) on 16q22 and 16p13, respectively. (**B**) and (**E**) Metaphase and interphase cells showing inversion, (**C**) metaphase cell showing deletion of CBFB probe region on 16p13, and (**F**) metaphase cell showing translocation

array CGH. Identification of such submicroscopic changes would help better understand the complex molecular architecture of AML genome for further identification of patient sub-groups to monitor prognosis.

Recurrent Chromosomal Aberrations in AML

Many AML subtypes studied with sufficient number have shown recurrent chromosome abnormalities. Cytogenetically unrelated aberrant clones are detected in only 1% of the cases. This implies that the disease phenotype in AML is brought on by the clonal expansion of a single renegade cell. Even though the karyotype differs from case to case, there is strong evidence that the total distribution of changes is highly non-random and that cellular rearrangements are of fundamental importance in leukemogenesis. A brief description of several newly identified translocations in addition to those described earlier [22] is given in the following section.

Chromosome 1

t(1;3)(p36;q21)

t(1;3)(p36;q21) may be a variant rearrangement of inv(3)(q21q26) and t(3;3)(q21;q26) due to the shared involvement of 3q21, the presence of dysmegakaryocytopoietic features, and the lack of FAB preference [5]. In most cases, t(1;3) is found in adults, some of which present with myelodysplastic syndromes (MDS) [33].

t(1;7)(q10;p10)

Many of patients with t(1;7) presented with secondary MDS or therapy-related relapse. About half of these presented it as the sole abnormality. Most t(1;7) cases are classified under AML M4. Aberrations involving t(1;7) are usually unbalanced, leading to the trisomy 1q and monosomy 7q and 7q− [60]. Poor clinical outcome in patients with MDS/AML with der(1;7) is shown to be associated with many risk factors [24].

t(1;11)(p32;q23) and t(1;11)(q21;q23)

In this set of translocations, the MLL gene (11q23) is fused to AF1p (1p32) [4] and AF1Q (1p21) [8]. The AF1q messenger RNA (mRNA) is highly expressed in the thymus but not in peripheral lymphoid tissues. In contrast to its restricted distribution in normal hematopoietic tissue, AF1q was expressed in all leukemic cell lines tested [57].

t(1;22)(p13;q13)

t(1;22) is more common in children, particularly in AML M7 cases than in adults and found as the only cytogenetic abnormality in the majority of cases. Patients exhibit extensive infiltration of abdominal organs by leukemic cells resulting in hepatosplenomegaly [12, 36]. They are also thrombocytopenic and have prominent bone marrow fibrosis. It is associated with a poor prognosis.

Chromosome 3

inv(3)(q21q26), ins(3;3)(q26;q21q26), t(3;3)(q21;q26)

A rare abnormality detected in about 2% of AML or MDS with inv(3)(q21q26) and t(3:3)(q21;q26) resulting in the activation of EVI1 (ecotropic viral integration site 1 isoform a) gene [53].

t(3;5)(q21-25;q31-35)

A novel fusion gene involving genes NPM1 nucleophosmin (nucleolar phos-phoprotein B23, numatrin) and MLF1 (myeloid leukemia factor 1) has been identified in three AML cases with t(3;5) [61].

t(3;8)(q26;q24)

Recently identified recurrent translocation in five cases with therapy-related AML or MDS. Associated abnormalities includes trisomy 13 (one case) and monosomy 7 (two cases) [29].

t(3;21)(q26;q22)

This aberration, like BCR (breakpoint cluster region)-ABL (c-abl oncogene 1, receptor tyrosine kinase), is more frequently found in CML (chronic myelo-genous leukemia) than in AML. In AML, these changes are frequently second-ary aberrations related to genotoxic exposure, especially the use of topoisome-rase II but are rarely found [45].

Chromosome 6p

t(6;9)(p23;q34)

Patients with the t(6;9) abnormality usually have an increased proportion of basophilic cells in the bone marrow, an otherwise rare finding in AML. This aberration is found in 2% of AML cases with abnormal karyotype. t(6;9) is often found as a single aberration in many cases examined. It is not limited to any FAB group but may be found more frequently in M2 and M4. It also seems that this anomaly is commonly found in younger patients. The breakpoint specificity appears to be high. On chromosome 9, the breakpoint is clustered within the intron of the CAN gene [51], Soekarman et al. 1992). Similarly, the partner 6p23 is found in the intron of the DEK (DEK oncogene) gene. How-ever, the function of the DEK-CAN product is not known and appears to have a nuclear localization. The t(6;9) appears to have a bad prognosis [17, 58].

t(6;11)(q27;q23)

The t(6;11)(q27;23) is one of the most common translocations observed in patients with acute myeloid leukemia (AML). The translocation breakpoint involves the MLL gene, which is the human homolog of the *Drosophila* trithorax gene at 11q23 and the AF6 gene at 6q27 [55]. As with most other MLL aberrations, prognosis is generally bad [54].

Chromosome 7

t(7;11)(p15;p15)

This recurrent anomaly is found mainly in oriental Asians [27] and is frequently the sole aberration associated with AML M2 and M4. Translocation involves HOXA9 (Homeobox A9) and NUP98 (nucleoporin 98 kDa) genes on chromosomes 7 and 11, respectively, and the fusion protein may promote leukemogenesis by inhibiting the function of HOXA9 in nucleocytoplasmic transport [6].

t(7;21)(p22;q22)

Most recently described t(7;21)(p22;q22) is a cryptic translocation not detected by karyotype analysis due to the location of the breakpoints at the telomeric ends on chromosomes 7p and 22q. By a combination of molecular methods, this translocation was identified to involve RUNX1 (runt-related transcription factor 1 isoform) and USP42 (Homo sapiens mRNA for ubiquitin-specific protease 42 (USP42 gene) gene. The role of this fusion gene in ubiquitin pathway may be a pathogenic factor in the development of AML [38].

Chromosome 8

t(8;9)(p21-23;p23-24)

A novel t(8;9)(p21-23;p23-24) involving PCM1 (pericentriolar material 1) and JAK2 (Janus kinase 2) genes has been identified as recurrent translocation in individuals presented many types of hematological malignancies with varying clinical outcome [43].

t(8;16)(p11;p13)

The recurrent translocation t(8;16)(p11;p13) is a cytogenetic hallmark for the M4/M5 subtype of acute myeloid leukemia involving fusion of MOZ (MYST histone acetyltransferase MYST3) gene at 8p11 with CBP (CREB-binding protein isoform b-CREBBP) at 16p13 [44].

t(8;21)(q22;q22)

t(8;21)(q22;q22) is one of the best known and most common recurrent chromosomal aberrations in AML and is strongly associated with childhood leukemia, particularly in AML M2, with well-defined and specific morphological features. The translocation t(8;21)(q22;q22) involves the AML1 (Homo sapiens AML1 mRNA for hypothetical protein) (21q22) and ETO (runt-related transcription factor 1; translocated to, 1) (8q22) genes. Occurrence is between 5 and

12% of AML and a third of FAB M2 cases although presentation in some M1 and M4 cases has also been reported.

t(8;21) is usually associated with a good response to chemotherapy and a high remission rate with long-term disease-free survival [20]. It is noteworthy that 50% of patients positive for t(8;21) eventually suffer disease relapse [28, 43]. Whether this outcome is determined by the presence of additional genetic alterations is unclear. Recent work has suggested that KIT mutations in codon D816, present in 20% of t(8;21) patients, are strongly associated with poor prognosis in pediatric t(8;21) AML [43, 48]. The KIT-D816 mutations confer a poor prognosis to AML1-ETO-positive AML and should therefore be included in the diagnostic workup. Patients with KIT-D816-positive/AML1-ETO-positive AML might benefit from early intensification of treatment or combination of conventional chemotherapy with KIT PTK inhibitors [47]. A large number of patients also demonstrate additional chromosome abnormalities: loss of sex chromosome and del(9)(q22) but no adverse outcome have been noted for either additional abnormality.

Chromosome 9

t(9;22)(q34;q11)

t(9;22)(q34;q11) is found in 3% of all AML patients with abnormal karyotype and in 15% AML patients overall [40]. Unlike CML, it is known to occur during remission of disease and is associated with AML M1 and M2 [21].

Chromosomes 10 and 11

Chromosomal aberrations involving chromosome 11 usually results in a poor prognosis. This consistency distinguishes it, aside from being one of the largest and more significant clusters of chromosomal aberrations in AML. Chromosome 11 changes have no FAB preference, and many involve alterations in the MLL gene (11q23). The MLL gene on 11q23 is involved in a number of translocations with different partner chromosomes. The most common translocations observed in childhood AML are the t(9;11)(p21;q23) and the t(11;19)(q23;p13.1); other translocations of 11q23 involve at least 50 different partners chromosomes. A partial tandem duplication of MLL gene has also been reported in the majority of adult patients whose leukemic blast cells have a +11 and in some with normal karyotype. Molecular studies have shown that MLL is rearranged more frequently than is revealed by conventional cytogenetic studies [30]. Changes in chromosome 10 are included in this group as well as so far, recurrent changes in chromosome 10 described in the literature are usually in association with chromosome 11.

t(11;17)(q23;q21) and t(11;17)(q23;q25)

In the translocation t(11;17)(q23;q21), a fusion product involving promyelocytic leukemia zinc finger (PLZF) gene and RARalpha on 17q21 have been reported. RARalpha gene is involved in both t(15;17) and t(11;17) suggests the importance of the modified RARalpha in AML but the function of this particular fusion product remains unknown [50]. Kang et al. [25] described two additional cases of AML, one with t(11;17(q23;q21) and second with t(11;17)(q23;q25) without the involvement of RARalpha at 17q.

t(11;19)(q23;p13)

There are two related translocations involving t(11;19) producing two different fusion products. In the first, t(11;19)(q23;p13.3), the MLL gene is fused to the MLLT1 (myeloid/lymphoid or mixed-lineage leukemia) gene. In the other, t(11;19)(q23;p13.1), the partner is ELL (eleven–nineteen lysine-rich leukemia gene) [56]. This product is mostly associated with M4, M5, and occasionally M1, M2.

t(11;20)(p15;q11)

Recurrent t(11;20)(p15;q11) reported in one case of polycythemia vera and a few cases of de novo AML M2 and therapy-related myelodysplastic syndrome (t-MDS) [41].

Chromosome 15

t(15;17)(q22;q21)

t(15;17)(q22;q21) is commonly associated with AML M3 and M3v ("hypergranular promyelocytic leukemia"). Its detection is implicative of a good clinical outcome. The prognostic value of M3 AML/t(15;17) is inferior to t(8;21) and inv(16) and superior to the poor prognostic group (AML with abnormalities of the chromosomes 5 and 7). AML M3 patients are, however, increasingly treated in independent protocols, rendering such comparison difficult. The sensitivity of M3 cells to all-trans retinoic acid led to the discovery that the retinoic acid receptor alpha (RARalpha) gene on 17q21 fuses with a zinc finger binding transcription factor on 15q22 (promyelocytic leukemia or PML) gene, thus giving rise to a PML-RARalpha fusion gene product. Rare cases lacking the classical t(15;17) have been described either having complex variant translocations involving both chromosomes 15 and 17 with additional chromosome(s), expressing in all studied cases, the PML/RARalpha transcript, or cases where neither chromosome 15 nor chromosome 17 are apparently involved, but with submicroscopic insertion of RARalpha into PML leading to expression of the PML/RARalpha transcript; these latter cases are considered as cryptic or masked t(15;17). Morphological analysis showed no major difference between the t(15;17) positive control group and the PML/RARalpha positive patients without t(15;17).

Chromosome 16

inv/del(16)(p13q22)/del(16)(q22)/t(16;16)(p13;q22)

inv(16)(p13q22) patients are usually grouped in AML M4 or M4EO and is characterized by the presence of an abnormal eosinophilic component. Occasionally, this abnormality has been seen in other myeloid malignancies, including AML M2, M4 without eosinophilia, M5, and MDS. A convergent study has revealed that patients with M4 AML with inv(16) and t(16;16) achieved higher complete remission (CR) rates. Conversely, del(16q) does not have a better outcome than other M4 AML or MDS. inv(16) and t(16;16) both result in the fusion of the CBFbeta gene at 16q22 to the smooth muscle myosin heavy chain (MYH11) at 16p13 [13]. CBFbeta/MYH11 is usually demonstrated by molecular studies. Thus, at diagnosis, the use of FISH and RT-PCR methods are important when evaluating inv(16) rather than G-band karyotyping. It is also noteworthy that CML patients with t(16;16)(p13;q22 and inv(16)(p13;q22) suggest an opposite prognosis, possessing both prominent extramedullary disease and poor response to treatment [37]. This may underlie significant mechanistic differentiation between chronic and acute leukemic progression.

Chromosome 17

i(17)(q10)

Among the various types of isochromosomes, i(17q) is the most frequently observed isochromosome in different cancer types including acute myeloid leukemia and MDS with varying frequency next to i(11q) [31]. There is no obvious FAB preference. Patients with this anomaly are predicted to have a poor outcome, characterized by rapid progression to AML, poor response to chemotherapy, and short survival after transformation [2].

Clinical Relevance of Chromosome Abnormalities in AML

Clinically the recurrent chromosomal aberrations serve as markers for diagnosis and patient management. The consistency of these translocations with specific disease type and presence of complex structural abnormalities within a disease type presents as a valuable tool for differential diagnosis and identification of clinical subsets with distinct clinical behavior. For example, recent studies show that KIT mutations are strongly associated with a poor prognosis in pediatric t(8;21) AML [49] despite t(8;21) being typically indicative of good clinical outcome. Here, the inclusion of molecular genetic data from mutation studies is important in disease analysis.

In many instances described here, physical visualization is insufficient to detect the presence of chromosomal aberrations. Typical aberrations such as inv(16) and t(16;16) which result in the formation of CBFB/MYH11 can also involve small deletions of chromosomal fragments. These are usually picked up by molecular methods such as FISH. Without these molecular methods, the invisibility of these aberrations may have led to inappropriate treatment and reduced treatment efficacy.

Within a few decades, a great number of cytogenetic aberrations in AML have been discovered and characterized. Yet, what is known now is that only some of the common abnormalities with diagnostic and clinical relevance. Many other less common aberrations are poorly characterized, and their involvement in leukemogenesis, or their diagnostic/clinical value is unknown. Furthermore, new recurrent aberrations are constantly added in the literature. There is a need to continue correlating recurrent chromosome aberrations with response rates, response duration, survival, and cure in AML patients treated with current and novel induction and post-induction regimens [4]. Certainly, routine use of cytogenetic testing prior to induction therapy is critical to stratify patients [59].

In recent years, new platforms such as end sequence profiling [42], array CGH [9] and oligonucleotide microarray [7] allow to detect micro-deletions, duplications amplification, and rearrangements associated with copy number changes not detectable physically through the microscope to be discovered and analyzed. The combined input from these platforms can potentially provide added support to cytogenetic data in revealing prognoses and guiding treatment regimes and perhaps lend further insight into the mechanistic roles of the translocations as well.

Conclusions/Future Directions

In the banding era, the number of recurrent chromosome translocations identified in AML were rearrangements and translocations involving large chromosomal segments. Discovery of new translocations involving breakpoints at the terminal regions of chromosomes became possible with the advent of fluorescence in situ hybridization technique. Furthermore, FISH resolved several submicroscopic deletions and inversions associated with previously identified translocations. Finer delineation of subtle alterations is now feasible with high-density array comparative genomic hybridization. Thus, we are entering an exciting era when a vast number of recurrent chromosomal alterations that can be used for better clinical stratification, disease management, and prognosis will be available for all AML patients.

Acknowledgments I thank Cancer Genetics Inc, New Jersey, USA for the inv (16) probe FISH images. This work was supported by grants from National Institutes of Health, NIH SBIR phase I grant 1R43CA091532-01 and Agency for Science Technology and Research (A-STAR) Singapore. Wilson GOH Wen Bin, for his assistance in the preparation of this article

References

1. Arrighi FE, Hsu TC. Localization of heterochromatin in human chromosomes. *Cytogenetics*. 1971;10:81–86.
2. Becher R, Carbonell F, Bartram CR. Isochromosome 17q in Ph1-negative leukemia: a clinical, cytogenetic, and molecular study. *Blood*. 1990;75:1679–1683.
3. Berger R, Bernheim A, Daniel MT, Flandrin G. t(15;17) in a promyelocytic form of chronic myeloid leukemia blastic crisis. *Cancer Genet Cytogenet*. 1983;8:149–152.
4. Bernard OA, Mauchauffe M, Mecucci C, Van den Berghe H, Berger R. A novel gene, AF-1p, fused to HRX in t(1;11)(p32;q23), is not related to AF-4, AF-9 nor ENL. *Oncogene*. 1994;9:1039–1045.
5. Bloomfield CD, Garson OM, Volin L, Knuutila S, de la Chapelle A. t(1;3)(p36;q21) in acute nonlymphocytic leukemia: a new cytogenetic-clinicopathologic association. *Blood*. 1985;66:1409–1413.
6. Borrow J, Shearman AM, Stanton VP, Jr, et al. The t(7;11)(p15;p15) translocation in acute myeloid leukaemia fuses the genes for nucleoporin NUP98 and class I homeoprotein HOXA9. *Nat Genet*. 1996;12:159–167.
7. Brennan C, Zhang Y, Leo C, et al. High-resolution global profiling of genomic alterations with long oligonucleotide microarray. *Cancer Res*. 2004;64:4744–4748.
8. Busson-Le Coniat M, Salomon-Nguyen F, Hillion J, Bernard OA, Berger R. MLL-AF1q fusion resulting from t(1;11) in acute leukemia. *Leukemia*. 1999;13:302–306.
9. Casas S, Aventín A, Fuentes F, et al. Genetic diagnosis by comparative genomic hybridization in adult de novo acute myelocytic leukemia. *Cancer Genet Cytogenet*. 2004; 153:16–25.
10. Caspersson T, Zech L, Johansson C. Analysis of human metaphase chromosome set by aid of DNA-binding fluorescent agents. *Exp Cell Res*. 1970;62:490–492.
11. Caligiuri MA, Strout MP, Lawrence D, et al. Rearrangement of ALL1 (MLL) in acute myeloid leukemia with normal cytogenetics. *Cancer Res*. 1998;58:55–59.
12. Chan WC, Carroll A, Alvarado CS, et al. Acute megakaryoblastic leukemia in infants with t(1;22)(p13;q13) abnormality. *Am J Clin Pathol*. 1992;98:214–221.
13. Claxton DF, Liu P, Hsu HB, et al. Detection of fusion transcripts generated by the inversion 16 chromosome in acute myelogenous leukemia. *Blood*. 1994;83:1750–1756.
14. de la Fuente J, Merx K, Steer EJ, et al. ABL-BCR expression does not correlate with deletions on the derivative chromosome 9 or survival in chronic myeloid leukemia. *Blood*. 2001;98:2879–2880.
15. Dewald GW, Wyatt WA, Juneau AL, et al. Highly sensitive fluorescence in situ hybridization method to detect double BCR/ABL fusion and monitor response to therapy in chronic myeloid leukemia. *Blood*. 1998;91:3357–3365.
16. Gall JG, Pardue ML. Formation and detection of RNA-DNA hybrid molecules in cytological preparations. *Proc Natl Acad Sci USA*. 1969;63:378–383.
17. Garçon L, Libura M, Delabesse E, et al. DEK-CAN molecular monitoring of myeloid malignancies could aid therapeutic stratification. *Leukemia*. 2005;19:1338–1344.

18. Goodpasture C, Bloom SE, Hsu TC, Arrighi FE. Human nucleolus organizers: the satellites or the stalks? *Am J Hum Genet.* 1976;28:559–566.
19. Grignani F, Fagioli M, Alcalay M, et al. Acute promyelocytic leukemia: from genetics to treatment. *Blood.* 1994;83:10–25.
20. Grimwade D, Walker H, Oliver F, et al. The importance of diagnostic cytogenetics on outcome in AML: analysis of 1,612 patients entered into the MRC AML 10 trial. The Medical Research Council Adult and Children's Leukaemia Working Parties. *Blood.* 1998;92:2322–2333.
21. Han JY, Theil KS. The Philadelphia chromosome as a secondary abnormality in inv(3)(q21q26) acute myeloid leukemia at diagnosis: confirmation of p190 BCR-ABL mRNA by real-time quantitative polymerase chain reaction. *Cancer Genet Cytogenet.* 2006;165:70–74.
22. Heim S, Mitelman F. *Cancer Cytogenetics.* New York: John Wiley & Sons, Inc; 1995.
23. Hopman AH, Wiegant J, Tesser GI, Van Duijn P. A non-radioactive in situ hybridization method based on mercurated nucleic acid probes and sulfhydryl-hapten ligands. *Nucleic Acids Res.* 1986;14:6471–6488.
24. Hsiao HH, Sashida G, Ito Y, et al. Additional cytogenetic changes and previous genotoxic exposure predict unfavorable prognosis in myelodysplastic syndromes and acute myeloid leukemia with der(1;7)(q10;p10). *Cancer Genet Cytogenet.* 2006;165: 161–166.
25. Kang LC, Smith SV, Kaiser-Rogers K, Rao K, Dunphy CH. Two cases of acute myeloid leukemia with t(11;17) associated with varying morphology and immunophenotype: rearrangement of the MLL gene and a region proximal to the RARalpha gene. *Cancer Genet Cytogenet.* 2005;159:168–173.
26. Klaus M, Haferlach T, Schnittger S, Kern W, Hiddemann W, Schoch C. Cytogenetic profile in de novo acute myeloid leukemia with FAB subtypes M0, M1, and M2: a study based on 652 cases analyzed with morphology, cytogenetics, and fluorescence in situ hybridization. *Cancer Genet Cytogenet.* 2004;155:47–56.
27. Kwong YL, Liu HW, Chan LC. Racial predisposition to translocation (7;11) *Leukemia.* 1992;6:232.
28. Leroy H, de Botton S, Grardel-Duflos N, et al. Prognostic value of real-time quantitative PCR (RQ-PCR) in AML with t(8;21). *Leukemia.* 2005;19:367–372.
29. Lin P, Medeiros LJ, Yin CC, Abruzzo LV. Translocation (3;8)(q26;q24): a recurrent chromosomal abnormality in myelodysplastic syndrome and acute myeloid leukemia. *Cancer Genet Cytogenet.* 2006;166:82–85.
30. Mauritzson N, Albin M, Rylander L, et al. Pooled analysis of clinical and cytogenetic features in treatment-related and de novo adult acute myeloid leukemia and myelo-dysplastic syndromes based on a consecutive series of 761 patients analyzed 1976-1993 and on 5098 unselected cases reported in the literature 1974-2001. *Leukemia.* 2002;16: 2366–2378.
31. Mertens F, Johansson B, Mitelman F. Isochromosomes in neoplasia. *Genes Chromosomes Cancer.* 1994;10:221–230.
32. Mitelman F, Johansson B, Mertens F. Fusion genes and rearranged genes as a linear function of chromosome aberrations in cancer. *Nat Genet.* 2004;36:331–334.
33. Moir DJ. A new translocation, t(1;3) (p36;q21), in myelodysplastic disorders. *Blood.* 1984;64:553–555.
34. Mrózek K, Heerema NA, Bloomfield CD. Cytogenetics in acute leukemia. *Blood Rev.* 2004;18:115–136.
35. Mrózek K, Heinonen K, de la Chapelle A, Bloomfield CD. Clinical significance of cytogenetics in acute myeloid leukemia. *Semin Oncol.* 1997;24:17–31.
36. Oshima M, Fukushima T, Koike K, Hoshida C, Izumi I, Tsuchida M. Infant leukemia with t(1;22) presenting proliferation of erythroid and megakaryocytic cell lineages. *Rinsho Ketsueki.* 1999;40:230–235.

37. Patel BB, Mohamed AN, Schiffer CA. "Acute myelogenous leukemia like" translocations in CML blast crisis: two new cases of inv(16)/t(16;16) and a review of the literature. *Leuk Res.* 1999;30:225–232.
38. Paulsson K, Békássy AN, Olofsson T, Mitelman F, Johansson B, Panagopoulos I. A novel and cytogenetically cryptic t(7;21)(p22;q22) in acute myeloid leukemia results in fusion of RUNX1 with the ubiquitin-specific protease gene USP42. *Leukemia.* 2006;20:224–229.
39. Paulsson K, Fioretos T, Strömbeck B, Mauritzson N, Tanke HJ, Johansson B. Trisomy 8 as the sole chromosomal aberration in myelocytic malignancies: a multicolor and locus-specific fluorescence in situ hybridization study. *Cancer Genet Cytogenet.* 2003;140:66–69.
40. Ponzetto C, Guerrasio A, Rosso C, et al. ABL proteins in Philadelphia-positive acute leukaemias and chronic myelogenous leukaemia blast crises. *Br J Haematol.* 1990;76:39–44.
41. Potenza L, Sinigaglia B, Luppi M, et al. A t(11;20)(p15;q11) may identify a subset of nontherapy-related acute myelocytic leukemia. *Cancer Genet Cytogenet.* 2004;149: 164–168.
42. Raphael BJ, Pevzner PA. Reconstructing tumor amplisomes. *Bioinformatics.* 2004;20 Suppl 1:i265–i273.
43. Reiter A, Walz C, Watmore A, et al. The t(8;9)(p22;p24) is a recurrent abnormality in chronic and acute leukemia that fuses PCM1 to JAK2. *Cancer Res.* 2005;65:2662–2667.
44. Rozman M, Camós M, Colomer D, et al. Type I MOZ/CBP (MYST3/CREBBP) is the most common chimeric transcript in acute myeloid leukemia with t(8;16)(p11;p13) translocation. *Genes Chromosomes Cancer.* 2004;40:140–145.
45. Schneider NR, Bowman WP, Frenkel EP. Translocation (3;21)(q26;q22) in secondary leukemia. Report of two cases and literature review. *Ann Genet.* 1991;34:256–63.
46. Schessl C, Rawat VP, Cusan M, et al. The AML1-ETO fusion gene and the FLT3 length mutation collaborate in inducing acute leukemia in mice. *J Clin Invest.* 2005;115: 2159–2168.
47. Seabright M. The use of proteolytic enzymes for the mapping of structural rearrangements in the chromosomes of man. *Chromosoma.* 1972;36:204–210.
48. Schnittger S, Kohl TM, Haferlach T, et al. KIT-D816 mutations in AML1-ETO-positive AML are associated with impaired event-free and overall survival. *Blood.* 2006;107:1791–1799.
49. Shimada A, Taki T, Tabuchi K, et al. KIT mutations, and not FLT3 internal tandem duplication, are strongly associated with a poor prognosis in pediatric acute myeloid leukemia with t(8;21): a study of the Japanese Childhood AML Cooperative Study Group. *Blood.* 2006;107:1806–1809.
50. Sierra M, Hernández JM, García JL, et al. Hematological, immunophenotypic, and cytogenetic characteristics of acute myeloblastic leukemia with trisomy 11. *Cancer Genet Cytogenet.* 2005;160:68–72.
51. Soekarman D, von Lindern M, Daenen S, et al. The translocation (6;9) (p23;q34) shows consistent rearrangement of two genes and defines a myeloproliferative disorder with specific clinical features. *Blood.* 1992;79:2990–2997.
52. Soekarman D, von Lindern M, van der Plas DC, et al. Dek-can rearrangement in translocation (6;9)(p23;q34). *Leukemia.* 1992;6:489–494.
53. Suzukawa K, Parganas E, Gajjar A, et al. Identification of a breakpoint cluster region 3' of the ribophorin I gene at 3q21 associated with the transcriptional activation of the EVI1 gene in acute myelogenous leukemias with inv(3)(q21q26). *Blood.* 1994;84:2681–2688.
54. Takatsuki H, Yufu Y, Tachikawa Y, Uike N. Monitoring minimal residual disease in patients with MLL-AF6 fusion transcript-positive acute myeloid leukemia following allogeneic bone marrow transplantation. *Int J Hematol.* 2002;75:298–301.

55. Tanabe S, Zeleznik-Le NJ, Kobayashi H, et al. Analysis of the t(6;11)(q27;q23) in leukemia shows a consistent breakpoint in AF6 in three patients and in the ML-2 cell line. *Genes Chromosomes Cancer*. 1996;15:206–216.

56. Thirman MJ, Levitan DA, Kobayashi H, Simon MC, Rowley JD. Cloning of ELL, a gene that fuses to MLL in a t(11;19)(q23;p13.1) in acute myeloid leukemia. *Proc Natl Acad Sci USA*. 1994;91:12110–12114.

57. Tse W, Zhu W, Chen HS, Cohen A. A novel gene, AF1q, fused to MLL in t(1;11) (q21;q23), is specifically expressed in leukemic and immature hematopoietic cells. *Blood*. 1995;85:650–656.

58. von Lindern M, Fornerod M, van Baal S, et al. The translocation (6;9), associated with a specific subtype of acute myeloid leukemia, results in the fusion of two genes, dek and can, and the expression of a chimeric, leukemia-specific dek-can mRNA. *Mol Cell Biol*. 1992;12:1687–1697.

59. Weltermann A, Fonatsch C, Haas OA, et al. Impact of cytogenetics on the prognosis of adults with de novo AML in first relapse. *Leukemia*. 2004;18:293–302.

60. Willem P, Pinto M, Bernstein R. Translocation t(1;7) revisited. Report of three further cases and review. *Cancer Genet Cytogenet*. 1988;36:45–54.

61. Yoneda-Kato N, Look AT, Kirstein MN, et al. The t(3;5)(q25.1;q34) of myelodysplastic syndrome and acute myeloid leukemia produces a novel fusion gene, NPM-MLF1. *Oncogene*. 1996;18:265–275.

Chromosomal Deletions in AML

Lalitha Nagarajan

Abstract Several, acquired, non-random chromosomal deletions have been characterized in acute myelogenous leukemia (AML). While the deletion limits vary among patients, there are consistent regions of overlap among the deleted segments between patients. Furthermore, chromosomal deletions are achieved frequently by unbalanced translocations between two and more chromosomes resulting loss of candidate leukemia suppressor loci from the affected chromosomes. Most deletions occurring as sole anomalies are associated with good–intermediate clinical outcome, but complex cytogenetic anomalies signify an aggressive clinical course. Thanks to the exciting development in microarray, siRNA technologies, a number of candidate AML suppressor genes localizing to the critical regions of overlap within the deletions have been identified recently. Most of the candidate genes do not function by the classical "two hits," namely loss of an allele unmasking inactivating mutations in the remaining allele. Gene dosage, epigenetic silencing, and uniparental disomy appear to be common mechanisms of gene inactivation in AML. While several of the newly discovered candidate genes lead to new pathways, a few of them affect previously known leukemogenic targets. Thus the investments made over the years on leukemia suppressor gene discovery are beginning to yield reasonable results at the present time. Future beholds promise for targeted therapy of these poorly characterized AMLs, as we uncover the mutations driving their clonal evolution.

Introduction

Acute myelogenous leukemia (AML) can be broadly classified into three categories based on cytogenetics: (i) diploid, (ii) reciprocal translocations, and (iii) complex cytogenetic anomalies. Among these, recurrent loss of material within specific chromosomal arm(s) is associated with the complex cytogenetic

L. Nagarajan (✉)
Department of Genetics, MD Anderson Cancer Center, Houston, TX 77030, USA
e-mail: lnagaraj@mdanderson.org

L. Nagarajan (ed.), *Acute Myelogenous Leukemia*,
Cancer Treatment and Research 145, DOI 10.1007/978-0-387-69259-3_4,
© Springer Science+Business Media, LLC 2010

subset. The complex anomalies also signify poor prognosis in myelodysplasia (MDS) which can be viewed as a preleukemic condition in this subset [7].

According to the Knudson's two hit hypothesis physical loss of a tumor suppressor by chromosomal deletion is accompanied by inactivating mutations in the remaining chromosome [13]. Based on this paradigm, deletions associated with AML were hypothesized to be chromosomal regions harboring suppressor gene(s), with the remaining allele in the "normal chromosome" inactivated by intragenic mutations. Inactivation of the remaining normal allele may be achieved by gross deletion, point mutation, or loss of the entire wild-type chromosome and duplication of the remaining mutated chromosome. The last mechanism, due to mitotic disjunction and duplication results in two copies (disomy) from one of the parental (uniparental) chromosomes. Interestingly, uniparental disomy (UPD) for several novel loci and not the other modes of inactivation appears to be a recurrent theme in myeloid neoplasms.

There has been a long and hard search for AML suppressors from the deleted chromosomes. Recent progress discussed in the following sections suggests that gene dosage, epigenetic silencing of candidate suppressor genes may contribute to the pathogenesis rather than the conventional "two hits." These newly characterized mechanisms are likely to yield novel diagnostic and therapeutic tools.

The deletion break points for each chromosomal arm vary between patients, there are regions of overlap. Thus, there is a critical region of overlap (CRO) (Fig. 1). The target AML suppressors were hypothesized to reside within the CRO. To precisely delineate the regions of loss, techniques were developed to define deletion limits by microsatellite polymorphism-based allelotyping and fluorescence in situ hybridization (FISH). Most recently, single nucleotide polymorphism (SNP)-based microarray and comparative genomic hybridization technologies have yielded a comprehensive portrait of genome-wide losses and gains.

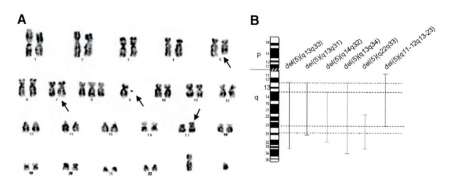

Fig. 1 Chromosomal deletions, suppressor gene discovery in AML. (A) Complex cytogenetic alterations in AML. This near-diploid metaphse shows deletions of 5q, 7q, monosomy 9, and additional material on 17p. Fluorescence in situ hybridization revealed the 17 p + material to be 5q. (B) Common deletion patterns in 5q. Note that there are at least two critical regions of overlap between the different deletions

At present, at least nine different chromosomal deletions have been identified in AML. The non-random, clonal nature of these changes suggests a causal role for these aberrations in AML pathogenesis. While deletions of 5q, 7q, 11q, 17p, 20q, are more common, loss of 1p, 4q, 9q, 16p are also reported (Table 1). Typically, the deletions are found within a clone that is near diploid in chromosomal content.

Table 1 Common chromosomal deletions in AML

Chromosome	Additional mutations	Prognosis	Candidate genes
1p	Multiple	Poor	Unknown
4q	Sole anomaly	Good	TET2
5q	Two or more anomalies	Poor	Multiple
7q	Two or more anomalies	Poor	Unknown
9q	Sole anomaly, CEBPA mutation	Good	Unknown
11q	Multiple other anomalies	Poor	CBL
16q	Sole anomaly	Unknown	Unknown
17p	Multiple other abnormalities	Poor	TRP53
20q	Sole or multiple anomalies	Good (sole), poor (complex)	ASXL1

As investigations on 4q, 5q, 7q, 11q, 17p, and 20q deletions have resulted in candidate gene discoveries or finer delineation of the critical region of loss, current developments on these are presented in the following sections.

Chromosome 4q

Translocations of 4q24 and submicroscopic deletions of 4q were originally identified in both the lymphoid and the myeloid cells of AML patients raising the possibility of a stem cell involvement [26]. In myeloproliferative neoplasms, a candidate gene of unknown function, TET2, harboring inactivating mutations has been found recently. In a limited survey of AML, TET2 mutations were detected in AML arising from MDS (a case each of diploid, complex cytogenetics, post-CMML and APL) [23]. Thus inactivation of TET2 appears to cooperate with a variety of mitogenic events. The cellular processes affected by this mutation are unknown at the present time.

Chromosome 5q

Of all the deletions associated with AML that of the long arm of chromosome 5 (5q⁻) is the most extensively characterized. In 1974, at a time Philadelphia chromosome was the only other anomaly associated with leukemia, Van den Berghe and colleagues identified interstitial deletions of the long arm chromosome 5 (5q⁻), to be the sole acquired anomaly in a group of patients with refractory anemia (RA) [25]. Subsequently, deletions of 5q were found to be recurring anomalies in MDS and AML. Interstitial deletions of the long arm of chromosomes 5 and 7 were also the first non-random anomalies to be recognized in therapy induced AML (t-AML) [14]. However, unlike RA, these anomalies were associated with poor prognosis in both de novo and t-AML. Frequently, in AML, they were one of several other cytogenetic anomalies whereas in RA the 5q-chromosome is the sole abnormality. Typically these are large deletions resulting in loss of 60% of 5q, which translates to 1.6% of haploid genome.

A number of candidate leukemia suppressors have been identified from four different region of loss on 5q. The candidate genes have been identified by multiple approaches: (i) a positional cloning approach by delineating the deletion limits in a large number of patients. The resulting critical region of overlap or minimally deleted region, served as the starting point for gene discovery. This approach resulted in two distinct critical regions within 5q31, the proximal MDS/AML locus and the distal RA locus. Genes within the deleted segments were systematically evaluated for homozygous deletions or point mutations. However, none of the genes fitted the classical "two hit" criterion due to absence of inactivating mutations in the remaining allele. These findings led to the possibility that the 5q leukemia suppressor functions by gene silencing of the remaining allele or gene dosage. Gene dosage effect would imply that a 50% decrease in the expression confers a clonal advantage.

To date, nine candidate suppressor genes have been identified. Thus, complete and dosage sensitive loss of function of *SSBP2, RIL, APC, SMAD5, EGR1, CTNNA, RPS14, SPARC,* and *NPM* (nucleophosmin) appears to contribute to leukemogenesis [3, 2, 19, 11, 12, 18, 6, 15, 9]. *SMAD5, EGR1, CTNNA* localize to the AML locus, whereas *RPS14* and *SPARC* map to the RA interval. While *SSBP2, RIL,* and *CTNNA* may be silenced by epigenetic mechanisms, loss of a single *EGR1, RPS14,* or *NPM* allele has phenotypic consequences. In vitro, reduction in the expression of *RPS14,* encoding a ribosomal protein results in a pronounced erythroid and megakaryocytic differentiation of hematopoietic stem/progenitor cells. In contrast, ENU-treated *Egr1* hemizygous mice show a myeloproliferative defect. Although *Npm* localizes outside the critical region of overlap, *Npm* heterozygous mice exhibit features of trilineage dysplasia. As most deletions of 5q are large, potential contribution of all these genes including NPM and APC cannot be ruled out.

In short, both gene dosage and epigenetic silencing of both alleles of candidate gene(s) may play a role in AML pathogenesis. Given the large size of the deletion, interplay between one or more candidate suppressor is likely. If the synergistic interaction between two candidate genes contributes significantly to loss of growth arrest, block in differentiation, the clonal advantage for retaining the 5q⁻ anomaly can be explained readily.

Chromosome 7q

Complete and partial loss of chromosome 7 and trisomy 8 constitutes the most common chromosomal anomalies in AML. Since 22% of the cytogenetically characterized AML cases harbor complete or partial deletions of chromosome 7, the same loci may be subtly altered in other poor prognosis patients. Interestingly, -7 or 7q- may be the sole anomaly in 8.5% of AML cases. Two distinct critical regions of loss at 7q22 and 31 have been delineated [17]. In AML cell lines, chromosome 7 also undergoes UPD resulting in segmental duplication of one of the chromosomes with loss of a potential wild-type copy [16]. Interestingly, a similar genotype has been found in primary blasts from MDS patients [10].

Chromosome 11q

Although 11q deletions have long been recognized as markers for poor prognosis, the candidate genes were unknown [7]. Recent microarray-based analyses revealed uniparental disomy for a mutated allele of the *CBL* gene [5]. *CBL* encodes an E3 ubiquitin ligase with a potential to downregulate FLT3-induced proliferative response. CBL mutations appear to co-segregate with inv (16) or t(8;21) suggesting a strong cooperative selection in core-binding factor-associated AML [20]. Lack of CBL activity will remove the negative regulation of FLT3 activation, resulting in an unrestrained mitogenic signal. Future studies will reveal whether *CBL* mutation accounts for all cases of deletion 11q or there are other regulatory genes that contribute to AML development.

Chromosome 17p

While the discovery of tumor suppressor genes on chromosomes 5 and 7 has been challenging, studies of chromosome 17p have been rewarding because of the *TP53* suppressor gene on this chromosome. But, as stated earlier, 17p deletions are frequently undetected by conventional cytogenetic methods, and hence this entity was not recognized for a long time. It was the recognition by Fenaux et al. of a strong correlation between 17p deletion and loss of the *TP53* gene and a form of dysgranulopoiesis with pseudo-Pelger–Huet hypolobulation and small vacuoles in the granulocytes that pinpointed the significance of 17p deletions. A screen of patients with MDS/AML who had the

"17p- syndrome" (i.e., cytogenetically detectable deletions of chromosome 17) then provided clear evidence of the loss of a *TP53* allele in 14/17 cases [22]. Furthermore, the loss of 17p was associated with an unbalanced translocation with chromosome 5 in 9/17 cases. Interestingly, Christiansen et al. noted a similar correlation between deletions of chromosome 5, a complex karyotype, and a *TP53* mutation [4]. Upon screening a series of 77 patients with t-MDS ($n = 55$) or t-AML ($n = 22$), 21 cases had mutations in the *TP53* gene. Of these, six patients had a cytogenetic loss of 17p13. Based on single-stranded DNA conformation polymorphism (SSCP) analysis, these investigators concluded that there was a loss of the wild-type *TP53* allele in nine more cases but no cytogenetic loss. Cytogenetic analysis of these nine patients revealed numerous other anomalies. Such observations have led to the suggestion that the normal *TP53* allele is lost with duplication of the homologous chromosome 17 with the mutated allele [4]. The most extensive analysis of the *TP53* gene in AML was conducted by Schoch et al., who reviewed 565 cases of AML with complex anomalies [21]. They were able to sequence 49 cases of complex anomalies for the *TP53* mutation and found that 27 cases (55%) had loss of a normal chromosome and a mutation in the remaining allele, whereas nearly a third (16 cases) had a mutation without a detectable chromosome deletion. Once again, it is likely that these latter cases involved duplications of the mutated chromosome 17 that were cytogenetically undetectable [23, 24] .

Chromosome 20q

Deletions of 20q in AML were one of the anomalies to be recognized several years ago [24]. Using the conventional allelotyping techniques, critical regions of loss were defined [1]. No strong candidate genes have emerged thus far except for truncating mutations in the *ASX1L* gene in 16% of MDS [8]. Although *ASX1L* may not reside within the critical region of loss, detection of mutations in this polycomb-associated gene indicates inactivation of a critical epigenetic pathway.

Mechanisms of Deletions

As chromosomal deletions are found as sole anomalies in a diploid clone as well as in cells with three or more alterations, it is unlikely that a single mechanism governs chromosomal deletions in AML. The complex anomalies, however, are derived from two-or three-way unbalanced translocations between chromosomes. While some of these may be unstable in the early stages of leukemogenesis, with evolution, clones harboring anomalies with survival, growth advantages are selected for. As the unbalanced translocations result in loss of chromosomal material between partner chromosomes, a selective advantage may be conferred by loss or haploinsufficiency for critical genes. Frequently, certain recurrent partners appear to be preferred (e.g., 5q, 7q, 17p) suggesting a functional cooperation between losses of genes from these chromosomes. As mentioned earlier, dicentric (5;17) or derivative chromosome 17 (with chromosome 5 material on 17p) and iso 17q are recurring anomalies in both de novo MDS and t-MDS implicating the loss of suppressor genes from these chromosomes in the preleukemic phase.

Conclusions

Chromosomal deletions represent an important genetic mechanism in AML. The deletions, achieved by diverse mechanisms, also signify variable outcome. While deletions detected as sole anomalies signify good prognosis, complex or numerous anomalies indicate poor response. Dismal outcome is predicted when a clone harboring a single deletion relapses with multiple additional abnormalities. Commonly co-segregating deletions (e.g., 5q, 7q, 17p) suggest a functional cooperation between the loss of suppressor elements. The sequence of specific chromosomal loss during AML pathogenesis may also be varied. For example, the loss of genes in 5q may contribute to early stages of evolution, inactivation of TP53 resulting in a highly aggressive clinical course is likely to be a late event. With the advent of highly sensitive tools to detect subtle deletions and duplications, gene discovery is gaining momentum. Evaluation of candidate genes by functional assays in vitro by RNAi-based approaches and in vivo by targeted deletion in murine models in the coming years will improve our understanding of critical genetic alterations underlying AML pathogenesis.

Acknowledgment Studies on the SSBP2 gene in the author's laboratory were supported by HL-074449.

References

1. Bench AJ, et al. Chromosome 20 deletions in myeloid malignancies: reduction of the common deleted region, generation of a PAC/BAC contig and identification of candidate genes. UK Cancer Cytogenetics Group (UKCCG). *Oncogene*. 2000;19(34):3902–3913.
2. Boumber YA, et al. RIL, a LIM gene on 5q31, is silenced by methylation in cancer and sensitizes cancer cells to apoptosis. *Cancer Res*. 2007;67(5):1997–2005.
3. Castro P, et al. A novel, evolutionarily conserved gene family with putative sequence-specific single-stranded DNA-binding activity. *Genomics*. 2002;80(1):78–85.
4. Christiansen DH, Andersen MK, Pedersen-Bjergaard J. Mutations with loss of heterozygosity of p53 are common in therapy-related myelodysplasia and acute myeloid leukemia after exposure to alkylating agents and significantly associated with deletion or loss of 5q, a complex karyotype, and a poor prognosis. *J Clin Oncol*. 2001;19(5):1405–1413.
5. Dunbar AJ, et al. 250 K single nucleotide polymorphism array karyotyping identifies acquired uniparental disomy and homozygous mutations, including novel missense substitutions of c-Cbl, in myeloid malignancies. *Cancer Res*. 2008;68(24):10349–10357.

6. Ebert BL, et al. Identification of RPS14 as a 5q- syndrome gene by RNA interference screen. *Nature*. 2008;451(7176):335–339.
7. Estey E, Dohner H. Acute myeloid leukaemia. *Lancet*. 2006;368(9550):1894–1907.
8. Gelsi-Boyer V, et al. Mutations of polycomb-associated gene ASXL1 in myelodysplastic syndromes and chronic myelomonocytic leukaemia. *Br J Haematol*. 2009;145(6):788–800.
9. Grisendi S, et al. Role of nucleophosmin in embryonic development and tumorigenesis. *Nature*. 2005;437(7055):147–153.
10. Heinrichs S, et al. Accurate detection of uniparental disomy and microdeletions by SNP array analysis in myelodysplastic syndromes with normal cytogenetics. *Leukemia*. 2009;23(9):1605–1613.
11. Hejlik DP, Nagarajan L. Deletion of 5q in myeloid leukemia cells HL-60: an L1 element-mediated instability. *Cancer Genet Cytogenet*. 2005;156(2):97–103.
12. Joslin JM, et al. Haploinsufficiency of EGR1, a candidate gene in the del(5q), leads to the development of myeloid disorders. *Blood*. 2007;110(2):719–726.
13. Knudson AG. Antioncogenes and human cancer. *Proc Natl Acad Sci USA*. 1993;90(23):10914–10921.
14. Le Beau MM, et al. Clinical and cytogenetic correlations in 63 patients with therapy-related myelodysplastic syndromes and acute nonlymphocytic leukemia: further evidence for characteristic abnormalities of chromosomes no. 5 and 7. *J Clin Oncol*. 1986;4(3):325–345.
15. Lehmann S, et al. Common deleted genes in the 5q- syndrome: thrombocytopenia and reduced erythroid colony formation in SPARC null mice. *Leukemia*. 2007;21(9):1931–1936.
16. Liang H, et al. Finer delineation and transcript map of the 7q31 locus deleted in myeloid neoplasms. *Cancer Genet Cytogenet*. 2005;162(2):151–159.
17. Liang JC, et al. Spectral karyotypic study of the HL-60 cell line: detection of complex rearrangements involving chromosomes 5, 7, and 16 and delineation of critical region of deletion on 5q31.1. *Cancer Genet Cytogenet*. 1999;113(2):105–109.
18. Liu TX, et al. Chromosome 5q deletion and epigenetic suppression of the gene encoding alpha-catenin (CTNNA1) in myeloid cell transformation. *Nat Med*. 2007;13(1):78–83.
19. Qian Z, et al. A critical role for Apc in hematopoietic stem and progenitor cell survival. *J Exp Med*. 2008;205(9):2163–2175.
20. Reindl C, et al. CBL exon 8/9 mutants activate the FLT3 pathway and cluster in core binding factor/11q deletion acute myeloid leukemia/myelodysplastic syndrome subtypes. *Clin Cancer Res*. 2009;15(7):2238–2247.
21. Schoch C, et al. Genomic gains and losses influence expression levels of genes located within the affected regions: a study on acute myeloid leukemias with trisomy 8, 11, or 13, monosomy 7, or deletion 5q. *Leukemia* 2005;19(7):1224–1228.
22. Soenen V, et al. 17p Deletion in acute myeloid leukemia and myelodysplastic syndrome. Analysis of breakpoints and deleted segments by fluorescence in situ. *Blood*. 1998;91(3):1008–1015.
23. Tefferi A, et al. TET2 mutations and their clinical correlates in polycythemia vera, essential thrombocythemia and myelofibrosis. *Leukemia*. 2009;23(7):1343–1345.
24. Testa JR, et al. Deletion of the long arm of chromosome 20 [del(20)(q11)] in myeloid disorders. *Blood*. 1978;52(5):868–877.
25. Van den Berghe H, et al. Distinct haematological disorder with deletion of long arm of no. 5 chromosome. *Nature*. 1974;251(5474):437–438.
26. Viguie F, et al. Common 4q24 deletion in four cases of hematopoietic malignancy: early stem cell involvement? *Leukemia*. 2005;19(8):1411–1415.

Genes Predictive of Outcome and Novel Molecular Classification Schemes in Adult Acute Myeloid Leukemia

Roel G.W. Verhaak and Peter J.M. Valk

Abstract The pretreatment karyotype of leukemic blasts is currently the key determinant in therapy decision making in acute myeloid leukemia (AML). The World Health Organization (WHO) has recognized this important information by including, besides clinical, cytological, cytochemical, and immunophenotypical features, recurrent cytogenetic abnormalities in its classification (Table 1). However, although the WHO defines important biologically and clinically relevant entities, the prognostic value of some of the well-defined cytogenetic subgroups is partially masked in the WHO classification. Moreover, in the recent past a number of novel molecular aberrations with marked prognostic value, which are not yet incorporated in the WHO classifications have been identified. These molecular abnormalities include mutations (e.g., in *FLT3*, c-*KIT*, and *NPM1*), partial duplications (e.g., of *MLL* and *FLT3*), and abnormal expression of pathogenetic genes (e.g., *EVI1*, *WT1*, *BCL2*, *MDR1*, *BAALC*, and *ERG*). In addition, novel molecular approaches in genomics, like monitoring the expression levels of thousands of genes in parallel using DNA microarray technology, open possibilities for further refinement of prognostication of AML. Gene expression profiling in AML is already well established and has proven to be valuable to recognize various cytogenetic subtypes, discover novel AML subclasses, and predict clinical outcome. The current advances made in molecular understanding of AML will ultimately lead to a further refinement of prognostics of AML.

P.J.M. Valk (✉)
Department of Hematology, Erasmus University Medical Center, Rotterdam,
The Netherlands
e-mail: p.valk@erasmusmc.nl

L. Nagarajan (ed.), *Acute Myelogenous Leukemia*,
Cancer Treatment and Research 145, DOI 10.1007/978-0-387-69259-3_5,
© Springer Science+Business Media, LLC 2010

AML Diagnostics

Since 1976 AML has been diagnosed according to the criteria of the French-American-British (FAB) group. This classification system is strictly based on morphological and cytochemical examination of the leukemic bone marrow and peripheral blood. The FAB classes M0, M6, and M7 appear to have some prognostic information in multivariate analyses; however, these represent only a small fraction of AML patients. The FAB classification in general does not carry any prognostic value, apart from two categories linked to chromosomal abnormalities with marked prognostic significance (FAB M3 and FAB M4Eo).

Table 1 The WHO classification of AML

Acute myeloid leukemia with recurrent genetic abnormalities

- Acute myeloid leukemia with t(8;21)(q22;q22) (AML1-ETO)
- Acute myeloid leukemia with abnormal bone marrow eosinophils and inv(16)(p13q22) or t(16;16)(p13;q22) (CBFβ-MYH11)
- Acute promyelocytic leukemia with t(15;17)(q22;q22) (PML-RARα) and variants
- Acute myeloid leukemia with 11q23 (MLL) abnormalities

Acute myeloid leukemia with multilineage dysplasia

- Following myeloid dysplastic syndrome (MDS) or MDS/myeloproliferative disorder (MPD)
- Without antecedent MDS or MDS/MPD, but with dysplasia in at least 50% of cells in two or more myeloid lineages

Acute myeloid leukemia and myelodysplastic syndromes, therapy related

- Alkylating agent/radiation-related type
- Topoisomerase II inhibitor-related type (some may be lymphoid)
- Others

Acute myeloid leukemia, not otherwise categorized

- Acute myeloid leukemia, minimally differentiated
- Acute myeloid leukemia without maturation
- Acute myeloid leukemia with maturation
- Acute myelomonocytic leukemia (AMML)
- Acute monoblastic/acute monocytic leukemia
- Acute erythroid leukemia (erythroid/myeloid and pure erythroleukemia)
- Acute megakaryoblastic leukemia
- Acute basophilic leukemia
- Acute panmyeloid with myelofibrosis
- Myeloid sarcoma

In the past two decades it has become evident that particular recurrent cytogenetic abnormalities are frequently present in distinct subsets of AML. In 1999 the WHO constructed a novel classification of hematologic malignancies [34]. The WHO classification of AML recognizes four major categories (Table 1). It distinguishes besides clinical, morphologic, immunologic features, important recurrent cytogenetic abnormalities in AML [34]. Hence, current diagnostics of AML requires a multidisciplinary approach, which includes adequate clinical assessment, morphological and cytochemical examination, immunophenotyping, cytogenetics, and molecular genetics.

Cytogenetic Analyses and AML Prognostics

More traditional prognostic markers to risk-stratify de novo AML are age and white blood cell count [43, 47]. However, certain cytogenetic abnormalities are consistently associated with particular subsets of AML and appear to have marked prognostic value. In fact, karyotyping is currently the most important indicator for risk-stratification of AML. Approximately 40% of all AML patients are currently classified into distinct groups with variable prognosis based on the presence or absence of specific recurrent cytogenetic abnormalities (Table 2).

Table 2 Cytogenetic abnormalities in AML, their prevalence, and prognostic value

Cytogenetic abnormality	Genes involved	Prognosis	Frequency (%)
t(8;21)(q22;q22)	AML1-ETO	Favorable	8–12
inv(16)(p11;q22)	CBFβ-MYH11	Favorable	4–10
t(15;17)(q22;q11)	PML-RARα	Favorable	8–12
11q23 abnormalities	MLL	Unfavorable/standard	4–6
-5/5q-	Unknown	Unfavorable	2–4
-7/7q-	Unknown	Unfavorable/standard	6–8
t(6;9)(q34;q11)	DEK-CAN	Unfavorable/standard	1
t(9;22)(q34;q11)	BCR-ABL	Unfavorable	<1
3q26 abnormalities	EVI1	Unfavorable	1–3
Complex karyotype	Unknown	Unfavorable	4–6

For instance, the chromosomal inversion of chromosome 16 [inv(16)] and balanced translocation of chromosomes 8 and 21 [t(8;21)], expressing the fusion oncoproteins CBFB/MYH11 and AML1/ETO, respectively, characterize AML with relatively favorable treatment outcome [47, 46]. In addition, patients with acute promyelocytic leukemia (APL), an AML subtype characterized by the accumulation of t(15;17)-containing promyelocytes, are particularly sensitive to all trans-retinoic acid (ATRA) and hence have a particularly good treatment outcome. ATRA induces differentiation of the t(15;17)-containing promyelocytes, because of the presence of the t(15;17)-encoded PML/RARA chimeric protein.

The chromosomal abnormalities t(8;21), inv(16) and t(15;17) are established molecular indicators for favorable treatment outcome in AML and patients with these abnormalities are classified in WHO subgroup 1 (Table 1). Together, these subgroups account for approximately 20% of all cases of AML. Roughly

75–80% of all patients with AML respond to therapy by reaching a complete remission and overall survival of these patients is approximately 45% after 5 year. In contrast, AML patients with an inv(16), t(8;21), or t(15;17) show a CR rate of 88–98% and an overall survival of 60–70% after 5 years [46, 48]. AML patients with 11q23 abnormalities are also placed into WHO class 1, while 11q23 abnormalities are sometimes defined as poor-risk markers [12, 67] and sometimes intermediate or standard risk [12, 30, 29]. For instance, t(9;11) is sometimes recognized as standard risk marker [12], whereas others treat all AML cases with 11q23 abnormalities as a single AML subclass [67, 30, 29]. The different partner chromosomes, which are fused to 11q23 in AML, such as chromosome 6, -9, -10, and -19, may in fact contribute to this observed difference in treatment outcome. Likewise, cytogenetic aberrations, e.g., t(6;9) and chromosome 7(q), are defined as poor-risk markers by some [12, 67], whereas others consider these cases standard risk [12, 30, 29]. AML with one of the latter chromosomal abnormalities are distributed over the WHO groups and AML cases with 11q23 abnormalities are included in WHO class 1 together with the favorable markers inv(16), t(8;21), and t(15;17), implying that the WHO classification in part dilutes the prognostic value of several of these recurrent cytogenetic aberrations. Thus, although the WHO classification defines important biological entities in AML, it does not take all prognostically relevant cytogenetic information known today into account.

Also, the largest cytogenetic subclass of AML, i.e., those patients with a normal karyotype, is categorized as standard risk, since these AML cases lack informative chromosomal markers. However, this group, accounting for approximately 40–45% of all AML patients, most probably contains a mixture of patients with favorable and unfavorable prognosis. Additional molecular markers are required to further refine the classification of these cases of AML.

Molecular Genetics and AML Prognostics

Molecular genetics has greatly improved our understanding of the pathophysiology of AML. In addition, in the past few years a number of novel molecular markers have been associated with AML prognostics (Table 3).

For instance, the presence of a partial tandem duplication of the mixed lineage leukemia gene (*MLL*-PTD) (approximately 8% of the cases) has been recognized as an unfavorable marker. Patients with a *MLL*-PTD generally have a shorter duration of remission in AML with a normal karyotype [65, 21, 15].

Another acquired molecular abnormality frequently detected in AML involves the fms-like tyrosine kinase-3 gene (*FLT3*), a hematopoietic growth factor receptor [27, 42, 50]. *FLT3* is mutated either as a result of an internal tandem duplication (ITD) (25–30% of the cases of AML) or by a base-pair substitution in the tyrosine kinase domain (TKD D835) (5–7% of the cases of AML) [27, 42]. The *FLT3* ITD as well as *FLT3* TKD mutations result in ligand-

Table 3 Molecular abnormalities in AML, their prevalence, and prognostic value

Marker gene	Abnormality	Prognosis	Frequency (%)
FLT3	Internal tandem duplication	Unfavorable	25
FLT3	Tyrosine kinase domain mutation	?	10–12
MLL	Partial tandem duplication	Unfavorable	8
CEBPA	Mutation	Favorable	5–10
TP53	Mutation	Unfavorable	3–5
c-*KIT*	Mutation	Unfavorable	3–5
N-*RAS*	Mutation	None	8–12
K-*RAS*	Mutation	None	3–5
MDR1	High expression	Unfavorable	~30
WT1	High expression	Unfavorable	~50
BCL2		Unfavorable	~50
BAALC	High expression	Unfavorable	50
EVI1	High expression	Unfavorable	5–10
ERG	High expression	Unfavorable	25

independent constitutive activation of the FLT3 receptor [27, 42]. Patients with a *FLT3* ITD mutation are associated with leukocytosis and generally have a poor treatment outcome [27, 42, 50, 24]. Interestingly, AML patients who also lack the *FLT3* wild-type allele or show increased *FLT3* ITD expression relative to the wild-type allele appear to have an even inferior prognosis [79, 70, 40, 4]. The prognostic significance of the *FLT3* TKD is controversial [6, 70, 80, 64]. Interestingly, it has been shown that AML patients with elevated levels of wild-type FLT3 also have constitutive activation of the receptor, which may be associated with poor response to treatment [56].

The CCAAT/enhancer binding protein alpha (CEBPA) is a transcription factor essential for granulocytic differentiation [57]. AML patients with biallelic mutations in *CEBPA* (approximately 7% of the cases) have been associated with good clinical outcome in the group of patients with standard-risk AML [74, 58, 25]. In contrast, low *CEBPA* mRNA expression in AML with intermediate-risk karyotypes may be associated with poor prognosis [74]. The favorable indication of *CEBPA* mutations was maintained after adjustment for cytogenetic risk, *FLT3* ITD, and *CEBPA* expression levels in multivariate analysis [74].

Mutations in the hematopoietic receptor c-KIT are present in a variety of diseases, such as systemic mastocytosis, gastrointestinal stromal tumors, and AML [41]. In AML mutations are present in the extracellular, i.e., exon 8, and intracellular, i.e., tyrosine kinase domain (TKD D816), juxtamembrane domain, and ITD mutations, in c-*KIT*, which are exclusively detected in corebinding factor (CBF) leukemias, i.e., AML with inv(16) and t(8;21) [7]. AML t(8;21) cases with c-KIT TKD(D816) mutation have a significantly higher incidence of relapse and a lower overall survival than AML with t(8;21) and wild-type c-*KIT* [61, 52, 13]. Poorer responses to treatment were also demonstrated in pediatric AML with t(8;21) and c-KIT TKD(D816) [66]. Mutations in the extracellular domain, i.e., exon 8 of c-KIT, are associated with relapse rate in AML with inv(16) [17]. Interestingly, treatment of AML with c-*KIT* exon 8 mutations with imatinib mesylate may be feasible [51, 14, 16]. In contrast, c-KIT TKD (D816) mutants are insensitive to imatinib [14, 78], but these AML patients may benefit from dasatinib [59]. Thus, treatment of CBF leukemias carrying c-KIT mutations may be improved by using these drugs. In fact, the appearance of c-KIT mutations only in CBF leukemias may suggest that even CBF leukemias overexpressing wild-type c-KIT may benefit from imatinib mesylate treatment [78].

Classical P53 mutations and mutations in the GTPases N-RAS and K-RAS occur at low frequency in AML (Table 3). RAS mutations do not show any association with clinical outcome in AML [9, 38, 54]. P53 mutations are indicative of poor response to therapy in AML, however, patients with P53 mutations often also carry abnormalities with proven poor indication, involving chromosomes 5 and 7 [49, 68].

Recently, Falini et al. identified specific 4 bp insertion mutations in the nucleophosmin (*NPM1*) gene in approximately 35% of all AML cases, which makes it the most common somatic gene mutation in AML [23]. These insertions induce a frame shift in the C-terminus of the NPM1 protein, resulting in aberrant cytoplasmic localization [23, 22]. *NPM1* mutations are significantly associated with age, i.e., more frequent in older patients [75, 18], high white blood cell counts, normal karyotypes, and *FLT3* ITD mutations [23, 75, 62, 20, 69]. *NPM1* mutations correlate negatively with the occurrence of inv(16), t(8;21) and t(15;17) as well as with mutations in *CEBPA* and N-*RAS* [23, 75, 62, 20, 69]. In the initial study, it was shown that *NPM1* mutations are associated with remission induction in AML [23]. More recently, patients with standard-risk AML with *NPM1* mutations, but without *FLT3* ITD mutations, were shown to have a significantly better overall and event-free survival [23, 75, 62, 20, 69]. Finally, in multivariate analysis *NPM1* mutations prove to be independent prognostic markers for favorable outcome [23, 75, 62, 20, 69].

Several individual genes have been identified as important prognostic marker genes in AML when abnormally expressed in the leukemic blast cells.

The proto-oncogene *EVI1*(ecotropic virus integration site 1) encodes a DNA binding protein, located on chromosome 3q26 [55]. *EVI1* is aberrantly expressed in cases of AML and myelodysplastic syndrome (MDS) carrying 3q26 rearrangements [55]. Interestingly, high expression of *EVI1* is also seen in approximately 8% of AML cases without 3q26 abnormalities [5]. In these cases of AML high *EVI1* expression is, like in AML with 3q26 abnormalities, an independent marker for poor prognosis [5]. The survival probability of high *EVI1* expressers was 0–8%, whereas those AML cases with low *EVI1* expression had a probability of 33% [5].

High expression of the tumor suppressor Wilms tumor 1 (WT1) has been proposed as a marker for inferior outcome [8], however, the true prognostic value of this marker is still unclear [8, 60, 26, 6, 81]. Interestingly, it has been shown that the expression of *WT1* significantly correlates with *BCL2* expression [37], which has also been proposed as poor prognostic marker in AML, when aberrantly expressed [36].

The multidrug resistance gene (*MDR1*) is frequently expressed in AML. MDR1 is a cellular drug efflux pump, which is thought to be one of the major causes of multidrug resistance. MDR1 expression is a disease-related unfavorable prognostic factor with significant impact on complete remission and overall survival in AML [31, 73].

BAALC (brain and acute leukemia, cytoplasmic), as a single gene, has been proposed to predict clinical outcome in AML [3]. Multivariate analyses showed high *BAALC* expression as independent risk factor for overall survival, event-free survival, and disease-free survival. Patients with high *BAALC* expression showed a significantly worse overall survival, i.e., median of 1.7 versus 5.8 years (survival at 3 years 39 versus 60%) [3], which was subsequently confirmed in a larger cohort with normal cytogenetics (survival at 3 years 36 versus 54%) [4]. The *BAALC* protein has been implicated to play a role in leukemia and normal expression of *BAALC* is restricted to hematopoietic progenitors.

Recently, Marcucci and colleagues demonstrated that overexpression of the ETS-related gene, *ERG*, predicted an adverse clinical outcome in AML [45]. Patients with high *ERG* expression (upper 25%, Q4) had an increased relapse rate and median survival time of 1.2 years, whereas the median survival time of the low expressors (Q1-3) was not reached in a follow-up of 5 years, with estimated survival rates of 19 and 51%, respectively. High *ERG* expression appeared to be an independent unfavorable prognostic factor for AML patients with a normal karyotype [45]. Interestingly, *ERG* overexpression only predicted shorter survival in those AML patients with low *BAALC* expression.

A combination of the aforementioned genetic markers with prognostic impact in AML, such as mutations in *MLL*, *FLT3*, *NPM1*, c-*KIT*, and *CEBPA* as well as aberrant expression levels of *EVI1*, *WT1*, *BCL2*, *MDR1*, *BAALC*, and *ERG* should ultimately refine the classification of standard-risk AML. However, many of the recently identified novel molecular markers have to be thoroughly evaluated, as single markers as well as in multivariate analyses, before reliable incorporation of these markers into therapeutic decision making.

Gene Expression Profiling and AML Prognostics

Although, nonrandom clonal aberrations are identified in 40–50% of all AML and the numbers of molecular genetic abnormalities are growing, a large proportion of AML patients cannot adequately be classified because of the lack of prognostically significant molecular abnormalities. Gene expression profiling, i.e., measuring the abundance of mRNA transcripts of thousands of genes concurrently using microarrays, has proven to be valuable in the classification of hematologic malignancies, including AML. Bone marrow and blood of patients with leukemia are ideal tissues to accurately determine gene expression profiles, since homogenous malignant cell populations are easily purified from these leukemic specimens. In recent years, gene expression profiling has been successfully introduced in clinical AML research. By gene expression profiling, cytogenetically as well as molecularly well-defined AML subclasses were predicted [77, 63, 38, 72, 10] and novel subtypes of AML were discovered [19, 72, 10, 76]. Attempts to predict response to therapy [35] as well as outcome [10] based on gene expression profiling have also been successful.

Prediction of Known Classes by Gene Expression Profiling

In pioneer studies AML was accurately distinguished from ALL and MLL by gene expression profiling [28, 2]. The diagnostic application was subsequently successfully demonstrated. In several studies it was shown that the established prognostically relevant recurrent chromosomal abnormalities such as inv(16), t(8;21), t(15;17), but also acquired mutations in *CEBPA* and *NPM1*, were predictable with very high accuracy [75, 63, 19, 39, 72, 10, 32, 33]. *NPM1* mutations, which carry prognostic information, are highly associated with a homeobox gene-specific expression signature in AML [75, 1] and are consequently predicted with high accuracy [75].

The majority of gene expression profiling studies focused on a limited number of subtypes of AML, generally comparing them to samples with a normal karyotype. Ideally these studies should be carried out using a representative AML patient-cohort, i.e., without excluding any subtype of AML, to ascertain that subtypes are predicted accurately [72, 10]. Interestingly, Haferlach et al. demonstrated recently that accurate prediction of AML with inv(16), t(8;21), t(15;17), a complex karyotype, 11q23 abnormalities and normal karyotype/other is feasible in a series of 937 bone marrow and peripheral blood samples from 892 patients with all clinically relevant leukemia subtypes [33]. There is a surprising concordance within the lists of discriminating genes predicting these specific subclasses of AML in the different gene expression profiling studies. These examples demonstrate the power and applicability of gene expression profiling, as a single assay to predict recurrent cytogenetic abnormalities [39].

Molecular genetic markers like t(11q23)/*MLL* abnormalities and *FLT3* mutations are also associated with characteristic gene expression patterns [19, 72, 10, 76, 53]. However, these signatures are less discriminative than those of AML with favorable cytogenetics and *CEBPA* mutations. While AML cases with *MLL* translocations exhibit a characteristic gene expression signature [72, 10, 2], no significant gene expression feature could be detected for cases with a PTD of the *MLL* gene [10]. Likewise, AML cases with *RAS* mutations displayed no apparent gene expression signature [72, 53]. These molecular abnormalities will, with the current level of gene expression technology, still require a combination of molecular diagnostic approaches.

Identification of Novel Subtypes of AML by Gene Expression Profiling

More challenging than predicting known classes is the discovery of novel subtypes of AML by using gene expression profiling followed by unsupervised cluster analyses, i.e., grouping of samples based on similarities in expression profiles. However, the number of studies dealing with this issue is still limited.

Bullinger and colleagues used unsupervised hierarchical cluster analyses to define two novel molecular subclasses of AML, with predominantly normal karyotypes and significant differences in survival times [10]. An outcome predictor consisting of 133 genes was constructed, which predicted overall survival and appeared a strong independent prognostic factor in multivariate analyses.

In a study of a representative cohort of 285 AML patients, 16 clusters of AML patients were revealed by gene expression profiling after unsupervised clustering [72] (Fig. 1). In this analysis novel clusters were characterized by high frequencies of certain molecular lesions or mutations, but also included patients without these molecular lesions. However, the heterogeneous nature of AML resulted in relatively small subsets, preventing statistically significant survival analyses [72].

Interestingly, after unsupervised clustering, AML samples with 11q23 abnormalities, *CEBPA* mutations and *EVI1* overexpression aggregated into two separate clusters, suggesting that a specific combination of acquired molecular lesions results in a distinct molecular signature. In addition, similar prognostic markers are sometimes recognized as poor-risk marker or standard-risk marker depending on the clinical study [12, 67, 30, 29, 44]. The separation of AML patients with a similar molecular or cytogenetic marker into two subgroups based on gene expression profiling may in fact reflect the differences in the prognostic values.

AML cases with *FLT3* ITD or *NPM1* mutations did not form one distinct cluster, but appeared to cluster within the 16 AML subtypes, suggesting that these cases may form separate disease entities within one AML subtype [72, 71, 11]. For instance, APL cases with a *FLT3* ITD, and concordant high WBC, cluster together within t(15;17) APL [71, 11]. In the future, these subtypes will include increased numbers of AML cases and *FLT3* ITD or *NPM1* mutation-specific signatures may then be formed. This would result in further refinement within the AML subtypes.

One of the 16 subclasses of AML comprised cases with no apparent cytogenetic marker as well as cases with a variety of known adverse cytogenetic markers, such as monosomies 5 and 7 and translocation t(9;22) [72]. Moreover, approximately half of the patients in this AML subtype had high expression of *EVI1* [5]. Patients of this unique cluster, which includes approximately 10% of all AML patients, showed a poor treatment outcome. In fact, the AML cells of these patients have signatures comparable to CD34$^+$ cell samples [72], suggesting that the cells may be resistant to chemotherapy, like CD34$^+$ progenitors. By

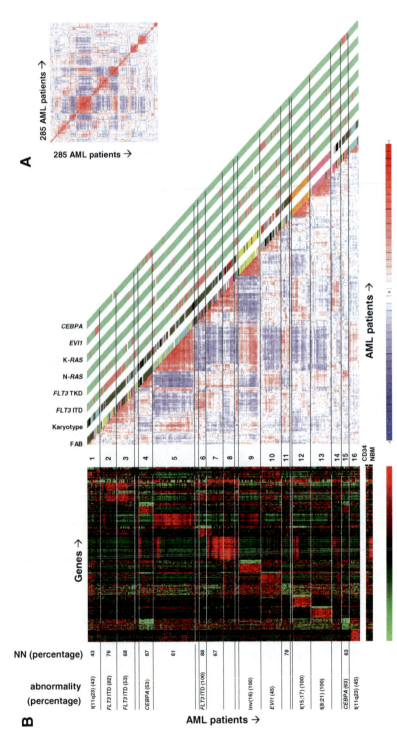

Fig. 1 (continued)

a supervised approach, i.e., analysis taking into account prior knowledge, Heuser and colleagues recently identified a characteristic gene expression signature that separated AML patients with good and poor response to induction therapy [35]. Interestingly, a substantial number of genes in this signature, including CD34, are expressed in normal hematopoietic stem/progenitor cells, which suggests that this signature identifies slow-dividing stem/progenitor cells that are resistant to induction chemotherapy [35].

These initial studies are illustrative of the applicability and power of gene expression profiling in the discovery of novel AML subtypes.

Outcome Prediction Using Genes Expression Profiling

With regard to the heterogeneity of AML patients, survival and survival time are likely to be noisy surrogates for the determination of molecular signatures that predict outcome. Based on a semi-supervised method, Bullinger and colleagues established an outcome class predictor in AML which added prognostic information over known prognostic factors as determined by multivariate proportional hazards analysis [10]. When applied to AML samples with normal karyotype, this predictor also defined good and poor outcome classes and recent analyses showed that gene expression-based outcome prediction is also feasible in larger more heterogeneous data sets [11].

◀————————————————————————————————————

Fig. 1 (continued) Unsupervised cluster analyses of 285 cases of primary AML. (**A**) Correlation view of 285 AML patients (2856 probe sets) [72]. The correlation visualization tool displays pair-wise correlations between the samples. The cells in the visualization are colored by Pearson's correlation coefficient values with deeper colors indicating higher positive (*red*) or negative (*blue*) correlations. The *scale bar* indicates 100% correlation (*red*) toward 100% anti-correlation (*blue*). One hundred percent anti-correlation would indicate that genes with high expression in one sample would always have low expression in the other sample and vice versa. The *red* diagonal displays the comparison of an AML patient with itself, i.e., 100% correlation. In order to reveal correlation patterns, a matrix ordering method is applied to rearrange the samples. The ordering algorithm starts with the most correlated sample pair and, through an iterative process, sorts all the samples into correlated blocks. Each sample is joined to a block in an ordered manner so that a correlation trend is formed within a block with the most correlated samples at the center. The blocks are then positioned along the diagonal of the plot in a similar ordered manner. (**B**) Adapted correlation view (2856 probe sets) of 285 AML patients (*right panel*) and the expression levels of the top 40 genes defining the 16 individual clusters of patients (*left panel*). FAB classification and karyotype based on cytogenetics are depicted in the columns along the original diagonal of the correlation view (FAB M0, *red;* M1, *green;* M2, *purple;* M3, *orange;* M4, *yellow;* M5, *blue;* M6, grey; karyotype: *normal-green*, inv(16), *yellow;* t(8;21), *purple;* t(15;17), *orange;* t(11q23)/ *MLL* abnormalities, *blue;* 7(q) abnormalities, *red;* +8, *pink;* complex, *black;* other, *grey*). *FLT3* ITD, *FLT3* TKD, N-*RAS*, K-*RAS,* and *CEBPA* mutations and *EVI1* overexpression are depicted in the same set of columns (*red bar*: positive and *green bar*: negative). (Reprinted with permission from Valk et al. [72], Copyright 2004, Massachusetts Medical Society)

Resistance to the first cycle of chemotherapy has been shown to be an independent prognostic factor for complete remission after the second cycle of chemotherapy as well as overall survival. Heuser et al. established a gene expression signature based on response to treatment, which separated an independent cohort of AML patients into two groups with different response rates (43.5% versus 66.7%) [35]. This may suggest that resistance to treatment may be predicted by gene expression profiling before the start of treatment.

The ultimate goal of gene expression profiling research in AML would be to establish a complete classification into good and poor outcome. So far, this has failed, due to the heterogeneity of AML. We have attempted to classify our cohort of 285 patients [72] into AML with good or poor treatment outcome, by comparing patients that achieved complete continuous remission with patients that experienced a relapse. However, this resulted in unacceptably high misclassification error of 40%, using a training and validation data set. A similar result was obtained when the same experiment was performed on samples with a normal karyotype only. Our results indicate that AML is currently too heterogeneous to accurately predict prognosis with only one classifier. It is more likely that useful prognostic classifiers will be defined to predict the outcome of AML cases within different subtypes of AML [71].

Conclusions and Perspectives Regarding Prognostics of AML

In previous years, a substantial number of novel molecular abnormalities with prognostic value in AML have been identified. Recently, gene expression profiling has been firmly established as an important tool to classify AML. Both these approaches will contribute to a comprehensive molecular classification of AML and will further support individualization of AML treatment based on characteristic molecular aberrations and expression patterns to enable a risk-adapted successful treatment of patients. In fact, while risk-stratification of AML patients based on gene expression profiling is currently still in an experimental phase, several molecular abnormalities, like *MLL*-PTD, *FLT3* ITD, *CEBPA*, and *NPM1* mutations have been validated and should be incorporated in routine molecular screening. However, although several of these candidate markers possess prognostic value, an absolute necessity for proper clinical implementation of all other candidate prognostic markers is adequate validation of the observations in larger cohorts and independent studies. One could envisage that single marker genes with prognostic value based on abnormal expression will be included in the gene expression signatures determined by genome-wide microarray analyses.

Since many of the previously characterized molecular abnormalities with predictable outcome do not display strong discriminative expression signatures, additional molecular diagnostics are still required in parallel. However, with the increasing quality and size of the microarrays as well as the number of genes included, it may ultimately be feasible to apply gene expression profiling as single comprehensive assay in AML prognostics.

The possibilities of whole genome genetic analyses are evolving rapidly. The integration of these technologies, e.g., array comparative genomic hybridization for assaying genomic DNA gains and losses, CpG island microarrays for measuring aberrant DNA methylation, and (semi-) high-throughput DNA sequencing, combined with sophisticated novel bioinformatical tools will ultimately lead toward a comprehensive AML prognostic classification. Importantly, these genome-wide analyses will not only be beneficial for prognostics but also give important insights into the true pathophysiology of AML.

References

1. Alcalay M, Tiacci E, Bergomas R, et al. Acute myeloid leukemia bearing cytoplasmic nucleophosmin (NPMc + AML) shows a distinct gene expression profile characterized by up-regulation of genes involved in stem cell maintenance. *Blood*. 2005
2. Armstrong SA, Staunton JE, Silverman LB, et al. MLL translocations specify a distinct gene expression profile that distinguishes a unique leukemia. *Nat Genet*. 2002;30:41–47.
3. Baldus CD, Tanner SM, Ruppert AS, et al. BAALC expression predicts clinical outcome of de novo acute myeloid leukemia patients with normal cytogenetics: a Cancer and Leukemia Group B Study. *Blood*. 2003;102:1613–1618.
4. Baldus CD, Thiede C, Soucek S, et al. BAALC expression and FLT3 internal tandem duplication mutations in acute myeloid leukemia patients with normal cytogenetics: prognostic implications. *J Clin Oncol*. 2006;24:790–797.
5. Barjesteh van Waalwijk van Doorn-Khosrovani S, et al. High EVI1 expression predicts poor survival in acute myeloid leukemia: a study of 319 de novo AML patients. *Blood*. 2003;101:837–845.
6. Barragan E, Cervera J, Bolufer P, et al. Prognostic implications of Wilms' tumor gene (WT1) expression in patients with de novo acute myeloid leukemia. *Haematologica*. 2004;89:926–933.
7. Beghini A, Peterlongo P, Ripamonti CB, et al. C-kit mutations in core binding factor leukemias. *Blood*. 2000;95:726–727.
8. Bergmann L, Miething C, Maurer U, et al. High levels of Wilms' tumor gene (wt1) mRNA in acute myeloid leukemias are associated with a worse long-term outcome. *Blood*. 1997;90:1217–1225.
9. Bowen DT, Frew ME, Hills R, et al. RAS mutation in acute myeloid leukemia is associated with distinct cytogenetic subgroups but does not influence outcome in patients younger than 60 years. *Blood*. 2005;106:2113–2119.
10. Bullinger L, Dohner K, Bair E, et al. Use of gene-expression profiling to identify prognostic subclasses in adult acute myeloid leukemia. *N Engl J Med*. 2004;350:1605–1616.
11. Bullinger L, Valk PJ. Gene expression profiling in acute myeloid leukemia. *J Clin Oncol*. 2005;23:6296–6305.
12. Byrd JC, Mrozek K, Dodge RK, et al. Pretreatment cytogenetic abnormalities are predictive of induction success, cumulative incidence of relapse, and overall survival in adult patients with de novo acute myeloid leukemia: results from Cancer and Leukemia Group B (CALGB 8461). *Blood*. 2002;100:4325–4336.
13. Thiede C, Koch S, Creutzig E, et al. Prevalence and prognostic impact of NPM1 mutations in 1485 adult patients with acute myeloid leukemia (AML). *Blood*. 2006.

14. Cairoli R, Beghini A, Morello E, et al. Imatinib mesylate in the treatment of Core Binding Factor leukemias with KIT mutations. A report of three cases. *Leuk Res.* 2005;29:397–400.
15. Caligiuri MA, Strout MP, Lawrence D, et al. Rearrangement of ALL1 (MLL) in acute myeloid leukemia with normal cytogenetics. *Cancer Res.* 1998;58:55–59.
16. Cammenga J, Horn S, Bergholz U, et al. Extracellular KIT receptor mutants, commonly found in core binding factor AML, are constitutively active and respond to imatinib mesylate. *Blood.* 2005;106:3958–3961.
17. Care RS, Valk PJ, Goodeve AC, et al. Incidence and prognosis of c-KIT and FLT3 mutations in core binding factor (CBF) acute myeloid leukaemias. *Br J Haematol.* 2003; 121:775–777.
18. Cazzaniga G, Dell'oro MG, Mecucci C, et al. Nucleophosmin mutations in childhood acute myelogenous leukemia with normal karyotype. *Blood.* 2005
19. Debernardi S, Lillington DM, Chaplin T, et al. Genome-wide analysis of acute myeloid leukemia with normal karyotype reveals a unique pattern of homeobox gene expression distinct from those with translocation-mediated fusion events. *Genes Chromosomes Cancer.* 2003;37:149–158.
20. Dohner K, Schlenk RF, Habdank M, et al. Mutant nucleophosmin (NPM1) predicts favorable prognosis in younger adults with acute myeloid leukemia and normal cytogenetics: interaction with other gene mutations. *Blood.* 2005;106: 3740–3746.
21. Dohner K, Tobis K, Ulrich R, et al. Prognostic significance of partial tandem duplications of the MLL gene in adult patients 16 to 60 years old with acute myeloid leukemia and normal cytogenetics: a study of the Acute Myeloid Leukemia Study Group Ulm. *J Clin Oncol.* 2002;20:3254–3261.
22. Falini B, Bolli N, Shan J, et al. Both carboxy-terminus NES motif and mutated tryptophan(s) are crucial for aberrant nuclear export of nucleophosmin leukemic mutants in NPMc+ AML. *Blood.* 2006
23. Falini B, Mecucci C, Tiacci E, et al. Cytoplasmic nucleophosmin in acute myelogenous leukemia with a normal karyotype. *N Engl J Med.* 2005;352:254–266.
24. Frohling S, Schlenk RF, Breitruck J, et al. Prognostic significance of activating FLT3 mutations in younger adults (16 to 60 years) with acute myeloid leukemia and normal cytogenetics: a study of the AML Study Group Ulm. *Blood.* 2002;100:4372–4380.
25. Frohling S, Schlenk RF, Stolze I, et al. CEBPA mutations in younger adults with acute myeloid leukemia and normal cytogenetics: prognostic relevance and analysis of cooperating mutations. *J Clin Oncol.* 2004;22:624–633.
26. Garg M, Moore H, Tobal K, et al. Prognostic significance of quantitative analysis of WT1 gene transcripts by competitive reverse transcription polymerase chain reaction in acute leukaemia. *Br J Haematol.* 2003;123:49–59.
27. Gilliland DG, Griffin JD. The roles of FLT3 in hematopoiesis and leukemia. *Blood.* 2002;100:1532–1542.
28. Golub TR, Slonim DK, Tamayo P, et al. Molecular classification of cancer: class discovery and class prediction by gene expression monitoring. *Science.* 1999;286:531–537.
29. Grimwade D, Walker H, Harrison G, et al. The predictive value of hierarchical cytogenetic classification in older adults with acute myeloid leukemia (AML): analysis of 1065 patients entered into the United Kingdom Medical Research Council AML11 trial. *Blood.* 2001;98:1312–1320.
30. Grimwade D, Walker H, Oliver F, et al. The importance of diagnostic cytogenetics on outcome in AML: analysis of 1,612 patients entered into the MRC AML 10 trial. The Medical Research Council Adult and Children's Leukaemia Working Parties. *Blood.* 1998;92:2322–2333.
31. Schnittger S, Kohl TM, Haferlach T, et al. KIT-D816 mutations in AML1-ETO-positive AML are associated with impaired event-free and overall survival. *Blood.* 2006;107:1791–1799.

32. Gutierrez NC, Lopez-Perez R, Hernandez JM, et al. Gene expression profile reveals deregulation of genes with relevant functions in the different subclasses of acute myeloid leukemia. *Leukemia*. 2005;19:402–409.
33. Haferlach T, Kohlmann A, Schnittger S, et al. Global approach to the diagnosis of leukemia using gene expression profiling. *Blood*. 2005;106:1189–1198.
34. Harris NL, Jaffe ES, Diebold J, et al. World Health Organization classification of neoplastic diseases of the hematopoietic and lymphoid tissues: report of the Clinical Advisory Committee meeting-Airlie House, Virginia, November 1997. *J Clin Oncol*. 1999;17:3835–3849.
35. Heuser M, Wingen LU, Steinemann D, et al. Gene-expression profiles and their association with drug resistance in adult acute myeloid leukemia. *Haematologica*. 2005;90:1484–1492.
36. Karakas T, Maurer U, Weidmann E, et al. High expression of bcl-2 mRNA as a determinant of poor prognosis in acute myeloid leukemia. *Ann Oncol*. 1998;9: 159–165.
37. Karakas T, Miething CC, Maurer U, et al. The coexpression of the apoptosis-related genes bcl-2 and wt1 in predicting survival in adult acute myeloid leukemia. *Leukemia*. 2002;16:846–854.
38. Kiyoi H, Naoe T, Nakano Y, et al. Prognostic implication of FLT3 and N-RAS gene mutations in acute myeloid leukemia. *Blood*. 1999;93:3074–3080.
39. Kohlmann A, Schoch C, Schnittger S, et al. Molecular characterization of acute leukemias by use of microarray technology. *Genes Chromosomes Cancer*. 2003;37: 396–405.
40. Kottaridis PD, Gale RE, Frew ME, et al. The presence of a FLT3 internal tandem duplication in patients with acute myeloid leukemia (AML) adds important prognostic information to cytogenetic risk group and response to the first cycle of chemotherapy: analysis of 854 patients from the United Kingdom Medical Research Council AML 10 and 12 trials. *Blood*. 2001;98:1752–1759.
41. Lennartsson J, Jelacic T, Linnekin D, et al. Normal and oncogenic forms of the receptor tyrosine kinase kit. *Stem Cells*. 2005;23:16–43.
42. Levis M, Small D. FLT3: ITDoes matter in leukemia. *Leukemia*. 2003;17:1738–1752.
43. Lowenberg B. Prognostic factors in acute myeloid leukaemia. *Best Pract Res Clin Haematol*. 2001;14:65–75.
44. Lowenberg B, Downing JR, Burnett A. Acute myeloid leukemia. *N Engl J Med*. 1999;341:1051–1062.
45. Marcucci G, Baldus CD, Ruppert AS, et al. Overexpression of the ETS-related gene, ERG, predicts a worse outcome in acute myeloid leukemia with normal karyotype: a Cancer and Leukemia Group B study. *J Clin Oncol*. 2005;23:9234–9242.
46. Mrozek K, Heerema NA, Bloomfield CD. Cytogenetics in acute leukemia. *Blood Rev*. 2004;18:115–136.
47. Mrozek K, Heinonen K, Bloomfield CD. Clinical importance of cytogenetics in acute myeloid leukaemia. *Best Pract Res Clin Haematol*. 2001;14:19–47.
48. Mrozek K, Heinonen K, Bloomfield CD. Prognostic value of cytogenetic findings in adults with acute myeloid leukemia. *Int J Hematol*. 2000;72:261–271.
49. Nakano Y, Naoe T, Kiyoi H, et al. Prognostic value of p53 gene mutations and the product expression in de novo acute myeloid leukemia. *Eur J Haematol*. 2000;65:23–31.
50. Nakao M, Yokota S, Iwai T, et al. Internal tandem duplication of the flt3 gene found in acute myeloid leukemia. *Leukemia*. 1996;10:1911–1918.
51. Nanri T, Matsuno N, Kawakita T, et al. Imatinib mesylate for refractory acute myeloblastic leukemia harboring inv(16) and a C-KIT exon 8 mutation. *Leukemia*. 2005;19: 1673–1675.
52. Nanri T, Matsuno N, Kawakita T, et al. Mutations in the receptor tyrosine kinase pathway are associated with clinical outcome in patients with acute myeloblastic leukemia harboring t(8;21)(q22;q22). *Leukemia*. 2005;19:1361–1366.

53. Neben K, Schnittger S, Brors B, et al. Distinct gene expression patterns associated with FLT3- and NRAS-activating mutations in acute myeloid leukemia with normal karyotype. *Oncogene*. 2005;24:1580–1588.
54. Neubauer A, Dodge RK, George SL, et al. Prognostic importance of mutations in the ras proto-oncogenes in de novo acute myeloid leukemia. *Blood*. 1994;83:1603–1611.
55. Nucifora G, Laricchia-Robbio L, Senyuk V. EVI1 and hematopoietic disorders: history and perspectives. *Gene*. 2006;368:1–11.
56. Ozeki K, Kiyoi H, Hirose Y, et al. Biologic and clinical significance of the FLT3 transcript level in acute myeloid leukemia. *Blood*. 2004;103:1901–1908.
57. Pabst T, Mueller BU, Zhang P, et al. Dominant-negative mutations of CEBPA, encoding CCAAT/enhancer binding protein-alpha (C/EBPalpha), in acute myeloid leukemia. *Nat Genet*. 2001;27:263–270.
58. Preudhomme C, Sagot C, Boissel N, et al. Favorable prognostic significance of CEBPA mutations in patients with de novo acute myeloid leukemia: a study from the Acute Leukemia French Association (ALFA). *Blood*. 2002;100:2717–2723.
59. Schittenhelm MM, Shiraga S, Schroeder A, et al. Dasatinib (BMS-354825), a dual SRC/ABL kinase inhibitor, inhibits the kinase activity of wild-type, juxtamembrane, and activation loop mutant KIT isoforms associated with human malignancies. *Cancer Res*. 2006;66:473–481.
60. Schmid D, Heinze G, Linnerth B, et al. Prognostic significance of WT1 gene expression at diagnosis in adult de novo acute myeloid leukemia. *Leukemia*. 1997;11:639–643.
61. Guerci A, Merlin JL, Missoum N, et al. Predictive value for treatment outcome in acute myeloid leukemia of cellular daunorubicin accumulation and P-glycoprotein expression simultaneously determined by flow cytometry. *Blood*. 1995;85:2147–2153.
62. Schnittger S, Schoch C, Kern W, et al. Nucleophosmin gene mutations are predictors of favorable prognosis in acute myelogenous leukemia with a normal karyotype. *Blood*. 2005;106:3733–3739.
63. Schoch C, Kohlmann A, Schnittger S, et al. Acute myeloid leukemias with reciprocal rearrangements can be distinguished by specific gene expression profiles. *Proc Natl Acad Sci U S A*. 2002;99:10008–10013.
64. Sheikhha MH, Awan A, Tobal K, et al. Prognostic significance of FLT3 ITD and D835 mutations in AML patients. *Hematol J*. 2003;4:41–46.
65. Shiah HS, Kuo YY, Tang JL, et al. Clinical and biological implications of partial tandem duplication of the MLL gene in acute myeloid leukemia without chromosomal abnormalities at 11q23. *Leukemia*. 2002;16:196–202.
66. Shimada A, Taki T, Tabuchi K, et al. KIT mutations, and not FLT3 internal tandem duplication, are strongly associated with a poor prognosis in pediatric acute myeloid leukemia with t(8;21): a study of the Japanese Childhood AML Cooperative Study Group. *Blood*. 2006;107:1806–1809.
67. Slovak ML, Kopecky KJ, Cassileth PA, et al. Karyotypic analysis predicts outcome of preremission and postremission therapy in adult acute myeloid leukemia: a Southwest Oncology Group/Eastern Cooperative Oncology Group Study. *Blood*. 2000;96: 4075–4083.
68. Stirewalt DL, Kopecky KJ, Meshinchi S, et al. FLT3, RAS, and TP53 mutations in elderly patients with acute myeloid leukemia. *Blood*. 2001;97:3589–3595.
69. Thiede C, Koch S, Creutzig E, et al. Prevalence and prognostic impact of NPM1 mutations in 1485 adult patients with acute myeloid leukemia (AML). *Blood*. 2006.
70. Thiede C, Steudel C, Mohr B, et al. Analysis of FLT3-activating mutations in 979 patients with acute myelogenous leukemia: association with FAB subtypes and identification of subgroups with poor prognosis. *Blood*. 2002;99:4326–4335.
71. Valk PJ, Delwel R, Lowenberg B. Gene expression profiling in acute myeloid leukemia. *Curr Opin Hematol*. 2005;12:76–81.
72. Valk PJ, Verhaak RG, Beijen MA, et al. Prognostically useful gene-expression profiles in acute myeloid leukemia. *N Engl J Med*. 2004;350:1617–1628.

73. van den Heuvel-Eibrink MM, van der Holt B, te Boekhorst PA, et al. MDR 1 expression is an independent prognostic factor for response and survival in de novo acute myeloid leukaemia. *Br J Haematol.* 1997;99:76–83.

74. van Waalwijk van Doorn-Khosrovani SB, Erpelinck C, et al. Biallelic mutations in the CEBPA gene and low CEBPA expression levels as prognostic markers in intermediate-risk AML. *Hematol J.* 2003;4:31–40.

75. Verhaak RG, Goudswaard CS, van Putten W, et al. Mutations in nucleophosmin NPM1 in acute myeloid leukemia (AML): association with other gene abnormalities and previously established gene expression signatures and their favorable prognostic significance. *Blood.* 2005

76. Vey N, Mozziconacci MJ, Groulet-Martinec A, et al. Identification of new classes among acute myelogenous leukaemias with normal karyotype using gene expression profiling. *Oncogene.* 2004;23:9381–9391.

77. Virtaneva K, Wright FA, Tanner SM, et al. Expression profiling reveals fundamental biological differences in acute myeloid leukemia with isolated trisomy 8 and normal cytogenetics. *Proc Natl Acad Sci U S A.* 2001;98:1124–1129.

78. Wang YY, Zhou GB, Yin T, et al. AML1-ETO and C-KIT mutation/overexpression in t(8;21) leukemia: implication in stepwise leukemogenesis and response to Gleevec. *Proc Natl Acad Sci U S A.* 2005;102:1104–1109.

79. Whitman SP, Archer KJ, Feng L, et al. Absence of the wild-type allele predicts poor prognosis in adult de novo acute myeloid leukemia with normal cytogenetics and the internal tandem duplication of FLT3: a cancer and leukemia group B study. *Cancer Res.* 2001;61:7233–7239.

80. Yanada M, Matsuo K, Suzuki T, et al. Prognostic significance of FLT3 internal tandem duplication and tyrosine kinase domain mutations for acute myeloid leukemia: a meta-analysis. *Leukemia.* 2005;19:1345–1349.

81. Yanada M, Terakura S, Yokozawa T, et al. Multiplex real-time RT-PCR for prospective evaluation of WT1 and fusion gene transcripts in newly diagnosed de novo acute myeloid leukemia. *Leuk Lymphoma.* 2004;45:1803–1808.

Receptor Tyrosine Kinase Alterations in AML – Biology and Therapy

Derek L. Stirewalt and Soheil Meshinchi

Abstract Acute myeloid leukemia (AML) is the most common form of leukemia in adults, and despite some recent progress in understanding the biology of the disease, AML remains the leading cause of leukemia-related deaths in adults and children. AML is a complex and heterogeneous disease, often involving multiple genetic defects that promote leukemic transformation and drug resistance. The cooperativity model suggests that an initial genetic event leads to maturational arrest in a myeloid progenitor cell, and subsequent genetic events induce proliferation and block apoptosis. Together, these genetic abnormalities lead to clonal expansion and frank leukemia. The purpose of this chapter is to review the biology of receptor tyrosine kinases (RTKs) in AML, exploring how RTKs are being used as novel prognostic factors and potential therapeutic targets.

Introduction

Acute myeloid leukemia (AML) is the most common form of leukemia in adults, and despite some recent progress in understanding the biology of the disease, AML remains the leading cause of leukemia-related deaths in adults and children [132] AML is a complex and heterogeneous disease, often involving multiple genetic defects that promote leukemic transformation and drug resistance. The co-operativity model suggests that an initial genetic event leads to maturational arrest in a myeloid progenitor cell, and subsequent genetic events induce proliferation and block apoptosis. Together, these genetic abnormalities lead to clonal expansion and frank leukemia [64, 170, 127, 176].

Mutations in receptor tyrosine kinases (RTKs) or their downstream effectors are extremely common in AML, with estimated 40–60% of AML patients

D.L. Stirewalt (✉)
Clinical Research Division, From Fred Hutchinson Cancer Research Center, Seattle, WA, USA
e-mail: dstirewa@fhcrc.org

L. Nagarajan (ed.), *Acute Myelogenous Leukemia*,
Cancer Treatment and Research 145, DOI 10.1007/978-0-387-69259-3_6,
© Springer Science+Business Media, LLC 2010

harboring a mutation abnormality in RTKs [144, 89, 45, 49]. In addition, another 15–25% of AML patients will have a mutation in one of the downstream effectors in RTK pathways [144, 168, 113, 69, 27]. More than 50 different RTKs have been identified, which are classified into 20 subfamilies based on their structural and functional characteristics (reviewed in Ref. [89]) [14, 38, 120]. A meticulous examination of all the different RTKs has yet to be performed, but such studies are underway. These studies will most likely discover that an even greater percentage of AML patients harbor mutations or abnormalities in RTK pathways. Although several RTKs have been implicated in malignancies, the vast majority of RTK mutations in AML, thus far, have been found in the RTK subclass III family (a.k.a. the PDGFR family) [120, 3, 124]. The subclass III RTK family consists of FLT3, KIT, PDGFRA, PDGFRB, and CSF1R, and the majority of this review will examine the biology, prognosis, and potential therapeutic targets of these RTKs, focusing on FLT3 and KIT mutations, which are by far the most common RTKs known to be affected in AML. The role of CSFR1 (a.k.a. C-FMS) and PDGFR in myeloid malignancies will also be briefly discussed. In addition, we will assess the growing interest in small molecule inhibitors against RTKs as potential therapeutic targets for AML.

Receptor Tyrosine Kinase Activation and Downstream Effectors

RTKs play a critical role in myeloid proliferation, differentiation, and apoptosis [115, 167, 126, 44, 133, 97, 102, 99, 93]. Structurally, subclass III RTKs consist of an extracellular ligand-binding (E) domain, transmembrane (TM) dimerization domain, juxtamembrane (JM) domain, and an intracellular tyrosine kinase domain (Fig. 1A). In their inactive state, RTKs exist primarily as monomers, and multiple autoinhibitory or intrinsic repressive forces prevent dimerization (a.k.a. receptor activation) [52, 123]. While in the monomeric state, the RTK displays a "closed" conformation, which prevents easy phosphorylation of specific tyrosine residues within the intercellular domains and limits inappropriate activation (Fig. 1A). Activation begins when a ligand binds to the extracellular domain (or domains), causing a conformational change in the receptor. The new conformation reverses the intrinsic repulsive forces of the receptor, promoting dimerization with either itself or other membrane-bound RTKs (Fig. 1B). Together, these changes lead to an "open" confirmation of the receptor, which facilitates the transfer of a phosphate from ATP to the tyrosine substrate within the intracellular kinase domain (Fig. 1B). This activated conformational change and/or phosphorylated tyrosine also promote the docking of adaptor proteins (e.g., SHC), which also become activated. The activated adapter proteins then interact with downstream effectors, efficiently transmitting the extracellular signal to the appropriate intracellular pathways (Fig. 2). After activation, RTKs are rapidly internalized and degraded, such that within 20 min the signal will start to dissipate [178, 163]. This rapid degradation and

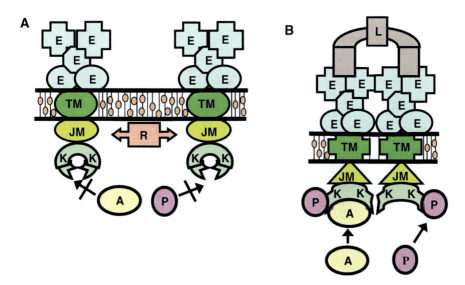

Fig. 1 Subclass III tyrosine kinase receptor structure and activation. (A) Inactive receptor. Subclass III RTKs consist of an extracellular ligand-binding domain (E), transmembrane dimerization domain (TM), juxtamembrane domain (JM), and an intracellular tyrosine kinase domain (K). In their inactive state, RTKs exist primarily as monomers, and multiple autoinhibitory or intrinsic repressive forces prevent dimerization (R). Note the "closed" conformation or the tyrosine kinase domain. (B) Activated receptor. RTKs become activated when a ligand (L) binds to their extracellular domain (E) causing a conformational change in the receptor. These changes lead to an "open" confirmation of the intracellular tyrosine kinase domain (K), which facilitates the transfer of phosphate (P) to the tyrosine kinase domain. This activated conformational change and/or phosphorylated tyrosine also promote the docking of adaptor proteins (A), which are then activated. The activated adapter proteins facilitate transmission of the extracellular signal to the appropriate intracellular pathways

downregulation of the receptor contributes to the tightly controlled signaling activity of RTKs.

Several different types of mutations (missense point mutations, deletions, insertions, and internal tandem duplications) in AML cells have been described in the different RTKs (mainly FLT3 and c-KIT, Fig. 3) [144, 89, 45, 49, 69, 101, 10]. Despite the different types of mutations, the RTK sequences tend to remain in frame, ensuring translation of the entire protein. These mutations have been found to be most frequent in JM domain or tyrosine kinase domain and will be described in more detail under the specific subclass RTKs; nevertheless, there are two general themes to these mutations. Those mutations involving the JM domain tend to be large insertions, which probably disrupt the intrinsic repulsive forces that naturally prevent dimerization. However, recently, small point mutations and deletions have been described in the JM domain of FLT3 [146]. Once these repulsive forces are disrupted, ligand-independent activation occurs. Whether the different types and sizes of

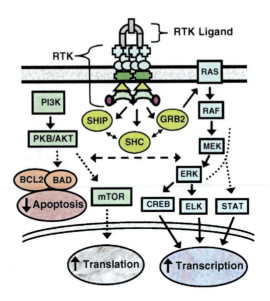

Fig. 2 Tyrosine kinase receptor pathway. Binding of a ligand (L) to tyrosine kinase receptor activates multiple downstream effectors, including the PI3K (phosphatidylinositol 3' kinase) and RAS pathways. *Solid* arrows are more direct interactions, while *broken* arrows represent associations that probably involve multiple steps between the proteins. The activated RTK interacts with multiple adapter proteins: SH2-containing sequence proteins (SHC), SH2-containing inositol phosphatase (SHIP), GRB2, and others, which connect the RTK to the PI3K and RAS pathways. RAS activation stimulates the MAPK kinase pathway: RAF, MAPK/ERK kinases (MEK), extracellular-signal-regulated kinase kinases (ERK), and 90-kDa ribosomal protein S6 kinase (RSK). These downstream effectors activate cyclic adenosine monophosphate-response element-binding protein (CREB), ELK, and signal transducer and activators of transcription (STATs), which lead to transcription of specific genes that promote proliferation. Activated PI3K stimulates protein kinase B (PKB/AKT) and other members of the PI3K pathway (e.g., rapamycin or mTOR), which promotes translation. In addition, activated PI3K induces phosphorylation of the pro-apoptotic BCL2 family protein (BAD), which blocks apoptosis by binding BCL2. Both pathways probably also interact with each other and other pathways/effectors such as BRCA1, p21WAF1, and p27KIP1

mutations in the JM domain have a unique biological and clinical significance is currently being investigated [145, 116]. The other domain of RTKs that is frequently mutated is that of the tyrosine kinase domain (or activation loop domain) [175, 1, 160, 42, 140, 67]. Missense point mutations in the tyrosine kinase domain (a.k.a. TKD mutations or FLT3/ALM) also constitutively activate the receptor. Similar to JM mutations, TKD mutations also activate downstream effectors, inappropriately stimulating pathways that are critical in the normal regulation of differentiation, proliferation, and apoptosis [175, 1].

In addition to RTK mutations, mutations in downstream effectors within RTK pathways may also play a critical role in leukemogenesis. For example, RAS, a component of many RTK pathways, is mutated in approximately

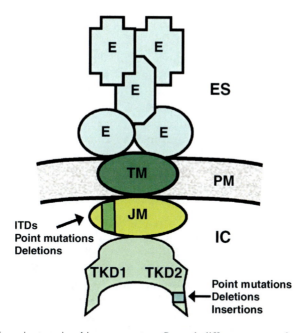

Fig. 3 Mutations in tyrosine kinase receptor. Several different types of mutations have been described in tyrosine kinase receptors. The vast majority of these mutations are either in the juxtamembrane (JM) domain or in the tyrosine kinase domain (TKD). Internal tandem duplications (*bright green*) are the most common type of mutations in JM domain (*star*), while point mutations (*bright blue*) are the most common type of mutations in the TKD. Other labels include the extracellular (E) and transmembrane (TM) domains, plasma membrane (PM), extracellular space (ES), and intracellular cytosol (IC)

20–25% of AML patients – usually either NRAS or KRAS [144, 89, 113, 35, 26, 104, 84, 17, 22, 119], and mutations in other members of the RTK pathways, such as PTNP11, have also been discovered in AML patients [27, 119, 138, 12]. Besides activating the RTK pathway, RTK mutations inappropriately activate other pathways such as JAK/STAT pathway [151, 54, 159, 13]. Recently, the JAK/STAT pathway has received attention due to novel small molecule inhibitors that make it a potential therapeutic target for several malignancies [95, 171, 174]. Whether it is the constitutive activation of the receptor by an RTK mutation [144, 89, 45, 49, 69, 101, 74, 2, 152, 130, 146, 175, 1, 19, 21, 9, 11, 24, 134, 129, 10, 160, 42, 140, 67, 143, 163, 117, 109], an autocrine/paracrine stimulation of the receptor by a ligand secreting tumor (e.g., VEGF) [161, 4, 48], or the activation of the downstream effectors [27, 119, 138, 12, 151, 54, 159, 13, 181, 96, 15, 76], inappropriate activation of RTK pathways directly contribute to pathogenesis of AML, progression of the disease, and its resistance to chemotherapy.

FLT3 Mutations

As FLT3 mutations have been implicated in the prognosis of AML, FLT3 mutations remain one of the most common genetic abnormalities in AML identified thus far [144, 89, 69, 101, 74, 2, 152, 130, 146, 175, 1, 145, 22, 157, 179, 59, 73, 37, 41]. As described above, FLT3 mutations occur within two specific regions within the FLT3 gene (juxtamembrane domain and tyrosine kinase domain). The most common type of a FLT3 mutation is that of internal tandem duplication (FLT3/ITD) in the JM domain, which occurs in 15–35% of AML patients [144, 89, 69, 74, 152, 130, 75, 90]. FLT3/ITDs are rare in infant AML, where only approximately 1% of children <1 year harbor a FLT3/ITD, but steadily increases with aging, such that 10–15% of pediatric and 20–35% of adult AML patients have FLT3/ITDs [144, 89, 69, 74, 152, 130, 75, 90]. FLT3/ ITDs cause ligand-independent dimerization and autophosphorylation of the receptor, leading to constitutive activation of the FLT3 and many downstream effectors (SHC, RAS, ERK, AKT, and STAT5) [151, 54, 88, 71, 70, 94]. Besides FLT3/ITDs, smaller insertions, deletions, and missense point mutations have recently been described in the JM of AML patients [146]. These mutations are relatively rare, occurring in less than 5% of patients [146]. Although the clinical significance of these "non-ITD" mutations in the JM (a.k.a. FLT3-JM-PM) are currently unknown, Reindl et al. recently found that FLT3-JM-PMs promoted ligand-independent dimerization, autophosphorylation, and constitutive activation of the receptor; however, the activation seemed to be "weaker" than compared to classic FLT3/ITD transduced cells, as indicated by a lower level of phosphorylation of the receptor and less activation of downstream STAT5 in cells transduced with FLT3-JM-PM [116].

FLT3/ITDs are associated with rapid disease progression and resistance to conventional therapy [69, 74, 152, 130, 75]. Initial studies demonstrated a strong prognostic significance for the presence of FLT3/ITD, such that patients with this mutation had an extremely poor clinical outcome compared to the patients without FLT3/ITD [69, 75]. In more recent studies using contemporary chemotherapy, the prognostic significance of FLT3/ITD has been less dramatic, with an overall survival of approximately 30% for the FLT3/ITD population compared to 45% to that of patients without FLT3/ITD [74, 152]. However, these studies also identified a subclass of patients with FTL3/ITD, in which the mutant ITD to wild-type allele (ITD allelic ratio or ITD-AR) correlates with clinical outcome. ITD-ARs vary significantly from patient to patient, and this difference may have clinical implications [152, 18, 173]. For example, some AML patients have a predominantly mutant ITD product with little or no normal product (high ITD-AR), whereas others have an equal or higher distribution of normal product (low ITD-AR). The ITD-AR has been used to identify patients with FLT3/ITD at a higher risk of relapse and poor outcome in a number of clinical trials [152, 18, 173]. Although ITD-AR may become a critical tool in the risk identification of FLT3/ITD-positive patients, the exact

ITD-AR threshold "cut-off" that identifies high-risk patients has not been established. Thiede et al. used ITD-AR threshold of 0.78 to define relapse risk in FLT3/ITD-positive patients [152]. Patients with low allelic ratio (AR $\leq$ 0.78) had an overall survival of nearly 60% compared to overall survival of 0% in patients with ITD-AR of >0.78 (p = 0.006) [152]. They also suggested that ITD allelic ratio may be a continuous variable, as changing the ITD-AR cut-off altered the clinical outcome. Similar findings were shown in a pediatric AML population from European cooperative studies. Using the ITD-AR median, Zwaan et al. demonstrated that patients with high ITD-AR (ITD-AR > median of 0.69) had a poor outcome, whereas the outcome for those with low ITD-AR (ITD-AR $\leq$ 0.69) was no different than patients without FLT3/ITD [173]. More recent analysis of ITD-AR in a cohort of 630 children treated in CCG-2941/2961 has revealed that ITD-AR of 0.4 can identify the highest proportion of FLT3/ITD-positive patients at high risk of relapse (personal communication by Meshinchi, submitted for publication). Underlying mechanisms for the allelic ratio variation is under study. We and others have presented data in support of loss of heterozygosity (LOH) in chromosome 13q12 as a possible mechanism for high ITD-AR in some of the AML patients [75, 18]. Other studies, however, have failed to demonstrate LOH in patients with high ITD-AR [152], suggesting that there may be additional factors responsible for variation in the ITD-AR.

Recently, we demonstrated that size of duplicated region in FLT3/ITD-positive AML may have prognostic significance. In a study of adult 151 AML patients treated, those AML patients with larger ITDs had a significantly higher relapse compared to those with smaller ITDs [145]. Ongoing studies in other populations are underway, but this data would reinforce the hypothesis that different sizes and types of mutations in the JM may have unique biological and clinical significance, adding yet another layer of complexity to the use of FLT3/ITDs as a prognostic marker [145, 116, 182].

FLT3 mutations in activation loop of the tyrosine kinase (FLT3/ALM) occur in approximately 5–10% of AML patients [89, 175, 1, 140] making FLT3/ALM the second most common type of FLT3 mutation. Like FLT3/ITDs, FLT3/ALM constitutively activates the FLT3 receptor and downstream effectors [175, 140, 67, 151, 54, 13, 73, 88, 71]. However, it remains to be determined whether FLT3/ITDs and FLT3/ALMs have similar biological consequences. Recent evidence would suggest that mutations in JM and TKD cause biologically different responses. When Grundler et al. transduced mice marrow with FLT3/ITDs, the mice developed the classic myeloproliferative syndrome, which had previously been described in other FLT3/ITD transduction murine models [147, 53]. However, mice transduced with FLT3/ALM developed an oligoclonal lymphoid disorder [147]. These data argue against the hypothesis that the biological consequences of FLT3/ITDs and FLT3/ALMs are the same. In addition, global RNA expression studies have found distinct expression patterns between FLT3/ITDs and FLT3/ALMs, which readily differentiate the two types of mutations, providing additional evidence that suggest biological differences between the two types of FLT3-activating mutations [65]. When one examines the clinical significance of FLT3/ALM, there appear to be clinical differences between patients with FLT3/ITDs and FLT3/AMLs. Unlike FLT3/ITDs, the frequency of FLT3/ALMs does not vary with age. In addition, available data suggest that patients with FLT3/ALM have a lower diagnostic white count, higher remission rates, lower relapse rate, and better overall survival than patients with FLT3/ITD [89, 152, 175, 78]. However, it must be noted that the frequency of FLT3/ALMs is considerably less than FLT3/ITDs, making it more difficult to obtain significant power to convincingly rule out a possible clinical significance for these mutations. Also, there have been fewer studies examining the clinical significance of FLT3/ ALMs, since these mutations were first identified in 2001, which is approximately 5 years after FLT3/ITDs were discovered [101, 175, 1, 140].

Together, the data suggest that FLT3/ITDs, FLT3/ALM, and FLT3/ JM/PM all promote constitutive activation of the receptor, but probably have biological differences that impact their prognostic significance. Therefore, in evaluating the prognostic relevance of FLT3-activating mutations, one must keep in mind not only the presence or absence of a mutation but also location, type, and allelic ratio of the mutation. The identification of a high-risk population in FLT3/ITD-positive patients is of great importance, as it may identify a significant number of patients who are destined for poor outcome and may benefit from alternative treatments such as early hematopoietic stem cell transplant. The ITD-AR or ITD size may provide additional tools to better risk-stratify AML patients. Besides understanding how these different FLT3 mutations respond to chemotherapy, it will be critical to determine how these different mutations impact the responsiveness to small molecule inhibitors. However, at this time, investigators are left with trying to better risk-stratify AML patients based on a variety of poorly understood surrogate FLT3 prognostic markers.

KIT Mutations

Activating mutations of KIT receptor gene have been reported in a variety of myeloid malignancies. Early studies implicated KIT mutations in the pathogenesis of mastocytosis [42, 87]; however, recent studies have found that these mutations may also be involved in the pathogenesis of AML, especially those with t(8;21) or inv(16) [170, 176, 89, 49, 27, 19, 24, 134, 129, 10, 85, 23]. Activating mutations in KIT receptor gene usually occur in either the JM domain, which regulates receptor dimerization, or the intracellular tyrosine kinase domain of the receptor gene. Mutations in both regions lead to constitutive activation of the KIT receptor [160, 42, 103, 68, 162].

KIT mutations have been reported in 3–15% of adult and pediatric AML [49, 24, 85] and in significantly higher proportion of those with t(8;21) or inv(16), where KIT mutations were observed in nearly 40–50% of AML involving the core-binding factor (CBF) [49]. Data suggest that KIT-activating mutations may co-operate with t(8;21) translocation to contribute to myeloid leukemogenesis, where KIT mutation leads to proliferative advantage in cells which have undergone maturation arrest due to t(8;21) translocations [170]. Although presence of KIT mutations may not display prognostic significance in AML patients at large, their presence may have prognostic and possible therapeutic implications in AML patients involving CBF [49, 19, 21, 134]. Cairoli et al. evaluated 67 patients with t(8;21) or inv(16) and demonstrated that 46% of the AML patients harbored a missense, insertion, deletion, or internal tandem duplication mutations in either exon 8, 11, or 17 of the KIT receptor gene [19, 164]. Missense mutations in the tyrosine kinase domain (TKD) of the receptor (D816) was the most common mutation observed, where 20/67 patients (29%) had D816 missense mutations. Correlation of the D816 mutation in KIT with clinical outcome demonstrated that those patients with D816 mutations had a significantly higher relapse rate and worse survival. A similar study reported an adverse outcome for pediatric AML patients, in which these investigators identified mutations in the KIT TKD in 17% of pediatric AML patients with t(8;21) [134]. In contrast, Care et al. found more prognostic significance for KIT mutations in exon 8 in CBF leukemias harboring Inv16, in which 24% of patients with Inv16 had a mutation in exon 8 [24]. Besides the potential prognostic implication of these mutations, KIT mutations may have therapeutic implications, as there are data to suggest that primary leukemic cells that harbor some forms of KIT mutations may be susceptible to pro-apoptotic effects of Imatinib Mesylate [77], thus providing a therapeutic modality for some AML patients with CBF abnormalities at high risk of relapse.

Other RTKs, c-FMS, and PDGFR

Activating mutations in CSF1R (a.k.a. c-FMS) were initially described in myeloid cell lines, MDS, and AML, having an estimated frequency of 2–10% in myeloid malignancies [50, 125, 165, 118, 153]; however, a later study in AML patients did not identify any CSF1R mutations in their AML population [32]. The true prevalence of CSF1R mutations remains to be defined, but the importance of this receptor for the development of leukemia should not be underestimated. CSF1R resides on chromosome 5q33-35 and aberrant regulation of the receptor through methylation or loss of heterozygosity may play a critical role in leukemogenesis [141, 16]. If disruption of the CSF1R occurs at an "inappropriate" time during myeloid differentiation, studies have found that it may predispose cells to malignant transformation [36].

With respect to mutations in the PDGFR1α (chromosome 4q11-13) and PDGFR1β (5q31-32) genes, large-scale mutation analyses of the two genes have not been conducted; however, translocations involving PDGFR1β and

TEL (a.k.a. ETV6, 12p13) have been identified in chronic myelomonocytic leukemia, activating PDGR1β and promoting malignant transformation [122, 50, 25, 155]. Recently, translocations involving PDGFR1α and FIP1L1 (4q12) have also been found in hypereosinophilic syndrome and chronic eosinophilic leukemia [154, 28, 30, 166, 51], suggesting a possible role for PDGFR1α in malignant transformation.

In addition to the activating mutations at the receptor level, activating mutations of the secondary mediators of the RTKs (e.g., RAS, BRAF genes) have also been reported in AML [144, 113, 27, 138, 96, 29, 60, 100, 121, 135, 86, 40]. Some of these activating mutations, such as those involving RAS, are common, occurring in 10–20% of AML patients. While the clinical significance of these mutations remains uncertain, ongoing studies are investigating how these mutations may co-operate with other genetic abnormalities to promote leukemogenesis and affect prognosis [89, 113, 69].

Small Molecule Inhibitors as Therapeutic Options

Recently, there has been a surge in the development of small molecule inhibitors for myeloid leukemia [111]. Imatinib Mesylate (Gleevec) has proven to be a major advancement for the treatment of chronic myeloid leukemia in chronic phase (CML-CP). Imatinib induces a high percentage of complete cytogenetic remission for patients with CML-CP, and many of these patients have maintained this remission for 3–5 years [33, 58]. The results for Imatinib have been less impressive in more advanced forms of CML (e.g., accelerated or blast phase), in which patients will sometimes obtain responses, but almost universally relapse with resistant disease [5]. Multiple factors, including mutations in the bcr-abl gene, genomic amplification of the bcr-abl, and/or other genetic abnormalities, may account for some of the disparity in the responses between CML-CP and its more advanced counterparts [34]. To counteract the resistance secondary to bcr-abl mutations, newer small molecule inhibitors such as BMS-354825 (a.k.a. Dasatinib) and AMN 107 have been developed that display activity against cells harboring the mutated bcr-abl gene [57, 106]. Although early clinical trials seem promising for these drugs, long-term studies will be necessary to ensure that resistance against these newer small molecule inhibitors does not develop. Unlike CML-CP, AML is a much more heterogeneous disease, as indicated by the variety of different cytogenetic abnormalities and mutations within AML cells. Therefore, the development of a "universal" small molecule inhibitor for AML will be more challenging, if not impossible. Yet, the RTK pathways offer the "potential" for such drug development, and we will briefly describe some of the novel tyrosine kinase inhibitors (TKIs) that have recently been developed that target RTK pathways.

FLT3 Inhibitors

The FLT3 pathway is an obvious target for TKIs. FLT3 mutations are one of the most common mutations in AML. Because FLT3 mutations constitutively activate the receptor's pathway and contribute to leukemogenesis, small molecule inhibitors that block their activation may have therapeutic benefits for many AML patients. In addition, an increased expression of the wild-type (WT) receptor may also play a role in leukemogenesis for some AML patients [128], suggesting that FLT3 inhibitors may be effective in more than just AML patients with the FLT3 mutations. Initial in vitro studies using non-specific TKI (herbimycin A, AG1296, and AG1295) found that these drugs blocked constitutive activation of FLT3/ITDs and preferentially killed cells harboring FLT3/ITDs [7, 180, 83, 158]. However, all these compounds are highly toxic in humans, initiating searches for more selective and less toxic drugs for clinical use. Through molecular screening, numerous compounds have now been identified (MLN518, PKC412, SU5416, SU5614, SU11248, CEP-701, CEP-5214), and we will briefly describe the progress and limitations of these compounds.

MLN518 (a.k.a. CT53518 from Millennium) has been found to inhibit the activation of FLT3/ITDs and growth potential of cells harboring these mutations [92]. Like many TKIs, MLN518 is not specific for the FLT3/ITD receptor, also inhibiting WT forms of FLT3, PDGF, and KIT. Heinrich et al. published their results of a phase I study for high-risk AML patients in 2002, which found that two of six patients had > 50% reduction in bone blasts [66]. A phase II study evaluated the efficacy of MLN518 in 18 FLT3/ITD-positive AML patients with relapsed or refractory disease that were unfit for conventional chemotherapy. In this study, 6 of 18 patients demonstrated an objective response, as measured by a decrease in the peripheral blood blast by a mean of 92% (range 85–100%) [55]. However, no complete responses (CRs) were achieved.

PKC412 (from Novartis) was initially developed as a VEGF receptor inhibitor, but studies found that this benzoylstaurosporine blocked FLT3, including FLT3 mutated receptors [128, 31]. Armstrong et al. also found that MLL cells over-expressing WT FLT3 were preferentially killed by PKC412 [128,172]. A phase II trial examined the efficacy of PKC412 as a single agent for relapsed or refractory AML patients with poor performance status [6]. The PKC412 was given orally at 75 mg three times a day. Initial results found the drug to be well tolerated, with the most common side effect being nausea. However, no CRs were obtained in this heavily pretreated population. Additional follow-up of this study was recently presented at the American Society of Hematology (ASH) conference. A total of 20 FLT/ITD-positive AML patients with mutant FLT3 with either relapsed/refractory AML or high-grade myelodysplasia were treated with single agent PKC412. The peripheral blood blasts decreased by 50% in 6 of the 20 patients, with 2 responders obtaining a blast percentage of <5%. Again, no CRs were observed in this high-risk population [150] but autophosphorylation of the mutant receptor was blocked in most of the responding patients, indicating an in vivo target response using the dose in the study. Given these results, PKC412 has been combined with daunorubicin and cytarabine for induction of AML patients [148, 46]. In a phase I trial examining this combination approach, investigators recently reported at ASH that those AML patients harboring FLT3/ITDs had a CR rate of 91% (10/11) compared to 53% (17/32) for those without FLT3/ITDs ($p = 0.033$) [46]. Although not a randomized study, these data suggest that combination therapies adding TKI with chemotherapy may improve remission induction for those AML patients harboring FLT3/ITDs, although its efficacy for improving survival remains to be defined.

Sugen has several drugs under development as potential TKIs (SU5416, SU5614, and SU11248) [38, 149, 105]. Similar to most of the other TKIs, these agents also block the activities of other tyrosine kinase receptors (KIT, PDGFR, and VEGF) [38, 149, 105]. Giles et al. treated 55 patients with refractory or relapsed AML with SU5416 at a dose of 145 mg/m^2 [177]. AML patients without FLT3 mutations were also included in the trial. Grade 3/4 toxicities were few, but included headaches (14%), dyspnea (14%), infusion-related reactions (11%), and thrombotic episodes (7%). As a single agent, only three patients (5%) obtain a partial response. Another phase I trial examined SU11248 in 32 patients with advanced AML. Again, the drug was relatively well tolerated, with major dose-limiting toxicity being fatigue [38, 47]. Although approximately 50% of the AML patients had a >50% reduction in their blast count percentage, complete remission was not achieved [38, 47].

CEP-701 and CEP-5214 (from Cephalon) are two indolocarbazole compounds that inhibit autophosphorylation of the FLT3-WT and FLT3-mutant receptors [39, 81]. Unlike some of the other TKIs, these drugs have less activity against other RTKs such as KIT, FMS, and PDGF, and most clinical studies have focused on CEP-701 (Lestaurtinib). Initial studies with CEP-701 demonstrated the ability of this agent to kill FLT3-mutated primary AML cells from patients, and subsequent murine studies have found that CEP-701 extends the survival of mice injected with BaF3 cells transformed using FLT3/ITD [39]. A phase I/II trial evaluated CEP-701 as a single agent for patients with refractory, relapsed, or poor risk AML expressing FLT3-activating mutations. Fourteen heavily pretreated AML patients with FLT3 mutations were treated with CEP-701. CEP-701 toxicities were minimal, with nausea, fatigue, and neutropenia being most commonly described. Five patients obtained objective clinical responses, which correlate closely with a block in the constitutive phosphorylation of the mutated receptor. Clinical responses included significant reductions in bone marrow and peripheral blood blasts; however, no CRs were observed [80, 137]. These investigators recently opened a study to examine if the addition of CEP-701 to conventional chemotherapy may improve clinical outcomes. A total of 48 AML patients with FLT3 mutations in first relapse were randomized to either receiving standard chemotherapy or standard chemotherapy with CEP-701. Of the 24 patients who received CEP-701, 5 achieved complete CR and another 5 obtained a CR with incomplete count recovery. For those patients receiving only standard chemotherapy, three achieved CR and an additional three obtained CR with incomplete count recovery. Accrual is continuing on this trial, and it is too early to know if the addition of CEP-701 will benefit these high-risk patients, but Levis et al. found that CEP-701 and chemotherapy killed a cell line harboring a FLT3/ITD in a synergistic fashion [136], suggesting that using CEP-701 in combination with chemotherapy may be beneficial.

Together, these results suggest that current TKIs probably are not extremely effective as single agents in heavily pretreated AML patients with FLT3 mutations. How are TKIs in previously untreated AML patients with FLT3 mutations remain to be determined? However, there is a push to use these agents in combination with more "standard" chemotherapy, believing that this offers the most potential for providing a therapeutic advantage with this class of drugs. Therefore, the efficacy of TKIs as a single agent in de novo AML patients may not ever be fully investigated. It will be critical to determine which AML patients with FLT3 mutations may have the highest likelihood of response with these novel drugs. As previously described, AML is a very heterogeneous disease event within the subgroup of patients with FLT3 mutations.

KIT and Other Small Molecule Inhibitors

There are few selective KIT inhibitors, and small molecules against KIT have not been extensively used for the treatment of AML patients. Many TKIs (e.g., SU5416, SU6668, Gleevec) have some activity against KIT [82, 56]. A recent study also found that both SU5416 and SU6668 inhibited KIT autophosphorylation and downstream effectors [56], and there has been at least one case report of an individual with refractory AML obtaining CR with SU5416 as monotherapy [139]. In addition, Gleevec has significant activity against WT-KIT and KIT harboring mutations in JM domain [82]. However, Gleevec has no activity against KIT with activating point mutations in the tyrosine kinase domain [82]; therefore, Gleevec's role in AML therapy may be limited to a highly selected group of AML patients, if at all. However, a newer TKI (BMS-354825 or Dasatinib) has been found to inhibit constitutive activation of KIT receptors with mutations in both the JMD and the TKD [106], suggesting that it may have a broader therapeutic application for AML with KIT mutations. To date, few large studies have actually targeted these drugs for AML patients with KIT mutations. This limited experience is probably due to lower prevalence of KIT mutations in AML as compared to FLT3 mutations, but given the recent discovery of the high rate of KIT mutations in AML involving CBF, additional studies using these drugs in AML patients with KIT mutations will certainly be developed [124].

In addition, small molecule inhibitors directed toward PDGF [91] or downstream effectors in the RTK pathway have been developed [43, 98, 156, 142, 108, 110, 169]. These drugs are currently in a variety of different stages of investigation. Because RAS mutations occur frequently in many different types of tumors [144, 113, 96, 29, 60, 100, 121, 135, 86, 40], RAS has been a natural target of the RTK pathway for many years, and farnesyltransferase inhibitors have led the way in trying to block the effect of constitutive activation of RAS in AML [114, 112, 107]. In early clinical trials examining farnesyltransferase inhibitors, the response rates as a single agent have been encouraging, ranging from 20 to 35%, but the efficacy of these compounds have not correlated with RAS mutations [107]. Details about the biology and clinical implications of farnesyltransferase inhibitors are beyond the scope of this review and have recently been extensively discussed by the leaders in this field [61, 131, 62, 63]. Other inhibitors have targeted downstream effectors such as MEK or RAF [Morgan, 2001 #1171;Rahmani, 2005 #2428;Sridhar, 2005 #2423;Ouyang, 2006 #2425;Wallace, 2006 #2427; Baines, 2000 #1356]. Although some studies suggest efficacy of these agents, major responses have been limited in AML patients.

Future Directions

There are many challenges ahead in the treatment of AML. One of the major challenges will be to begin to interrogate our understanding of the RTK pathways into the development of novel therapeutic approaches for the treatment of AML. As a start, it would be critical to be able to better risk-stratify AML patients harboring abnormalities in RTK pathways. In order to do this successfully, additional investigations will need to determine the exact frequency of mutations in RTK pathway and the clinical significance of these mutations. In addition, it will also be important to understand how the different types of mutations affect the RTK pathways. These investigations will be difficult, given the large number of RTKs, downstream effectors, and heterogeneity within AML.

Researchers are examining how inhibitors of RTK pathways may be useful in the treatment of AML. As single agents, current compounds unfortunately have not induced complete responses in AML patients, but the potential of combining these agents with standard chemotherapy regimens may be more beneficial. In addition, combinations of small molecule inhibitors blocking inappropriate RTK activation at several points along the pathway may also be more effective than using them as a single agent. The major limitation to such approaches may be toxicity, given that the RTK pathway is critical in the regulation of the normal function of the hematopoietic system.

Acknowledgments Support for the authors was provided by National Institutes of Health grants (K23 CA92405 and CA 114563). This work was supported by National Institute of Health grants no. K23 CA92405-01 and CA18029.

References

1. Abu-Duhier FM, Goodeve AC, Wilson GA, Care RS, Peake IR, Reilly JT. Identification of novel FLT-3 Asp835 mutations in adult acute myeloid leukaemia. *Br J Haematol.* 2001;113:983–988.
2. Abu-Duhier FM, Goodeve AC, Wilson GA, et al. FLT3 internal tandem duplication mutations in adult acute myeloid leukaemia define a high-risk group. *Br J Haematol.* 2000;111:190–195.
3. Agnes F, Shamoon B, Dina C, Rosnet O, Birnbaum D, Galibert F. Genomic structure of the downstream part of the human FLT3 gene: exon/intron structure conservation among genes encoding receptor tyrosine kinases (RTK) of subclass III. *Gene.* 1994;145:283–288.
4. Aguayo A, Estey E, Kantarjian H, et al. Cellular vascular endothelial growth factor is a predictor of outcome in patients with acute myeloid leukemia. *Blood.* 1999;94: 3717–3721.
5. Anstrom KJ, Reed SD, Allen AS, Glendenning GA, Schulman KA. Long-term survival estimates for imatinib versus interferon-alpha plus low-dose cytarabine for patients with newly diagnosed chronic-phase chronic myeloid leukemia. *Cancer.* 2004;101:2584–2592.
6. Armstrong SA, Kung AL, Mabon ME, et al. Inhibition of FLT3 in MLL. Validation of a therapeutic target identified by gene expression based classification. *Cancer Cell.* 2003;3:173–183.
7. Armstrong SA, Staunton JE, Silverman LB, et al. MLL translocations specify a distinct gene expression profile that distinguishes a unique leukemia. *Nat Genet.* 2002;30:41–47.
8. Beghini A, Larizza L, Cairoli R, Morra E. c-kit activating mutations and mast cell proliferation in human leukemia. *Blood.* 1998;92:701–702.
9. Beghini A, Peterlongo P, Ripamonti CB, et al. C-kit mutations in core binding factor leukemias. *Blood.* 2000;95:726–727.
10. Beghini A, Ripamonti CB, Cairoli R, et al. KIT activating mutations: incidence in adult and pediatric acute myeloid leukemia, and identification of an internal tandem duplication. *Haematologica.* 2004;89:920–925.
11. Beghini A, Ripamonti CB, Castorina P, et al. Trisomy 4 leading to duplication of a mutated KIT allele in acute myeloid leukemia with mast cell involvement. *Cancer Genet Cytogenet.* 2000;119:26–31.
12. Bentires-Alj M, Paez JG, David FS, et al. Activating mutations of the noonan syndrome-associated SHP2/PTPN11 gene in human solid tumors and adult acute myelogenous leukemia. *Cancer Res.* 2004;64:8816–8820.
13. Birkenkamp KU, Geugien M, Lemmink HH, Kruijer W, Vellenga E. Regulation of constitutive STAT5 phosphorylation in acute myeloid leukemia blasts. *Leukemia.* 2001; 15:1923–1931.
14. Blume-Jensen P, Hunter T. Oncogenic kinase signalling. *Nature.* 2001;411:355–365.
15. Bos JL. *Ras* oncogenes in human cancer: a review. *Cancer Res.* 1989;49:4682–4689.
16. Boultwood J, Rack K, Kelly S, et al. Loss of both CSF1R (FMS) alleles in patients with myelodysplasia and a chromosome 5 deletion. *Proc Natl Acad Sci USA.* 1991;88:6176–6180.
17. Bowen DT, Frew ME, Hills R, et al. RAS mutation in acute myeloid leukemia is associated with distinct cytogenetic subgroups but does not influence outcome in patients younger than 60 years. *Blood.* 2005;106:2113–2119.
18. Brandts CH, Sargin B, Rode M, et al. Constitutive activation of Akt by Flt3 internal tandem duplications is necessary for increased survival, proliferation, and myeloid transformation. *Cancer Res.* 2005;65:9643–9650.
19. Cairoli R, Beghini A, Grillo G, et al. Prognostic impact of c-KIT mutations in core binding factor leukemias. An Italian retrospective study. *Blood.* 2006. In press.
20. Cairoli R, Beghini A, Morello E, et al. Imatinib mesylate in the treatment of Core Binding Factor leukemias with KIT mutations. A report of three cases. *Leuk Res.* 2005;29:397–400.

21. Cairoli R, Grillo G, Beghini A, et al. C-Kit point mutations in core binding factor leukemias: correlation with white blood cell count and the white blood cell index. *Leukemia.* 2003;17:471–472.
22. Callens C, Chevret S, Cayuela JM, et al. Prognostic implication of FLT3 and Ras gene mutations in patients with acute promyelocytic leukemia (APL): a retrospective study from the European APL Group. *Leukemia.* 2005;19:1153–1160.
23. Cammenga J, Horn S, Bergholz U, et al. Extracellular KIT receptor mutants, commonly found in core binding factor AML, are constitutively active and respond to imatinib mesylate. *Blood.* 2005;106:3958–3961.
24. Care RS, Valk PJ, Goodeve AC, et al. Incidence and prognosis of c-KIT and FLT3 mutations in core binding factor (CBF) acute myeloid leukaemias. *Br J Haematol.* 2003; 121:775–777.
25. Carroll M, Tomasson MH, Barker GF, Golub TR, Gilliland DG. The TEL/platelet-derived growth factor beta receptor (PDGF beta R) fusion in chronic myelomono-cytic leukemia is a transforming protein that self-associates and activates PDGF beta R kinase-dependent signaling pathways. *Proc Natl Acad Sci USA.* 1996;93:14845–14850.
26. Casey G, Rudzki Z, Roberts M, Hutchins C, Juttner C. N-ras mutation in acute myeloid leukemia: incidence, prognostic significance and value as a marker of minimal residual disease. *Pathology.* 1993;25:57–62.
27. Christiansen DH, Andersen MK, Desta F, Pedersen-Bjergaard J. Mutations of genes in the receptor tyrosine kinase (RTK)/RAS-BRAF signal transduction pathway in therapy-related myelodysplasia and acute myeloid leukemia. *Leukemia.* 2005;19:2232–2240.
28. Cools J, DeAngelo DJ, Gotlib J, et al. A tyrosine kinase created by fusion of the PDGFRA and FIP1L1 genes as a therapeutic target of imatinib in idiopathic hypereosi-nophilic syndrome. *N Engl J Med.* 2003;348:1201–1214.
29. Cools J, Stover EH, Gilliland DG. Detection of the FIP1L1-PDGFRA fusion in idio-pathic hypereosinophilic syndrome and chronic eosinophilic leukemia. *Methods Mol Med.* 2006;125:177–187.
30. Cools J, Stover EH, Wlodarska I, Marynen P, Gilliland DG. The FIP1L1-PDGFRalpha kinase in hypereosinophilic syndrome and chronic eosinophilic leukemia. *Curr Opin Hematol.* 2004;11:51–57.
31. De Angelo DJ, Stone RM, Heaney ML, et al. Phase II evaluation of the tyrosine kinase inhibitor MLN518 in patients with acute myeloid leukemia (AML) bearing a FLT3 internal tandem duplication (ITD) mutation. *Blood.* 2004;104:1792a.
32. Dello Sbarba P, Pollard JW, Stanley ER. Alterations in CSF-1 receptor expression and protein tyrosine phosphorylation in autonomous mutants of a CSF-1 dependent macro-phage cell line. *Growth Factors.* 1991;5:75–85.
33. Druker BJ, Lydon NB. Lessons learned from the development of an abl tyrosine kinase inhibitor for chronic myelogenous leukemia. *J Clin Invest.* 2000;105:3–7.
34. Druker BJ, Sawyers CL, Kantarjian H, et al. Activity of a specific inhibitor of the BCR-ABL tyrosine kinase in the blast crisis of chronic myeloid leukemia and acute lymphoblastic leukemia with the Philadelphia chromosome. *N Engl J Med.* 2001;344:1038–1042.
35. Farr CJ, Saiki RK, Erlich HA, McCormick F, Marshall CJ. Analysis of RAS gene mutations in acute myeloid leukemia by polymerase chain reaction and oligonucleotide probes. *Proc Natl Acad Sci USA.* 1988;85:1629–1633.
36. Felgner J, Kreipe H, Heidorn K, et al. Increased methylation of the c-fms protooncogene in acute myelomonocytic leukemias. *Pathobiology.* 1991;59:293–298.
37. Fenski R, Flesch K, Serve S, et al. Constitutive activation of FLT3 in acute myeloid leukaemia and its consequences for growth of 32D cells. *Br J Haematol.* 2000;108:322–330.
38. Foran J, O'Farrell AM, Fiedler W, et al. An innovative single dose clinical study shows potent inhibition of FLT3 phosphorylation by SU11248 in vivo: a clinical and pharma-codynamic study in AML patients. *Blood.* 2002;100:2196a.

39. Foran J, Paquette R, Cooper M, et al. A phase I study of repeated oral dosing with SU11248 for the treatment of patients with acute myeloid leukemia who have failed, or are not eligible for conventional chemotherapy. *Blood.* 2002;100:2195a.

40. Forrester K, Almoguera C, Han K, Grizzle WE, Perucho M. Detection of high incidence of K-ras oncogenes during human colon tumorigenesis. *Nature.* 1987;327:298–303.

41. Frohling S, Schlenk RF, Breitruck J, et al. Prognostic significance of activating FLT3 mutations in younger adults (16 to 60 years) with acute myeloid leukemia and normal cytogenetics: a study of the AML Study Group Ulm. *Blood.* 2002;100:4372–4380.

42. Furitsu T, Tsujimura T, Tono T, et al. Identification of mutations in the coding sequence of the proto-oncogene c-kit in a human mast cell leukemia cell line causing ligand-independent activation of c-kit product. *J Clin Invest.* 1993;92:1736–1744.

43. Furuta T, Sakai T, Senga T, et al. Identification of potent and selective inhibitors of PDGF receptor autophosphorylation. *J Med Chem.* 2006;49:2186–2192.

44. Gabbianelli M, Pelosi E, Montesoro E, et al. Multi-level effects of flt3 ligand on human hematopoiesis: expansion of putative stem cells and proliferation of granulomonocytic progenitors/monocytic precursors. *Blood.* 1995;86:1661–1670.

45. Gari M, Goodeve A, Wilson G, et al. c-kit proto-oncogene exon 8 in-frame deletion plus insertion mutations in acute myeloid leukaemia. *Br J Haematol.* 1999;105:894–900.

46. Giles F, Schiffer C, Kantarjian H, et al. Phase 1 study of PKC412, an oral FLT3 kinase inhibitor, in sequential and concomitant combinations with daunorubicin and cytarabine (DA) induction and high-dose cytarabine consolidation in newly diagnosed patients with AML. *Blood.* 2004;104:262a.

47. Giles FJ, Stopeck AT, Silverman LR, et al. SU5416, a small molecule tyrosine kinase receptor inhibitor, has biologic activity in patients with refractory acute myeloid leukemia or myelodysplastic syndromes. *Blood.* 2003;102:795–801.

48. Gille H, Kowalski J, Yu L, et al. A repressor sequence in the juxtamembrane domain of Flt-1 (VEGFR-1) constitutively inhibits vascular endothelial growth factor-dependent phosphatidylinositol 3'-kinase activation and endothelial cell migration. *Embo J.* 2000; 19:4064–4073.

49. Goemans BF, Zwaan CM, Miller M, et al. Mutations in KIT and RAS are frequent events in pediatric core-binding factor acute myeloid leukemia. *Leukemia.* 2005;19: 1536–1542.

50. Golub TR, Barker GF, Lovett M, Gilliland DG. Fusion of PDGF receptor beta to a novel ets-like gene, tel, in chronic myelomonocytic leukemia with t(5;12) chromosomal translocation. *Cell.* 1994;77:307–316.

51. Gotlib J, Cools J, Malone JM, 3rd, Schrier SL, Gilliland DG, Coutre SE. The FIP1L1-PDGFRalpha fusion tyrosine kinase in hypereosinophilic syndrome and chronic eosinophilic leukaemia: implications for diagnosis, classification, and management. *Blood.* 2004;103:2879–2891.

52. Griffith J, Black J, Faerman C, et al. The structural basis for autoinhibition of FLT3 by the juxtamembrane domain. *Mol Cell.* 2004;13:169–178.

53. Grundler R, Miething C, Thiede C, Peschel C, Duyster J. FLT3-ITD and tyrosine kinase domain mutations induce 2 distinct phenotypes in a murine bone marrow transplantation model. *Blood.* 2005;105:4792–4799.

54. Hayakawa F, Towatari M, Kiyoi H, et al. Tandem-duplicated Flt3 constitutively activates STAT5 and MAP kinase and introduces autonomous cell growth in IL-3-dependent cell lines. *Oncogene.* 2000;19:624–631.

55. Heinrich MC, Druker BJ, Curtin PT, et al. A "first in man" study of the safety and PK/PD of an oral FLT3 inhibitor (MLN518) in patients with AML or high risk myelodyspsia. *Blood.* 2002;100:1305a.

56. Heinrich MC, Griffith DJ, Druker BJ, Wait CL, Ott KA, Zigler AJ. Inhibition of c-kit receptor tyrosine kinase activity by STI 571, a selective tyrosine kinase inhibitor. *Blood.* 2000;96:925–932.

57. Hochhaus A, Kreil S, Corbin AS, et al. Molecular and chromosomal mechanisms of resistance to imatinib (STI571) therapy. *Leukemia.* 2002;16:2190–2196.
58. Hughes TP, Kaeda J, Branford S, et al. Frequency of major molecular responses to imatinib or interferon alfa plus cytarabine in newly diagnosed chronic myeloid leukemia. *N Engl J Med.* 2003;349:1423–1432.
59. Iwai T, Yokota S, Nakao M, et al. Internal tandem duplication of the FLT3 gene and clinical evaluation in childhood acute myeloid leukemia. The Children's Cancer and Leukemia Study Group, Japan. *Leukemia.* 1999;13:38–43.
60. Janssen JW, Steenvoorden AC, Lyons J, et al. RAS gene mutations in acute and chronic myelocytic leukemias, chronic myeloproliferative disorders, and myelodysplastic syndromes. *Proc Natl Acad Sci USA.* 1987;84:9228–9232.
61. Karp JE, Lancet JE, Kaufmann SH, et al. Clinical and biologic activity of the farnesyltransferase inhibitor R115777 in adults with refractory and relapsed acute leukemias: a phase 1 clinical-laboratory correlative trial. *Blood.* 2001;97:3361–3369.
62. Karp JE, Lancet JE. Development of the farnesyltransferase inhibitor tipifarnib for therapy of hematologic malignancies. *Fut Oncol.* 2005;1:719–731.
63. Karp JE, Lancet JE. Targeting the process of farnesylation for therapy of hematologic malignancies. *Curr Mol Med.* 2005;5:643–652.
64. Kelly LM, Kutok JL, Williams IR, et al. PML/RARalpha and FLT3-ITD induce an APL-like disease in a mouse model. *Proc Natl Acad Sci USA.* 2002;99:8283–8288.
65. Kelly LM, Liu Q, Kutok JL, Williams IR, Boulton CL, Gilliland DG. FLT3 internal tandem duplication mutations associated with human acute myeloid leukemias induce myeloproliferative disease in a murine bone marrow transplant model. *Blood.* 2002;99: 310–318.
66. Kelly LM, Yu J, Boulton CL, et al. CT53518, a novel selective FLT3 antagonist for the treatment of acute myelogenous leukemia (AML). *Caner Cell.* 2002.
67. Kindler T, Breitenbuecher F, Kasper S, et al. Identification of a novel activating mutation (Y842C) within the activation loop of FLT3 in patients with acute myeloid leukemia (AML). *Blood.* 2005;105:335–340.
68. Kitayama H, Kanakura Y, Furitsu T, et al. Constitutively activating mutations of c-kit receptor tyrosine kinase confer factor-independent growth and tumorigenicity of factor-dependent hematopoietic cell lines. *Blood.* 1995;85:790–798.
69. Kiyoi H, Naoe T, Nakano Y, et al. Prognostic implication of FLT3 and N-RAS gene mutations in acute myeloid leukemia. *Blood.* 1999;93:3074–3080.
70. Kiyoi H, Ohno R, Ueda R, Saito H, Naoe T. Mechanism of constitutive activation of FLT3 with internal tandem duplication in the juxtamembrane domain. *Oncogene.* 2002;21:2555–2563.
71. Kiyoi H, Towatari M, Yokota S, et al. Internal tandem duplication of the FLT3 gene is a novel modality of elongation mutation which causes constitutive activation of the product. *Leukemia.* 1998;12:1333–1337.
72. Kohl TM, Schnittger S, Ellwart JW, Hiddemann W, Spiekermann K. KIT exon 8 mutations associated with core-binding factor (CBF)-acute myeloid leukemia (AML) cause hyperactivation of the receptor in response to stem cell factor. *Blood.* 2005;105:3319–3321.
73. Kondo M, Horibe K, Takahashi Y, et al. Prognostic value of internal tandem duplication of the FLT3 gene in childhood acute myelogenous leukemia. *Med Pediatr Oncol.* 1999;33:525–529.
74. Kottaridis PD, Gale RE, Frew ME, et al. The presence of a FLT3 internal tandem duplication in patients with acute myeloid leukemia (AML) adds important prognostic information to cytogenetic risk group and response to the first cycle of chemotherapy: analysis of 854 patients from the United Kingdom Medical Research Council AML 10 and 12 trials. *Blood.* 2001;98:1752–1759.
75. Kottaridis PD, Gale RE, Langabeer SE, Frew ME, Bowen DT, Linch DC. Studies of FLT3 mutations in paired presentation and relapse samples from patients with acute

myeloid leukemia: implications for the role of FLT3 mutations in leukemogenesis, minimal residual disease detection, and possible therapy with FLT3 inhibitors. *Blood.* 2002;100:2393–2398.

76. Kubo K, Naoe T, Kiyoi H, et al. Clonal analysis of multiple point mutations in the N-ras gene in patients with acute myeloid leukemia. *Jpn J Cancer Res.* 1993;84:379–387.

77. Kuchenbauer F, Feuring-Buske M, Buske C. AML1-ETO needs a partner: new insights into the pathogenesis of t(8;21) leukemia. *Cell Cycle.* 2005;4:1716–1718.

78. Lacayo NJ, Meshinchi S, Kinnunen P, et al. Gene Expression Profiles at Diagnosis in de novo Childhood AML Patients Identify FLT3 Mutations with Good Clinical Outcomes. *Blood.* 2004.

79. Lancet JE, Karp JE. Farnesyltransferase inhibitors in hematologic malignancies: new horizons in therapy. *Blood.* 2003;102:3880–3889.

80. Levis M, Allebach J, Fai-Tse K, et al. FLT3-targeted inhibitors kill FLT3-dependent modeled cells, leukemia-derived cell lines, and primary AML blasts in vitro and in vivo. *Blood.* 2001;89:721a.

81. Levis M, Allebach J, Tse KF, et al. A FLT3-targeted tyrosine kinase inhibitor is cytotoxic to leukemia cells in vitro and in vivo. *Blood.* 2002;99:3885–3891.

82. Levis M, Pham R, Smith BD, Small D. In vitro studies of a FLT3 inhibitor combined with chemotherapy: sequence of administration is important in order to achieve synergistic cytotoxic effects. *Blood.* 2004;104:1145–1150.

83. Levis M, Tse KF, Smith BD, Garrett E, Small D. A FLT3 tyrosine kinase inhibitor is selectively cytotoxic to acute myeloid leukemia blasts harboring FLT3 internal tandem duplication mutations. *Blood.* 2001;98:885–887.

84. Liang DC, Shih LY, Fu JF, et al. K-Ras mutations and N-Ras mutations in childhood acute leukemias with or without mixed-lineage leukemia gene rearrangements. *Cancer.* 2006;106:950–956.

85. Longley BJ, Jr., Metcalfe DD, Tharp M, et al. Activating and dominant inactivating c-KIT catalytic domain mutations in distinct clinical forms of human mastocytosis. *Proc Natl Acad Sci USA.* 1999;96:1609–1614.

86. McCoy MS, Toole JJ, Cunningham JM, Chang EH, Lowy DR, Weinberg RA. Characterization of a human colon/lung carcinoma oncogene. *Nature.* 1983;302:79–81.

87. Mead A, Linch D, Hills R, Wheatley K, Burnett A, Gale R. Favourable prognosis associated with FLT3 tyrosine kinase domain mutations in AML in contrast to the adverse outcome associated with internal tandem duplications. *Blood.* 2005;106:334a.

88. Meshinchi S, Alonzo TA, Gerbing R, Lang B, Radich J. FLT3 internal tandem duplication (FLT3/ITD) is a prognostic factor for poor outcome in pediatric AML: a CCG2961 study. *Blood.* 2003;102:335a.

89. Meshinchi S, Stirewalt DL, Alonzo TA, et al. Activating mutations of RTK/ras signal transduction pathway in pediatric acute myeloid leukemia. *Blood.* 2003;102:1474–1479.

90. Meshinchi S, Woods WG, Stirewalt DL, et al. Prevalence and prognostic significance of FLT3 internal tandem duplication in pediatric acute myeloid leukemia. *Blood.* 2001;97:89–94.

91. Mesters RM, Padro T, Bieker R, et al. Stable remission after administration of the receptor tyrosine kinase inhibitor SU5416 in a patient with refractory acute myeloid leukemia. *Blood.* 2001;98:241–243.

92. Minami Y, Kiyoi H, Yamamoto Y, et al. Selective apoptosis of tandemly duplicated FLT3-transformed leukemia cells by Hsp90 inhibitors. *Leukemia.* 2002;16:1535–1540.

93. Minami Y, Yamamoto K, Kiyoi H, Ueda R, Saito H, Naoe T. Different antiapoptotic pathways between wild-type and mutated FLT3: insights into therapeutic targets in leukemia. *Blood.* 2003;102:2969–2975.

94. Mizuki M, Fenski R, Halfter H, et al. Flt3 mutations from patients with acute myeloid leukemia induce transformation of 32D cells mediated by the Ras and STAT5 pathways. *Blood.* 2000;96:3907–3914.

95. Mizuki M, Schwable J, Steur C, et al. Suppression of myeloid transcription factors and induction of STAT response genes by AML-specific Flt3 mutations. *Blood.* 2003;101: 3164–3173.

96. Mizuki M, Schwaeble J, Steur C, et al. Suppression of myeloid transcription factors and induction of STAT response genes by AML-specific Flt3 mutations. *Blood.* 2003;101(8):3164–3173.

97. Moore TA, Zlotnik A. Differential effects of Flk-2/Flt-3 ligand and stem cell factor on murine thymic progenitor cells. *J Immunol.* 1997;158:4187–4192.

98. Morgan MA, Dolp O, Reuter CW. Cell-cycle-dependent activation of mitogen-activated protein kinase kinase (MEK-1/2) in myeloid leukemia cell lines and induction of growth inhibition and apoptosis by inhibitors of RAS signaling. *Blood.* 2001;97:1823–1834.

99. Murray LJ, Young JC, Osborne LJ, Luens KM, Scollay R, Hill BL. Thrombopoietin, flt3, and kit ligands together suppress apoptosis of human mobilized CD34 + cells and recruit primitive CD34 + Thy-1 + cells into rapid division. *Exp Hematol.* 1999;27:1019–1028.

100. Nakagawa T, Saitoh S, Imoto S, et al. Multiple point mutation of N-ras and K-ras oncogenes in myelodysplastic syndrome and acute myelogenous leukemia. *Oncology.* 1992;49:114–122.

101. Nakao M, Yokota S, Iwai T, et al. Internal tandem duplication of the flt3 gene found in acute myeloid leukemia. *Leukemia.* 1996;10:1911–1918.

102. Namikawa R, Muench MO, de Vries JE, Roncarolo MG. The FLK2/FLT3 ligand synergizes with interleukin-7 in promoting stromal-cell-independent expansion and differentiation of human fetal pro-B cells in vitro. *Blood.* 1996;87:1881–1890.

103. Nanri T, Matsuno N, Kawakita T, et al. Mutations in the receptor tyrosine kinase pathway are associated with clinical outcome in patients with acute myeloblastic leukemia harboring t(8;21)(q22;q22). *Leukemia.* 2005;19:1361–1366.

104. Neubauer A, Dodge RK, George SL, et al. Prognostic importance of mutations in the ras proto-oncogenes in de novo acute myeloid leukemia. *Blood.* 1994;83:1603–1611.

105. O'Farrell AM, Abrams TJ, Yuen HA, et al. SU11248 is a novel FLT3 tyrosine kinase inhibitor with potent activity in vitro and in vivo. *Blood.* 2003;101:3597–3605.

106. O'Hare T, Walters DK, Stoffregen EP, et al. In vitro activity of Bcr-Abl inhibitors AMN107 and BMS-354825 against clinically relevant imatinib-resistant Abl kinase domain mutants. *Cancer Res.* 2005;65:4500–4505.

107. Oliff A. Farnesyltransferase inhibitors: targeting the molecular basis of cancer. *Biochimica et Biophysica Acta.* 1999;1423:C19–30.

108. Ouyang B, Knauf JA, Smith EP, et al. Inhibitors of Raf kinase activity block growth of thyroid cancer cells with RET/PTC or BRAF mutations in vitro and in vivo. *Clin Cancer Res.* 2006;12:1785–1793.

109. Padua RA, Guinn BA, Al-Sabah AI, et al. RAS, FMS and p53 mutations and poor clinical outcome in myelodysplasias: a 10-year follow-up. *Leukemia.* 1998;12:887–892.

110. Panka DJ, Wang W, Atkins MB, Mier JW. The Raf inhibitor BAY 43-9006 (Sorafenib) induces caspase-independent apoptosis in melanoma cells. *Cancer Res.* 2006;66:1611–1619.

111. Pellegata NS, Sessa F, Renault B, et al. K-ras and p53 gene mutations in pancreatic cancer: ductal and nonductal tumors progress through different genetic lesions. *Cancer Res.* 1994;54:1556–1560.

112. Prendergast GC, Davide JP, deSolms SJ, et al. Farnesyltransferase inhibition causes morphological reversion of ras-transformed cells by a complex mechanism that involves regulation of the actin cytoskeleton. *Mol Cell Biol.* 1994;14:4193–4202.

113. Radich JP, Kopecky KJ, Willman CL, et al. N-ras mutations in adult de novo acute myelogenous leukemia: prevalence and clinical significance. *Blood.* 1990;76: 801–807.

114. Rahmani M, Davis EM, Bauer C, Dent P, Grant S. Apoptosis induced by the kinase inhibitor BAY 43-9006 in human leukemia cells involves down-regulation of Mcl-1 through inhibition of translation. *J Biol Chem.* 2005;280:35217–35227.

115. Ray RJ, Paige CJ, Furlonger C, Lyman SD, Rottapel R. Flt3 ligand supports the differentiation of early B cell progenitors in the presence of interleukin-11 and interleukin-7. *Eur J Immunol.* 1996;26:1504–1510.

116. Reindl C, Bagrintseva K, Vempati S, et al. Point mutations found in the juxtamembrane domain of FLT3 define a new class of activating mutations in AML. *Blood.* 2006.

117. Ridge SA, Worwood M, Oscier D, Jacobs A, Padua RA. FMS mutations in myelodysplastic, leukemic, and normal subjects. *Proc Natl Acad Sci USA.* 1990;87:1377–1380.

118. Ridge SA, Worwood M, Oscier D, Jacobs A, Padua RA. FMS mutations in myelodysplastic, leukemic, and normal subjects. *Proc Natl Acad Sci USA.* 1990;87:1377–1380.

119. Ritter M, Kim TD, Lisske P, Thiede C, Schaich M, Neubauer A. Prognostic significance of N-RAS and K-RAS mutations in 232 patients with acute myeloid leukemia. *Haematologica.* 2004;89:1397–1399.

120. Robinson DR, Wu YM, Lin SF. The protein tyrosine kinase family of the human genome. *Oncogene.* 2000;19:5548–5557.

121. Rodenhuis S, van de Wetering ML, Mooi WJ, Evers SG, van Zandwijk N, Bos JL. Mutational activation of the K-ras oncogene. A possible pathogenetic factor in adenocarcinoma of the lung. *N Engl J Med.* 1987;317:929–935.

122. Rosenbauer F, Wagner K, Kutok JL, et al. Acute myeloid leukemia induced by graded reduction of a lineage-specific transcription factor, PU.1. *Nat Genet.* 2004;36:624–630.

123. Roskoski R, Jr. Structure and regulation of Kit protein-tyrosine kinase–the stem cell factor receptor. *Biochem Biophys Res Commun.* 2005;338:1307–1315.

124. Rosnet O, Birnbaum D. Hematopoietic receptors of class III receptor-type tyrosine kinases. *Crit Rev Oncog.* 1993;4:595–613.

125. Roussel MF, Downing JR, Rettenmier CW, Sherr CJ. A point mutation in the extracellular domain of the human CSF-1 receptor (c-fms proto-oncogene product) activates its transforming potential. *Cell.* 1988;55:979–988.

126. Rusten LS, Lyman SD, Veiby OP, Jacobsen SE. The FLT3 ligand is a direct and potent stimulator of the growth of primitive and committed human CD34+ bone marrow progenitor cells in vitro. *Blood.* 1996;87:1317–1325.

127. Schessl C, Rawat VP, Cusan M, et al. The AML1-ETO fusion gene and the FLT3 length mutation collaborate in inducing acute leukemia in mice. *J Clin Invest.* 2005;115: 2159–2168.

128. Schittenhelm MM, Shiraga S, Schroeder A, et al. Dasatinib (BMS-354825), a dual SRC/ABL kinase inhibitor, inhibits the kinase activity of wild-type, juxtamembrane, and activation loop mutant KIT isoforms associated with human malignancies. *Cancer Res.* 2006;66:473–481.

129. Schnittger S, Kohl TM, Haferlach T, et al. KIT-D816 mutations in AML1-ETO-positive AML are associated with impaired event-free and overall survival. *Blood.* 2006;107: 1791–1799.

130. Schnittger S, Schoch C, Dugas M, et al. Analysis of FLT3 length mutations in 1003 patients with acute myeloid leukemia: correlation to cytogenetics, FAB subtype, and prognosis in the AMLCG study and usefulness as a marker for the detection of minimal residual disease. *Blood.* 2002;100:59–66.

131. Sebti SM, Der CJ. Opinion: searching for the elusive targets of farnesyltransferase inhibitors. *Nat Rev Cancer.* 2003;3:945–951.

132. SEER Cancer Statistics Review 1975-2001, Vol. 2004. National Cancer Institute; 2004. http://seer.cancer.gov/csr/1975_2004/. Accessed on 15 July, 2009.

133. Shah AJ, Smogorzewska EM, Hannum C, Crooks GM. Flt3 ligand induces proliferation of quiescent human bone marrow CD34+CD38– cells and maintains progenitor cells in vitro. *Blood.* 1996;87:3563–3570.

134. Shimada A, Taki T, Tabuchi K, et al. KIT mutations, and not FLT3 internal tandem duplication, are strongly associated with a poor prognosis in pediatric acute myeloid leukemia with t(8;21): a study of the Japanese Childhood AML Cooperative Study Group. *Blood.* 2006;107:1806–1809.

135. Slebos RJ, Kibbelaar RE, Dalesio O, et al. K-ras oncogene activation as a prognostic marker in adenocarcinoma of the lung. *N Engl J Med.* 1990;323:561–565.

136. Smith BD, Levis M, Beran M, et al. Single-agent CEP-701, a novel FLT3 inhibitor, shows biologic and clinical activity in patients with relapsed or refractory acute myeloid leukemia. *Blood.* 2004;103:3669–3676.

137. Smith BD, Levis M, Brown P, et al. Single agent CEP-701, a novel FLT-3 inhibitor, shows initial response in patients with refractory acute myeloid leukemia. *Blood.* 2002;100:314a.

138. Smith ML, Snadden J, Neat M, et al. Mutation of BRAF is uncommon in AML FAB type M1 and M2. *Leukemia.* 2003;17:274–275.

139. Smolich BD, Yuen HA, West KA, Giles FJ, Albitar M, Cherrington JM. The antiangiogenic protein kinase inhibitors SU5416 and SU6668 inhibit the SCF receptor (c-kit) in a human myeloid leukemia cell line and in acute myeloid leukemia blasts. *Blood.* 2001;97:1413–1421.

140. Spiekermann K, Bagrintseva K, Schoch C, Haferlach T, Hiddemann W, Schnittger S. A new and recurrent activating length mutation in exon 20 of the FLT3 gene in acute myeloid leukemia. *Blood.* 2002;100:3423–3425.

141. Springall F, O'Mara S, Shounan Y, Todd A, Ford D, Iland H. c-fms point mutations in acute myeloid leukemia: fact or fiction? *Leukemia.* 1993;7:978–985.

142. Sridhar SS, Hedley D, Siu LL. Raf kinase as a target for anticancer therapeutics. *Mol Cancer Ther.* 2005;4:677–685.

143. Stephenson SA, Slomka S, Douglas EL, Hewett PJ, Hardingham JE. Receptor protein tyrosine kinase EphB4 is up-regulated in colon cancer. *BMC Mol Biol.* 2001;2:15.

144. Stirewalt DL, Kopecky KJ, Meshinchi S, et al. FLT3, RAS, and TP53 mutations in elderly patients with acute myeloid leukemia. *Blood.* 2001;97:3589–3595.

145. Stirewalt DL, Kopecky KJ, Meshinchi S, et al. Size of FLT3 internal tandem duplication has prognostic significance in patients with acute myeloid leukemia. *Blood.* 2006;107: 3724–3726.

146. Stirewalt DL, Meshinchi S, Kussick SJ, et al. Novel FLT3 point mutations within exon 14 found in patients with acute myeloid leukaemia. *Br J Haematol.* 2004;124:481–484.

147. Stirewalt DL, Radich JP. The role of FLT3 in haematopoietic malignancies. *Nat Rev Cancer.* 2003;3:650–665.

148. Stone RM, DeAngelo DJ, Klimek V, et al. Patients with acute myeloid leukemia and an activating mutation in FLT3 respond to a small-molecule FLT3 tyrosine kinase inhibitor, PKC412. *Blood.* 2005;105:54–60.

149. Stone RM, Fischer T, Paquette R, et al. Phase 1B study of PKC412, an oral FLT3 kinase inhibitor, in sequential and simultaneous combinations with daunorubicin and cytarabine (DA) induction and high-dose consolidation in newly diagnosed patients with AML. *Blood.* 2005;106:404a.

150. Stone RM, Klimek V, J. DD, et al. PKC412, an oral FLT3 inhibitor, has activity in mutant FLT3 acute myeloid leukemia (AML): a phase II clinical trial. *Blood.* 2002;100:316a.

151. Tartaglia M, Martinelli S, Iavarone I, et al. Somatic PTPN11 mutations in childhood acute myeloid leukaemia. *Br J Haematol.* 2005;129:333–339.

152. Thiede C, Steudel C, Mohr B, et al. Analysis of FLT3-activating mutations in 979 patients with acute myelogenous leukemia: association with FAB subtypes and identification of subgroups with poor prognosis. *Blood.* 2002;99:4326–4335.

153. Tobal K, Pagliuca A, Bhatt B, Bailey N, Layton DM, Mufti GJ. Mutation of the human FMS gene (M-CSF receptor) in myelodysplastic syndromes and acute myeloid leukemia. *Leukemia.* 1990;4:486–489.

154. Tomasson MH, Sternberg DW, Williams IR, et al. Fatal myeloproliferation, induced in mice by TEL/PDGFbetaR expression, depends on PDGFbetaR tyrosines 579/581. *J Clin Invest.* 2000;105:423–432.
155. Tomasson MH, Williams IR, Hasserjian R, et al. TEL/PDGFbetaR induces hematologic malignancies in mice that respond to a specific tyrosine kinase inhibitor. *Blood.* 1999;93:1707–1714.
156. Tong FK, Chow S, Hedley D. Pharmacodynamic monitoring of BAY 43-9006 (Sorafenib) in phase I clinical trials involving solid tumor and AML/MDS patients, using flow cytometry to monitor activation of the ERK pathway in peripheral blood cells. *Cytometry B Clin Cytom.* 2006;70(3):107–114.
157. Towatari M, Iida H, Tanimoto M, Iwata H, Hamaguchi M, Saito H. Constitutive activation of mitogen-activated protein kinase pathway in acute leukemia cells. *Leukemia.* 1997;11:479–484.
158. Tse KF, Allebach J, Levis M, Smith BD, Bohmer FD, Small D. Inhibition of the transforming activity of FLT3 internal tandem duplication mutants from AML patients by a tyrosine kinase inhibitor. *Leukemia.* 2002;16:2027–2036.
159. Tse KF, Mukherjee G, Small D. Constitutive activation of FLT3 stimulates multiple intracellular signal transducers and results in transformation. *Leukemia.* 2000;14: 1766–1776.
160. Tsujimura T, Furitsu T, Morimoto M, et al. Ligand-independent activation of c-kit receptor tyrosine kinase in a murine mastocytoma cell line P-815 generated by a point mutation. *Blood.* 1994;83:2619–2626.
161. Tsujimura T, Kanakura Y, Kitamura Y. Mechanisms of constitutive activation of c-kit receptor tyrosine kinase. *Leukemia.* 1997;11 Suppl 3:396–398.
162. Tsujimura T, Morimoto M, Hashimoto K, et al. Constitutive activation of c-kit in FMA3 murine mastocytoma cells caused by deletion of seven amino acids at the juxtamembrane domain. *Blood.* 1996;87:273–283.
163. Turner AM, Lin NL, Issarachai S, Lyman SD, Broudy VC. FLT3 receptor expression on the surface of normal and malignant human hematopoietic cells. *Blood.* 1996;88: 3383–3390.
164. Ueda S, Ikeda H, Mizuki M, et al. Constitutive activation of c-kit by the juxtamembrane but not the catalytic domain mutations is inhibited selectively by tyrosine kinase inhibitors STI571 and AG1296. *Int J Hematol.* 2002;76:427–435.
165. van der Geer P, Hunter T. Identification of tyrosine 706 in the kinase insert as the major colony-stimulating factor 1 (CSF-1)-stimulated autophosphorylation site in the CSF-1 receptor in a murine macrophage cell line. *Mol Cell Biol.* 1990;10:2991–3002.
166. Vandenberghe P, Wlodarska I, Michaux L, et al. Clinical and molecular features of FIP1L1-PDFGRA (+) chronic eosinophilic leukemias. *Leukemia.* 2004;18:734–742.
167. Veiby OP, Lyman SD, Jacobsen SE. Combined signaling through interleukin-7 receptors and flt3 but not c-kit potently and selectively promotes B-cell commitment and differentiation from uncommitted murine bone marrow progenitor cells. *Blood.* 1996;88: 1256–1265.
168. Vogelstein B, Civin CI, Preisinger AC, et al. RAS gene mutations in childhood acute myeloid leukemia: a Pediatric Oncology Group study. *Genes Chromosomes Cancer.* 1990;2:159–162.
169. Wallace EM, Lyssikatos J, Blake JF, et al. Potent and selective mitogen-activated protein kinase kinase (MEK) 1,2 inhibitors. 1. 4-(4-bromo-2-fluorophenylamino)-1-methylpyridin-2(1H)-ones. *J Med Chem.* 2006;49:441–444.
170. Wang YY, Zhou GB, Yin T, et al. AML1-ETO and C-KIT mutation/overexpression in t(8;21) leukemia: implication in stepwise leukemogenesis and response to Gleevec. *Proc Natl Acad Sci USA.* 2005;102:1104–1109.
171. Weiner HL, Zagzag D. Growth factor receptor tyrosine kinases: cell adhesion kinase family suggests a novel signaling mechanism in cancer. *Cancer Invest.* 2000;18:544–554.

172. Weisberg E, Boulton C, Kelly LM, et al. Inhibition of mutant FLT3 receptors in leukemia cells by the small molecule tyrosine kinase inhibitor PKC412. *Cancer Cell.* 2002;1:433–443.

173. Whitman SP, Archer KJ, Feng L, et al. Absence of the wild-type allele predicts poor prognosis in adult de novo acute myeloid leukemia with normal cytogenetics and the internal tandem duplication of FLT3: a cancer and leukemia group B study. *Cancer Res.* 2001;61:7233–7239.

174. Wiener JR, Kassim SK, Yu Y, Mills GB, Bast RC, Jr. Transfection of human ovarian cancer cells with the HER-2/neu receptor tyrosine kinase induces a selective increase in PTP-H1, PTP-1B, PTP-alpha expression. *Gynecol Oncol.* 1996;61:233–240.

175. Yamamoto Y, Kiyoi H, Nakano Y, et al. Activating mutation of D835 within the activation loop of FLT3 in human hematologic malignancies. *Blood.* 2001;97:2434–2439.

176. Yamashita N, Osato M, Huang L, et al. Haploinsufficiency of Runx1/AML1 promotes myeloid features and leukaemogenesis in BXH2 mice. *Br J Haematol.* 2005;131:495–507.

177. Yee KW, O'Farrell AM, Smolich BD, et al. SU5416 and SU5614 inhibit kinase activity of wild-type and mutant FLT3 receptor tyrosine kinase. *Blood.* 2002;100:2941–2949.

178. Yee NS, Langen H, Besmer P. Mechanism of kit ligand, phorbol ester, and calcium-induced down-regulation of c-kit receptors in mast cells. *J Biol Chem.* 1993;268:14189–14201.

179. Yokota S, Kiyoi H, Nakao M, et al. Internal tandem duplication of the FLT3 gene is preferentially seen in acute myeloid leukemia and myelodysplastic syndrome among various hematological malignancies. A study on a large series of patients and cell lines. *Leukemia.* 1997;11:1605–1609.

180. Zhao M, Kiyoi H, Yamamoto Y, et al. In vivo treatment of mutant FLT3-transformed murine leukemia with a tyrosine kinase inhibitor. *Leukemia.* 2000;14:374–378.

181. Zheng R, Levis M, Piloto O, et al. FLT3 ligand causes autocrine signaling in acute myeloid leukemia cells. *Blood.* 2004;103:267–274.

182. Zwaan CM, Meshinchi S, Radich JP, et al. FLT3 internal tandem duplication in 234 children with acute myeloid leukemia: prognostic significance and relation to cellular drug resistance. *Blood.* 2003;102:2387–2394.

Lineage-Specific Transcription Factor Aberrations in AML

Beatrice U. Mueller and Thomas Pabst

Abstract Transcription factors play a key role in the commitment of hemato-poietic stem cells to differentiate into specific lineages [78]. This is particularly important in that a block in terminal differentiation is the key contributing factor in acute leukemias. This general theme of the role of transcription factors in differentiation may also extend to other tissues, both in terms of normal development and cancer. Consistent with the role of transcription factors in hematopoietic lineage commitment is the frequent finding of aberrations in transcription factors in AML patients. Here, we intend to review recent findings on aberrations in lineage-restricted transcription factors as observed in patients with acute myeloid leukemia (AML).

Leukemic Stem Cells as a Model for Cancer Development

Growing evidence indicates that only a minor subpopulation of cells of a neoplasm is maintaining the uncontrolled production of cancerous daughter cells [56, 26, 30]. According to this model, a relatively small number of such tumor stem cells give rise to the bulk of tumor cells such as leukemic blasts. Leukemic stem cells (LSC) seem to be the cellular source of a hierarchy of daughter cells, which greatly differ in their potency of unlimited proliferation and maintaining the disease [26, 30]. LSCs apparently share important stem cell functions with normal hematopoietic stem cells (HSCs), such as self-renewal, initial differentiation, and survival. It is therefore assumed that a similar set of critical genes controls both LSCs and HSCs. These genes comprise the polycomb family member Bmi-1 [36], the ubiquitous transcription factor

Declaration: All authors declare that they have no interest in a company – or a competitor of a company – whose product was analyzed in the present work. All authors agree with the manuscript in its present form.

B.U. Mueller (✉)
Department of Internal Medicine, University Hospital, 3010 Bern, Switzerland
e-mail: beatrice.mueller@insel.ch

L. Nagarajan (ed.), *Acute Myelogenous Leukemia*,
Cancer Treatment and Research 145, DOI 10.1007/978-0-387-69259-3_7,
© Springer Science+Business Media, LLC 2010

jun-B [57], and the Wing-less type (Wnt) pathway [27]. However, the molecular pathways underlying the transformation of HSCs into LSCs are poorly understood.

The hallmark of AML is the terminal differentiation block of hematopoietic cells of the myeloid lineage, while proliferation and self-renewal are preserved [18]. Thus, it is assumed that the molecular events underlying LSC development in AML must be critical enough to block terminal differentiation while still allowing basic stem cell function. In normal HSCs, lineage-specific transcription factors have been identified as potent regulators of these functional programs.

The Role of Transcription Factors for Normal Hematopoietic Stem Cell Function

There is a large literature indicating that a small group of transcription factors with mostly lineage restricted expression patterns plays a crucial role in controlling normal hematopoiesis [50]. Among the best characterized examples are the transcription factors AML1/RUNX1, SCL/Tal-1, c-myb, PU.1, GATA1, and CCAAT/enhancer binding protein α (CEBPA). Knock-out of the genes encoding these factors in mice displayed profound hematopoietic defects as reviewed elsewhere [78] and summarized in Table 1. Moreover, those transcription factors were shown to regulate broad ranges of important target genes, hereby directly programming hematopoietic precursors to differentiate along a complex developmental pathway [77].

A number of these transcription factors have been linked to HSC functions. AML1/RUNX1 and SCL/Tal-1 have been demonstrated to be indispensable for the specification of fetal liver HSCs [49, 70]. However, recent studies using conditional knockout mice showed that AML1/RUNX1 and SCL/Tal-1 are less critical for the maintenance of HSCs in the adult organism [21, 28], indicating that these nuclear factors may be required for the generation of HSCs from more pluripotent cells in the embryo but not for HSC self-renewal.

Several reports link the transcription factor PU.1 to HSC function. Expression of PU.1 is detected in HSCs and increases progressively during differentiation into neutrophils or monocytes [1, 48]. Previous studies showed that, upon transplantation into congenic mice, PU.1$^{-/-}$ fetal liver HSCs poorly engraft the bone marrow and are incapable of long-term reconstitution [66]. Furthermore, although PU.1$^{-/-}$-derived cells contribute to erythrocyte development in chimeric mouse studies, this effect is only transient, strongly suggesting a defect at the level of the HSCs [66]. More recent approaches using both conventional and conditional PU.1 knockout mice suggest that PU.1$^{-/-}$ HSCs possess impaired self-renewal capacity and lack the ability to differentiate into the earliest lymphoid and myeloid progenitor stages [32, 11].

Table 1 Transcription factors involved in normal hematopoiesis and the association of alterations in these factors with human AML

Factor	Expression	Gene-target products	Knockouts	AML	Mutation
AML1	HSCs and most others	M-CSF receptor and many others	Lack all definitive hematopoiesis	M0 M2	Often biallelic mutations t(8;21); t(3;21)
PU.1	All progenitors, suppressed in erythroid and T cells	Receptors for GM-CSF, G-CSF, M-CSF; and many others	Loss of B cells & macrophages; delayed development of T cells and granulocytes	M0,4,5,6 M2 t(8;21) M3 t(15;17) FLT3-ITD	Mutations Suppression Suppression Suppression
CEBPA	HSCs, CMPs, and GMPs, not in MEPs and lymphoid cells	Receptors for G-CSF, IL-6, E2F, c-myc; primary granule proteins and many others	Loss of granulocytic maturation; block at the CMP to GMP stage	M1,2,(4) M2 t(8;21) t(3;21) M4 inv(16) FLT3-ITD	Mutations Suppression Suppression Suppression Suppression
GATA1	HSCs, CMPs, MEPs, not in GMPs or lymphoid cells	Erythropoietin receptor and many others	Lack definitive erythroid cells	M7 assoc. with Down's syndrome	Mutations

HSC, hematopoietic stem cells; CMP, common myeloid progenitor; CLP, common lymphoid progenitor; GMP, granulocyte/macrophage progenitor; MEP, megakaryocyte/erythrocyte progenitor; IL-6, interleukin-6; GM-CSF, granulocyte-monocyte colony-stimulating factor.

CEBPA, a crucial regulator of granulopoiesis, has also been implicated in the control of HSCs. CEBPA-deficient HSCs display enhanced competitive repopulation activity in murine transplantation models, possibly by a mechanism that includes altered cell divisions through increased Bmi-1 expression [88, 29].

Differing Expression Levels of Lineage-Restricted Transcription Factors Regulate Differentiation Fates

The pattern of lower level expression in HSCs and lineage-restricted up- or downregulation in more mature cells is characteristic for a number of hematopoietic transcription factors, including PU.1, CEBPA, GATA-1, and c-myb [1]. This hallmark expression pattern suggests a stage-specific difference in the need for those proteins during differentiation and emphasizes a requirement for basic transcription factor functions at the HSC level.

PU.1 was the first hematopoietic transcription factor for which a causal association of expression levels and differentiation fates has been shown. A model was developed by which different cellular concentrations of PU.1 direct distinct cell fates, with the highest levels required for macrophage and – to a lesser extent – granulocytic development and lower levels for B-cell lineage adoption [10, 12]. Another transcription factor with clear dosage sensitivity in hematopoiesis is c-myb. Whereas the null mutation of c-myb completely prevents progenitor cell development, low c-myb levels (5–10% of wild type) do allow initial HSC differentiation into progenitors, but they are insufficient for further lineage commitment and terminal differentiation [14]. Finally, GATA-1 appears to regulate erythroid, megakaryocytic, mast cell, and eosinophilic cell fates in a concentration-dependent manner [69, 42, 87].

Transcription Factors Play a Crucial Role in Human AML

Transcription factors are among the most frequently mutated or dysregulated genes in patients with AML [31]. The involvement of transcription factors in acute leukemias was first suggested by common somatically acquired chromosomal translocations [38]. In patients with AML, the most frequently found translocation products involving transcription factors are AML1/ETO [t(8;21)], core-binding factor β /myosin heavy polypeptide 11 (CBFβ/MYH11) [inv16], mixed lineage leukemia (MLL) gene fusions [t11q23], and promyelocytic leukemia/retinoic acid receptor α (PML/RARα) [t(15;17)] [31]. More recently, smaller mutations in the coding regions of several transcription factor genes have been identified in AML patients. Typically, such mutations represent the sole detectable genomic aberration in these AML patients, and they are limited to particular subtypes of AML. Since most of these mutations are heterozygous, they result in reduced – but not absent – transcription factor activity. The block of normal differentiation upon transcription factor reduction caused by such mutations might result in the accumulation of a progenitor pool from which LSCs can arise.

Transcription Factor Activity of the Myeloid Key Transcription Factor CEBPA Is Critically Suppressed in AML Patients

A rapidly growing literature indicates that the function of the granulocytic transcription factor CEBPA can be inhibited by a number of mechanisms in blasts of AML patients (Fig. 1).

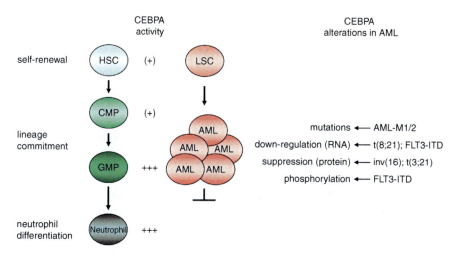

Fig. 1 Role of CEBPA in normal HSC renewal, early lineage decision, and terminal differentiation as opposed to the differentiation block in AML caused by CEBPA aberrations. In normal hematopoiesis, CEBPA is expressed at low levels in HSCs suggesting a role in limiting HSC self-renewal SCs. CEBPA is upregulated to highest levels in granulocyte monocyte progenitors (GMP) whereas it is not expressed in precursors of lymphoid cells and downregulated as common myeloid progenitors (CMP) differentiate to megakaryocyte and erythroid progenitors. In AML with suppressed CEBPA activity, the transition from CMPs to GMPs is blocked leading to the accumulation of myeloid blasts arrested at this particular stage. LSC: leukemic stem cells

CEBPA Mutations in AML Patients

Eleven different groups now have reported the presence of CEBPA mutations in AML patients, with the frequency approximating 5–14% [53, 19, 61, 74, 83, 17, 67, 5, 37, 35, 72, 46]. The mutations can be largely divided into two common types: First, carboxyl-terminal in-frame mutations disrupt the basic zipper region, thus affecting DNA binding as well as homo- and heterodimerization with other CEBP family members [53]. Often, this type of mutation is associated with a second mutation in the other allele. Second, amino-terminal frame shift mutations result in premature termination of the wild type 42 kDa form of the CEBPA protein while preserving the 30 kDa form [53, 61, 83, 46]. The latter inhibits the wild type CEBPA 42 kDa protein in a dominant negative manner [4].

Several clinical characteristics of patients with CEBPA mutations are becoming increasingly clear. CEBPA mutations are typically seen in AML-M1 and AML-M2 patients, and only rarely in AML-M4 patients [53, 61, 74, 35]. Approximately 70% of AML patients with CEBPA mutations have a normal karyotype [53, 61, 74, 83, 17, 5, 37, 35, 72]. CEBPA mutations are not observed in patients with common karyotype abnormalities – such as t(15;17), inv(16), or t(8;21) [53, 17, 35]. AML patients with CEBPA mutations have better relapse-free survival and/or overall survival [61, 74, 83, 17, 5, 37, 35, 72]. Why CEBPA mutations confer good prognosis is unclear [26]. There are contradictory studies on cooperating mutations as far as the effect of coexisting FLT3-ITD adversely affects the favorable prognosis of CEBPA mutations [39, 38, 47]. There appears to be an association of deletion 9q and CEBPA loss-of-function mutations suggesting that loss of a critical segment of 9q and disruption of CEBPA function possibly cooperate in the pathogenesis of del(9q) AML [16]. Using microarray analysis, AML patients with CEBPA mutations express a distinctive gene expression signature [82, 6]. There is increasing consensus that the favorable prognostic value of CEBPA mutations justifies screening for CEBPA mutations in AML-M1 and AML-M2 patients with intermediate risk cytogenetic result at diagnosis [35]. The combination of CEBPA mutations with other markers such as FLT3-ITD, BAALC, nucleophosmin, or EVI1 expression allows risk assessment at diagnosis particularly in AML patients with a normal karyotype [5, 72, 81, 39, 2, 15, 3].

Two families have recently been reported in whom three and four members affected by AML-carried germline heterozygous CEBPA mutations [67, 73]. Intriguingly, the site and type of CEBPA mutations in the two families are almost identical. The early onset of AML in these two families contrasts markedly with the usual age of onset of sporadic AML, reflecting predisposition. The observation that germline mutations in CEBPA predispose to AML and the fact that no additional chromosomal aberrations were detected in these cases of familial AML suggest that CEBPA mutations might be sufficient to cause myeloid leukemia. This is in contrast to the situation in non-conditional CEBPA knockout mice models, in which no overt AML is observed, and it provides – once more – a lesson on the difficulty to mimic a malignant human disease using mouse models [65].

Suppression of CEBPA Expression in AML Patients

CEBPA mRNA expression is exclusively suppressed in the presence of the AML1-ETO fusion protein both in vitro and in vivo [52]. AML1-ETO is thought to inhibit CEBPA mRNA expression through inhibition of autoregulation [52]. Gene expression analyses have not revealed any other subset of AML patients with consistently suppressed CEBPA mRNA [82, 6, 81]. However, CEBPA mRNA is repressed by FLT3/ITD signaling [89]. This repression can be overcome by treatment with a FLT3 inhibitor, CEP-701 [89]. Finally, a report studying hypermethylation of the CEBPA promoter indicates that 9% of AML-M2 patients had methylated CEBPA promoter [8].

Posttranscriptional Suppression of CEBPA in AML Patients

The arrest of differentiation is a hallmark of chronic myelogenous leukemia in myeloid blast crisis (CML-BC). Loss of CEBPA function represents an obvious candidate event for this block. Whereas CEBPA mutations are absent in CML-BC [54], CEBPA protein is not detectable in CML-BC cells [59]. In contrast, CEBPA mRNA is clearly present in CML-BC samples [59]. Expression of CEBPA was found to be suppressed posttranscriptionally by interaction of the poly(rC)-binding protein hnRNP E2 with CEBPA mRNA [59, 58]. Whether hnRNP E2 is involved in posttranscriptional CEBPA suppression in AML remains to be tested.

The ability of oncogenic proteins to regulate the rate of translation of specific mRNA subsets provides a rapid mechanism to modulate the levels of the corresponding proteins. Such a mechanism was recently demonstrated for CEBPA in AML with t(3;21) encoding the AML1-MDS1-EVI1 fusion gene (AME) [24, 23]. AME suppresses CEBPA protein in vitro and in AML patients [24]. In contrast, CEBPA mRNA levels remain unchanged. Interestingly, the RNA-binding protein calreticulin was strongly activated in AME patient samples. Calreticulin interacts with GCN repeats within CEBPA (and CEBPB) mRNAs [79]. GCN repeats within these mRNAs form stable stem loop structures [79]. The interaction of calreticulin with the stem loop structure of CEBPA mRNA leads to inhibition of translation of CEBP proteins [79]. In myeloid cells, inhibition of calreticulin by siRNA powerfully restores CEBPA levels [24, 23].

Posttranslational Inhibition of CEBPA in AML Patients

It was proposed that phosphorylation of CEBPA at serine 21 is mediated by extracellular signal-regulated kinases 1 and/or 2 (ERK1/2) [79]. This phosphorylation induces a conformational change in CEBPA such that the transactivation domains of two CEBPA molecules within a dimer move further apart [79]. Phosphorylation of CEBPA at serine 21 was shown to inhibit granulopoiesis [79].

Additional phosphorylation sites have been reported [64, 85]. Activated Ras acts on serine 248 of the CEBPA transactivation domain [64]. Interestingly,

PKC inhibitors block the activation of CEBPA by Ras thus impairing the ability of CEBPA to induce granulocytic differentiation [4]. Other posttranslational mechanisms of CEBPA modulation in AML remain to be tested, such as heterodimerization with other CEBP family members [55].

An unresolved issue is the ultimate clarification of how CEBPA is involved in the differentiation block of leukemic cells of patients with acute promyelocytic leukemia (APL). There is some evidence that induction of PML-RARα inhibits CEBPA activity [80]. Also, restoring CEBPA activity in leukemic cells from PML-RARα transgenic mice suppresses growth and induces partial differentiation in vitro [80]. In vivo, enhanced expression of CEBPA prolongs survival of such animals [80]. ATRA treatment of patients with APL powerfully induces CEBPA mRNA expression [45]. However, further work is needed to elucidate exactly how CEBPA is affected by the presence of PML-RARα.

Alterations of the AML1/RUNX1 Transcription Factor in AML

Traditionally, the AML1/RUNX1 gene was regarded as a translocation partner involved in all kind of chromosomal rearrangements. This view had to be modified when RUNX1 point mutations in sporadic and in familial myeloid leukemia were reported [51, 75]. Subsequent studies have identified point mutations in 9% of AML patients, with an obvious preference for the M0 subtype [62, 33, 40, 71]. Until recently, mutations were believed to cluster within the Runt domain of the RUNX1 gene. However, mutations in the carboxyl terminal region, outside the Runt domain, have been identified predominantly in MDS-AML [22].

Heterozygous RUNX1 Mutations and the Concept of Haploinsufficiency

The majority of RUNX1 mutations in leukemia patients are heterozygous, pointing to the concept of haploinsufficiency as the mode of action in these leukemias. This notion is supported by the finding of entire RUNX1 gene deletions in familial platelet disorder (FPD)-associated AML [75]. It is believed that the RUNX1 +/− status is likely to provide the cells with a growth advantage. However, clustering of RUNX1 mutations in the Runt domain suggests a dominant-negative mechanism in addition to genuine haploinsufficiency. It is interesting that families with RUNX1 mutations acting simply via haploinsufficiency have a lower incidence of leukemia than families with mutations acting in a dominant-negative fashion [75, 41]. Thus, low RUNX1 activity appears to correlate with high leukemogenicity.

Biallelic RUNX1 Mutations in the AML M0 Subtype

RUNX1 point mutations are most frequently found in the AML M0 subtype. In addition, the majority of AML M0 patients with RUNX1 mutations show the biallelic type of mutation whereas this type is hardly ever observed in other leukemia subtypes. Therefore, it is clear that RUNX1 point mutations of the biallelic type are tightly associated with the AML M0 subtype. Of note is the fact that the biallelic type mutation is strictly confined to the status with two copies of RUNX1. In contrast, biallelic mutation in trisomy 21 is associated with myeloid malignancies other than M0.

RUNX1 Mutations Evolving as Secondary Leukemia from Myelodysplastic Syndrome (MDS) or Therapy-Related MDS

RUNX1 point mutations are the most consistent genomic aberration found in MDS and subsequent overt leukemia. These RUNX1 mutations are characteristically distributed, as they are also detected in the carboxyl terminal region which is rarely targeted in other types of leukemias [22]. Even if the carboxyl terminal region has previously not been investigated in many studies with de novo AML, RUNX1 point mutations including the C-terminal moiety appear to be found predominantly in MDS-AML.

Acquired Trisomy 21 and Non-MO Myeloid Malignancy

Whereas RUNX1 point mutations are rarely observed in AML in congenital trisomy 21 patients, AML with acquired trisomy 21 is associated with a high frequency of RUNX1 point mutations [62]. Interestingly, trisomy 21 increases the copy number of the mutant allele. In other words, acquired (but not congenital) trisomy 21 is thought to be a secondary change to the RUNX1 +/− status.

RUNX1 Mutations in Familial Leukemia (FPD-AML)

Heterozygous germline mutation in the RUNX1 gene causes familial platelet disorder with predisposition to acute myeloid leukemia (FPD-AML). A total of 12 pedigrees have been identified so far. FPD alone, before the onset of AML, is not fatal but significantly inhibits blood clotting, which could be life threatening after surgery, injury, dental treatment, or childbirth. The incidence of leukemia among affected individuals varies from 20 to 50% (average 35%).

Hypomorphic PU.1 Function in AML

PU.1 Mutations in AML

Heterozygous mutations in the PU.1 gene were identified in 7% of patients with AML [44]. These mutations largely resulted in decreased ability of PU.1 to synergize with interacting proteins such as AML1 or c-Jun in the activation of target genes. In contrast, other studies found either no PU.1 mutations or that they occurred less frequently [13]. The exact reason for this discrepancy has yet to be revealed.

PU.1 Is Suppressed in Acute Promyelocytic Leukemia (APL)

PU.1 expression and/or function appears to be downregulated by several important oncogenic products [89, 45, 84, 43]. In particular, PU.1 downregulation is a critical effect of the PML-RARα protein (Fig. 2), whereas treatment with all-trans retinoic acid (ATRA) powerfully restores PU.1 expression [45]. Restoring PU.1 per se in APL blasts is sufficient to induce terminal neutrophil differentiation [45]. Consequently, and similar to the absence of CEBPA mutations in t(8;21) leukemia, these observations might explain the lack of frequent PU.1 mutations in AML.

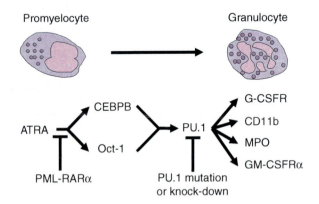

Fig. 2 PU.1-induced terminal neutrophil differentiation can be blocked in leukemia. Two pathways are depicted: (**a**) by the PML-RARα fusion gene product in human acute promyelocytic leukemia (APL) [45] and (**b**) by reduced function of PU.1 itself in AML patients with PU.1 mutations or in knockdown or radiation induced AML in mice [9]. ATRA restores PU.1 expression in APL by activation of CEBPB and Oct-1, thereby allowing induction of PU.1 target genes [9]

Hypomorphic PU.1 Function Acts Leukemogenic

Heterozygous mutations as well as downregulation of PU.1 by leukemogenic fusion proteins indicate a scenario in which critically lowered transcription factor activity acts leukemogenic. This is supported by a number of animal models harboring hypomorphic transcription factor function. First of all, PU.1 knockdown mice provided first definitive proof that graded downregulation of a single transcription factor to a level above nullizygosity is sufficient to induce myeloid transformation [63]. In this study, PU.1 knockdown was engineered by deletion of the –14 kb PU.1 URE, thus allowing 20% residual PU.1 expression in HSCs and myeloid progenitors. These animals had normal numbers of HSCs, but an increased myeloid progenitor compartment, and after a short preleukemic phase they frequently developed an aggressive AML [63].

Although it has been reported that null alleles of PU.1 can also lead to leukemia [11], clear experimental evidence for the advantage of hypomorphic PU.1 function over its complete disruption in provoking AML came from the analysis of γ-irradiated mice [9]. Radiation-induced myeloid leukemias regularly acquired a combination of a deletion on one copy of chromosome 2, which included the PU.1 gene locus, and a recurring single-point mutation in the ETS domain of the remaining PU.1 allele, which impaired DNA binding (Fig. 2). Most strikingly, no tumors were found which had both PU.1 alleles deleted nor were there any cases where the remaining PU.1 allele suffered a null mutation, suggesting a specific selection during transformation for those clones with preserved minimal PU.1 activity over those that suffered a complete loss of function.

Collectively, both observations on patient samples and analysis of experimentally generated myeloid leukemia in animal models support a concept in which LSC activity is generally associated with hypomorphic rather than with complete abrogation of PU.1 activity. The quantitative reduction to a critical functional dosage might meet a highly malignant threshold level which is dominant over the effect of true null alleles in inducing cancer. The involvement of most leukemia-affiliated transcription factors in the regulation of basic stem-cell functions of normal HSCs, such as self-renewal and initial differentiation, could explain the general requirement for such residual activity to meet key LSC functions, which might not be as efficiently satisfied by completely abolished transcription-factor activity.

Deregulated GATA1 Function in AML

Inherited mutations in the amino-terminal zinc finger domain of GATA1 are associated with congenital dyserythropoietic anemia and thrombocytopenia [47]. In contrast, somatically acquired GATA1 mutations (Fig. 3) are exclusively found in patients with congenital trisomy 21 who developed concomitant acute megakaryoblastic leukemia (AMKL) or transient myeloproliferative

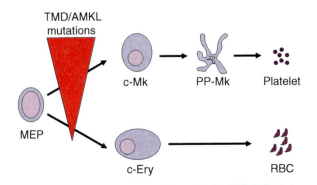

Fig. 3 Mutations in GATA-1 lead to distinct blocks in hematopoiesis. Acquired mutations in GATA1 – as observed in essentially all cases of transient myeloproliferative disorder (TMD) and Down's syndrome associated acute megakaryoblastic leukemia (AMKL) [47, 86] – lead to an early block in megakaryocytic development and the resulting proliferation of megakaryoblasts within the peripheral blood and either the fetal liver (TMD) or the bone marrow (AMKL). In addition, both of these disorders frequently exhibit dyserythropoiesis, consistent with the observation that these blasts express markers of both the erythroid and the megakaryocytic lineages. In contrast, experimentally induced mutations in the N-finger of GATA1 block megakaryocytic development prior to proplatelet formation, while erythroid development is hardly affected. These observations suggest that the megakaryocytic development is more sensitive to alterations in GATA1. MEP, megakaryocyte erythroid precursor; c-MK, committed megakaryocyte; c-Ery, committed erythroblast; PP, proplatelet producing megakaryocyte; RBC, red blood cell

disorder (TMD) [86]. Thus, GATA1 mutations are tightly linked to Down's syndrome (DS) and trisomy 21. It has been estimated that the incidence of AMKL in children with DS is 1 in 500, with a median age of presentation of 2 years [34]. This corresponds to a 500-fold increased risk of AMKL compared to children without DS [90].

In DS-associated leukemias, alterations in GATA1 include missense, non-sense, and splice site mutations, as well as short insertions or deletions that alter the correct reading frame of the GATA1 transcript [20, 25]. All of these mutations reside within exon 2 (or the intron immediately downstream). Similar to CEBPA mutations, expression of full length 47 kDa GATA1 is prevented, while expression of a short truncated 40 kDa isoform is preserved [86]. In contrast, however, to CEBPA mutations, the 40 kDa GATA1 form cannot act as a dominant negative on full-length GATA1 in the leukemic cells since GATA1 is an X-linked gene, and only one allele is expressed at any one time. This 40 kDa GATA1 lacks the amino-terminal activation domain but retains both zinc fingers and the entire carboxyl-terminus, and thus leads to a hypomorphic protein that is still able to bind to DNA and to friend-of-GATA (FOG), an important cofactor of GATA1 protein function, but has reduced transactivation potential [20, 7]. Thus, the principle of a strong association between leukemogenesis and hypomorphic instead of absent transcription factor function appears to apply also to GATA-1. This concept has recently been supported by the observation that GATA1 knockdown mice developed leukemia whereas mice completely lacking GATA1 did not [76, 68].

Concluding Remarks

Although there is now ample evidence for the dominant roles of transcription factors in both normal and leukemic stem-cell activities, there are still numerous open questions, in particular concerning the functional mechanisms. Most major insights into the biologic roles of transcription factors were derived from studying knockout mice. However, such gene disruption usually leads to the complete abrogation of expression. The discovery of small heterozygous mutations in key transcription factors in AML patients indicates that genes are not simply expressed in a turn on or off mode in malignancies, but are rather suppressed to meet very specific thresholds, which trigger distinct biologic functions.

A major challenge for the future is therefore to consider the impact of small gradations in the dosages on both normal and neoplastic developments. It is essential for our understanding of malignant events that we obtain a detailed view of the mechanisms that adjust the concentrations of especially those transcription factors with a role in human cancer.

Acknowledgments Research grants: This work was supported by grants from the Swiss National Science Foundation SF 3100A0-100445 to B.U.M. and SF 310000-109388 to TP.We apologize to all authors whose contribution to the field could not be cited due to limitations in space.

References

1. Akashi K, Traver D, Miyamoto T, Weissman IL. A clonogenic common myeloid progenitor that gives rise to all myeloid lineages. *Nature.* 2000;404:193–197.
2. Avivi I, Rowe JM. Prognostic factors in acute myeloid leukemia. *Curr Opin Hematol.* 2005;12:62–67.
3. Barjesteh van Waalwijk van Doorn-Khosrovani S, Erpelinck C, van Putten WL, et al. High EVI1 expression predicts poor survival in acute myeloid leukemia: a study of 319 de novo AML patients. *Blood.* 2003;101:837–845.
4. Behre G, Singh SM, Liu H, et al. Ras signaling enhances the ability of CEBPA to induce granulocytic differentiation by phosphorylation of serine 248. *J Biol Chem.* 2002;277:26293–26299.
5. Bienz M, Ludwig M, Mueller BU, et al. Risk assessment in patients with acute myeloid leukemia and a normal karyotype. *Clin Cancer Res.* 2005;11:1416–1425.
6. Bullinger L, Dohner K, Bair E, et al. Use of gene-expression profiling to identify prognostic subclasses in adult acute myeloid leukemia. *N Engl J Med.* 2004;350:1605–1616.
7. Calligaris R, Bottardi S, Cogoi S, Apezteguia I, Santoro C. Alternative translation initiation site usage results in two functionally distinct forms of the GATA-1 transcription factor. *Proc Natl Acad Sci USA.* 1995;92:11598–11602.

8. Chim CS, Wong ASY, Kwong YL. Infrequent hypermethylation of CEBPA promoter in acute myeloid leukaemia. *Br J Haemat.* 2002;119:988–990.
9. Cook WD, McCaw BJ, Herring CD, et al. PU.1 is a suppressor of myeloid leukemia, inactivated in mice by gene deletion and mutation of its DNA-binding domain. *Blood.* 2004;104:3437–3444.
10. Dahl R, Walsh JC, Lancki D, et al. Regulation of macrophage and neutrophil cell fates by the PU.1:C/EBPalpha ratio and granulocyte colony-stimulating factor. *Nat Immunol.* 2003;4:1029–1036.
11. Dakic A, Metcalf D, Di Rago L, et al. Pu.1 regulates the commitment of adult hematopoietic progenitors and restricts granulopoiesis. *J Exp Med.* 2005;201:1487–1502.
12. DeKoter RP, Singh H. Regulation of B lymphocyte and macrophage development by graded expression of PU.1 *Science.* 2000;288:1439–1441.
13. Dohner K, Tobis K, Bischof T, et al. Mutation analysis of the transcription factor PU.1 in younger adults (16 to 60 years) with acute myeloid leukemia: a study of the AML Study Group Ulm (AMLSG ULM). *Blood.* 2002;100:4680–4681.
14. Emambokus N, Vegiopoulos A, Haman B, et al. Progression through key stages of hematopoiesis is dependent on distinct threshold levels of c-myb. *EMBO J.* 2003;22;4478–4488.
15. Falini B, Mecucci C, Tiacci E, et al. Cytoplasmic nucleophosmin in acute myelogenous leukemia with a normal karyotype. *N Engl J Med.* 2005;352:254–266.
16. Fröhling S, Schlenk RF, Krauter J. Acute myeloid leukemia with deletion 9q within a noncomplex karyotype is associated with CEBPA loss-of-function mutations. *Genes Chrom Cancer.* 2005;42:427–432.
17. Fröhling S, Schlenk RF, Stolze I, et al. CEBPA mutations in younger adults with acute myeloid leukemia and normal cytogenetics: prognostic relevance and analysis of cooperating mutations. *J Clin Oncol.* 2004;22:624–633.
18. Gilliland DG, Tallman MS. Focus on acute leukemias. *Cancer Cell.* 2002;1:417–420.
19. Gombart AF, Hofmann WK, Kawano S, et al. Mutations in the gene encoding the transcription factor CCAAT/enhancer binding protein alpha in myelodysplastic syndromes and acute myeloid leukemias. *Blood.* 2002;99:1332–1340.
20. Gurbuxani S, Vyas P, Crispino JD. Recent insights into the mechanisms of myeloid leukemogenesis in Down syndrome. *Blood.* 2004;103:399–406.
21. Hall MA, Curtis DJ, Metcalf D. The critical regulator of embryonic hematopoiesis, SCL, is vital in the adult for megakaryopoiesis, erythropoiesis, and lineage choice in CFU-S12. *Proc Natl Acad Sci USA.* 2003;100:992–997.
22. Harada H, Harada Y, Niimi H, Kyo T, Kimura A, Inaba T. High incidence of somatic mutations in the AML1/RUNX1 gene in myelodysplastic syndrome and low blast percentage myeloid leukemia with myelodysplasia. *Blood.* 2004;103: 2316–2324.
23. Helbling D, Mueller BU, Timchenko NA, et al. CBFB-SMMHC is correlated with increased calreticulin expression and suppresses the granulocytic differentiation factor CEBPA in AML with inv(16). *Blood.* 2005;106:1369–1375.
24. Helbling D, Mueller BU, Timchenko NA, et al. The leukemic fusion gene AML1-MDS1-EVI1 suppresses CEBPA in acute myeloid leukemia by activation of calreticulin. *Proc Natl Acad Sci USA.* 2004;101:13312–13317.
25. Hitzler JK, Zipursky A. origins of leukaemia in children with Down's syndrome. *Nat Rev Cancer.* 2005;5:11–20.
26. Hope KJ, Jin L, Dick JE. Acute myeloid leukemia originates from a hierarchy of leukemic stem cell classes that differ in self-renewal capacity. *Nat Immunol.* 2004;5:738–743.
27. Huntly BJ, Shigematsu H, Deguchi K, et al. MOZ-TIF2, but not BCR-ABL, confers properties of leukemic stem cells to committed murine hematopoietic progenitors. *Cancer Cell.* 2004;6:587–596.

28. Ichikawa M, Asai T, Saito T, et al. AML-1 is required for megakaryocytic maturation and lymphocytic differentiation, but not for maintenance of hematopoietic stem cells in adult hematopoiesis. *Nat Med.* 2004;10:299–304.
29. Iwama A, Oguro H, Negishi M, et al. Enhanced self-renewal of hematopoietic stem cells mediated by the polycomb gene product Bmi-1. *Immunity.* 2004;21:843–851.
30. Jamieson CH, Ailles LE, Dylla SJ, et al. Granulocyte-macrophage progenitors as candidate leukemic stem cells in blast crisis CML. *N Engl J Med.* 2004;351:657–667.
31. Kelly LM, Gilliland DG. Genetics of myeloid leukemias. *Annu Rev Genomics Hum Genet.* 2002;3:179–198.
32. Kim HG, De Guzman CG, Swindle CS, et al. The ETS family transcription factor PU.1 is necessary for the maintenance of fetal liver hematopoietic stem cells. *Blood.* 2004;104:3894–3900.
33. Langabeer SE, Gale RE, Rollinson SJ, Morgan GJ, Linch DC. Mutations of the AML1 gene in acute myeloid leukemia of FAB types M0 and M7. *Genes Chromosomes Cancer.* 2002;34:24–32.
34. Lange B. The management of neoplastic disorders of haematopoiesis in children with Down's syndrome. *Br J Haematol.* 2000;110:512–524.
35. Leroy H, Roumier C, Huyghe P, et al. CEBPA point mutations in haematological malignancies. *Leukemia.* 2005;19:329–334.
36. Lessard J, Sauvageau G. Bmi-1 determines the proliferative capacity of normal and leukaemic stem cells. *Nature.* 2003;423:255–260.
37. Lin LI, Chen CY, Lin DT, et al. Characterization of CEBPA mutations in acute myeloid leukemia: most patients with CEBPA mutations have biallelic mutations and show distinct immunophenotype of the leukemic cells. *Clin Cancer Res.* 2005;11:1372–1379.
38. Look AT. Oncogenic transcription factors in the human acute leukemias. *Science.* 1997;278:1059–1064.
39. Marcucci G, Mrozek K, Bloomfield CD. Molecular heterogeneity and prognostic biomarkers in adults with acute myeloid leukemia and normal cytogenetics. *Curr Opinion Hematol.* 2005;12:68–75.
40. Matsuno N, Osato M, Yamashita N, et al. Dual mutations in the AML1 and FLT3 genes are associated with leukemogenesis in acute myeloblastic leukemia of the M0 subtype. *Leukemia.* 2003;17:2492–2499.
41. Michaud J, Wu F, Osato M, Cottles GM, et al. In vitro analyses of known and novel RUNX1/AML1 mutations in dominant familial platelet disorder with predisposition to acute myelogenous leukemia: implications for mechanisms of pathogenesis. *Blood.* 2002;99:1364–1372.
42. Migliaccio AR, Rana RA, Sanchez M, et al. GATA-1 as a regulator of mast cell differentiation revealed by the phenotype of the GATA-1 low mouse mutant. *J Exp Med.* 2003;197:281–296.
43. Mizuki M, Schwable J, Steur C, et al. Suppression of myeloid transcription factors and induction of STAT response genes by AML-specific Flt3 mutations. *Blood.* 2003;101:3164–3173.
44. Mueller BU, Pabst T, Osato M, et al. Heterozygous PU.1 mutations are associated with acute myeloid leukemia. *Blood.* 2002;100:998–1007.
45. Mueller BU, Pabst T, Petkovic V, et al. ATRA resolves the differentiation block in t(15;17) acute myeloid leukemia by restoring PU.1 expression through CEBP induction. *Blood.* 2006;107:3330–3338.
46. Nerlov C. C/EBPα mutations in acute myeloid leukaemias. *Nat Rev.* 2004;4:394–400.
47. Nichols KE, Crispino JD, Poncz M, et al. Familial dyserythropoietic anaemia and thrombocytopenia due to an inherited mutation in GATA1. *Nat Genet.* 2000;24:266–270.
48. Nutt SL, Metcalf D, D'Amico A, et al. Dynamic regulation of PU.1 expression in multipotent hematopoietic progenitors. *J Exp Med.* 2005;201:221–231.

49. Okuda T, van Deursen J, Hiebert SW, Grosveld G, Downing JR. AML1, the target of multiple chromosomal translocations in human leukemia, is essential for normal fetal liver hematopoiesis. *Cell*. 1996;84:321–330.
50. Orkin SH. Diversification of haematopoietic stem cells to specific lineages. *Nat Rev Genet*. 2000;1:57–64.
51. Osato M, Asou N, Abdalla E, et al. Biallelic and heterozygous point mutations in the runt domain of the AML1/PEBP2alphaB gene associated with myeloblastic leukemias. *Blood*. 1999;93:1817–1824.
52. Pabst T, Mueller BU, Harakawa N, et al. AML1-ETO downregulates the granulocytic differentiation factor CEBPA in t(8;21) myeloid leukemia. *Nat Med*. 2001;7:444–451.
53. Pabst T, Mueller BU, Zhang P, et al. Dominant-negative mutations of CEBPA, encoding CCAAT/enhancer binding protein-alpha (CEBPA), in acute myeloid leukemia. *Nat Genet*. 2001;27:263–270.
54. Pabst T, Stillner E, Neuberg D, et al. Mutations of the myeloid transcription factor CEBPA are not associated with the blast crisis of chronic myeloid leukemia. *Br J Haematol*. 2006. In press.
55. Parkin SE, Baer M, Copeland TD, et al. Regulation of CCAAT/enhancer binding protein (C/EBP) activator proteins by heterodimerization with C/EBP□ (Ig/EBP). *J Biol Chem*. 2002;277:23563–23572.
56. Passegue E, Jamieson CH, Ailles LE, Weissman IL. Normal and leukemic hematopoiesis: are leukemias a stem cell disorder or a reacquisition of stem cell characteristics? *Proc Natl Acad Sci USA*. 2003;100(suppl 1):11842–11849.
57. Passegue E, Wagner EF, Weissman IL. JunB deficiency leads to a myeloproliferative disorder arising from hematopoietic stem cells. *Cell*. 2004;119:431–443.
58. Perrotti D, Calabretta B. Translational regulation by the p210 BCR/ABL oncoprotein. *Oncogene*. 2004;23:3222–3229.
59. Perrotti D, Cesi V, Trotta R, et al. BCR-ABL suppresses CEBPA expression through inhibitory action of hnRNP E2. *Nat Genet*. 2002;30:48–58.
60. Perrotti D, Marcucci G, Caliguri MA. Loss of CEBPA and favorable prognosis of acute myeloid leukemias: a biological paradox. *J Clin Oncol*. 2004;22:582–584.
61. Preudhomme C, Sagot C, Boisset N, et al. Favorable prognostic significance of CEBPA mutations in patients with de novo acute myeloid leukemia: a study from the Acute Leukemia French Association (ALFA). *Blood*. 2002;100:2717–2723.
62. Preudhomme C, Warot-Loze D, Roumier C, et al. High incidence of biallelic point mutations in the Runt domain of the AML1/PEBP2 alpha B gene in Mo acute myeloid leukemia and in myeloid malignancies with acquired trisomy 21. *Blood*. 2000;96:2862–2869.
63. Rosenbauer F, Wagner K, Kutok JL, et al. Acute myeloid leukemia induced by graded reduction of a lineage-specific transcription factor PU.1. *Nat Genet*. 2004;36:624–630.
64. Ross SE, Radomska HS, Wu B, et al. Phosphorylation of C/EBPα inhibits granulopoiesis. *Mol Cell Biol*. 2004;24:675–686.
65. Schwieger M, Löhler J, Fischer M. A dominant-negative mutant of CEBPA, associated with acute myeloid leukemias, inhibits differentiation of myeloid and erythroid progenitors of man but not mouse. *Blood*. 2004;103:2744–2752.
66. Scott EW, Fisher RC, Olson MC, et al. PU.1 function in a cell-autonomous manner to control the differentiation of multipotential lymphoid-myeloid progenitors. *Immunity*. 1997;6:437–447.
67. Sellick GS, Spendlove HE, Catovsky D, et al. Further evidence that germline CEBPA mutations cause dominant inheritance of acute myeloid leukemia. *Leukemia*. 2005;19:1276–1278.
68. Shimizu R, Kuroha T, Ohneda O, et al. Leukemogenesis caused by incapacitated GATA-1 function. *Mol Cell Biol*. 2004;24:10814–10825.
69. Shivdasani RA, Fujiwara Y, McDevitt MA, Orkin SH. A lineage-selective knockout establishes the critical role of transcription factor GATA-1 in megakaryocyte growth and platelet development. *EMBO J*. 1997;16:3965–3973.

70. Shivdasani RA, Mayer EL, Orkin SH. Absence of blood formation in mice lacking the T-cell leukaemia oncoprotein tal-1/SCL. *Nature*. 1995;373:432–434.
71. Silva FP, Morolli B, Storlazzi CT, et al. Identification of RUNX1/AML1 as a classical tumor suppressor gene. *Oncogene*. 2003;22:538–547.
72. Smith ML, Arch R, Smith LL, et al. Development of a human acute myeloid leukaemia screening panel and consequent identification of novel gene mutation in FLT3 and CCND3. *Br J Haemat*. 2005;128:318–323.
73. Smith ML, Cavenagh JD, Lister TA, Fitzgibbon J. Mutation of CEBPA in familial acute myeloid leukemia. *N Engl J Med*. 2004;351:2403–2407.
74. Snaddon J, Smith ML, Neat M, et al. Mutations of CEBPA in acute myeloid leukemia FAB types M1 and M2. *Genes Chromosomes Cancer*. 2003;37:72–78.
75. Song WJ, Sullivan MG, Legare RD, et al. Haploinsufficiency of CBFA2 causes familial thrombocytopenia with propensity to develop acute myelogenous leukaemia. *Nat Genet*. 1999;23:166–75.
76. Takahashi S, Onodera K, Motohashi H, et al. Arrest in primitive erythroid cell development caused by promoter-specific disruption of the GATA-1 gene. *J Biol Chem*. 1997;272:12611–12615.
77. Tenen DG, Hromas R, Licht JD, Zhang DE. Transcription factors, normal myeloid development, and leukemia. *Blood*. 1997;90:489–519.
78. Tenen, DG. Transcription factors in myeloid differentiation and leukemia. *Nat Rev Cancer*. 2003;3:89–101.
79. Timchenko NA, Iakova P, Welm AL, et al. Calreticulin interacts with C/EBPα and C/EBPβ mRNAs and represses translation of C/EBP proteins. *Mol Cell Biol*. 2002;22:7242–7257.
80. Truong BTH, Lee YJ, Lodie TA, et al. CCAAT/enhancer binding proteins repress the leukemic phenotype of acute myeloid leukemia. *Blood*. 2003;101:1141–1148.
81. Valk PJM, Delwel R, Lowenberg B. Gene expression profiling in acute myeloid leukemia. *Curr Opinion Hemat*. 2005;12:76–81.
82. Valk PJM, Verhaak RGW, Beijen MA, et al. Prognostically useful gene-expression profiles in acute myeloid leukemia. *N Engl J Med*. 2004;350:1617–1628.
83. Van Waalwijk van Doorn-Khosrovani SB, Erpelnick C, Meijer J, et al. Biallelic mutations in the CEBPA gene and low CEBPA expression levels as prognostic markers in intermediate-risk AML. *Hematol J*. 2003;4:31–40.
84. Vangala RK, Heiss-Neumann MS, Rangatia JS, et al. The myeloid transcription factor PU.1 is inactivated by AML1-ETO in t(8;21) myeloid leukemia. *Blood*. 2003;101:270–277.
85. Wang GL, Iakova P, Wilde M, et al. Liver tumors escape negative control of proliferation via PI3K/Akt-mediated block of CEBPA growth inhibitory activity. *Genes Dev*. 2004;18:912–925.
86. Wechsler J, Greene M, McDevitt MA, et al. Acquired mutations in GATA1 in the megakaryoblastic leukemia of Down syndrome. *Nat Genet*. 2002;32:148–152.
87. Yu C, Cantor AB, Yang H, et al. Targeted deletion of a high-affinity GATA-binding site in the GATA-1 promoter leads to selective loss of the eosinophil lineage in vivo. *J Exp Med*. 2002;195:1387–1395.
88. Zhang P, Iwasaki-Arai J, Iwasaki H, et al. Enhancement of hematopoietic stem cell repopulating capacity and self-renewal in the absence of the transcription factor C/EBPα. *Immunity*. 2004;21:853–863.
89. Zheng R, Friedman AD, Levis M, et al. Internal tandem duplication mutation of FLT3 blocks myeloid differentiation through suppression of C/EBPα expression. *Blood*. 2004;103:1883–1890.
90. Zipursky A. Transient leukemia – a benign form of leukaemia in newborn infants with trisomy 21. *Br J Haematol*. 2003;120:930–968.

Proleukemic RUNX1 and CBFβ Mutations in the Pathogenesis of Acute Leukemia

Michael E. Engel and Scott W. Hiebert

Abstract The existence of non-random mutations in critical regulators of cell growth and differentiation is a recurring theme in cancer pathogenesis and provides the basis for our modern, molecular approach to the study and treatment of malignant diseases. Nowhere is this more true than in the study of leukemogenesis, where research has converged upon a critical group of genes involved in hematopoietic stem and progenitor cell self-renewal and fate specification. Prominent among these is the heterodimeric transcriptional regulator, RUNX1/CBFβ. RUNX1 is a site-specific DNA-binding protein whose consensus response element is found in the promoters of many hematopoietically relevant genes. CBFβ interacts with RUNX1, stabilizing its interaction with DNA to promote the actions of RUNX1/CBFβ in transcriptional control. Both the RUNX1 and the CBFβ genes participate in proleukemic chromosomal alterations. Together they contribute to approximately one-third of acute myelogenous leukemia (AML) and one-quarter of acute lymphoblastic leukemia (ALL) cases, making RUNX1 and CBFβ the most frequently affected genes known in the pathogenesis of acute leukemia. Investigating the mechanisms by which RUNX1, CBFβ, and their proleukemic fusion proteins influence leukemogenesis has contributed greatly to our understanding of both normal and malignant hematopoiesis. Here we present an overview of the structural features of RUNX1/CBFβ and their derivatives, their roles in transcriptional control, and their contributions to normal and malignant hematopoiesis.

M.E. Engel (✉)
Division of Pediatric Hematology/Oncology, Department of Pediatrics Monroe Carell Jr. Children's Hospital at Vanderbilt and the Vanderbilt-Ingram Cancer Center, Nashville, TN 37232-6310, USA
e-mail: mike.engel@vanderbilt.edu

L. Nagarajan (ed.), *Acute Myelogenous Leukemia*,
Cancer Treatment and Research 145, DOI 10.1007/978-0-387-69259-3_8,
© Springer Science+Business Media, LLC 2010

Introduction

Hematopoietic stem cells (HSCs) and progenitor cells combine the capacity for self-renewal with the ability to generate developmentally restricted progenitors responsible for producing all mature circulating cells. Normal hematopoiesis requires the interplay of an appropriate microenvironment, secreted growth, and differentiation factors, their transmembrane receptors, the signaling machinery mobilized by them, and the nuclear factors that receive and reconcile multiple inputs to regulate distinct patterns of gene expression. Through a balance between cell proliferation and differentiation, hematopoietic stem and progenitor cells maintain appropriate numbers of circulating formed elements in the face of continual and disparate hematopoietic challenges. The systems that govern this balance between cell proliferation and fate specification must be simultaneously flexible and economical. For this reason, critical hematopoietic regulators often function in both cell proliferation and cell fate specification, and these processes have an inverse relationship. Given this duality, it is not surprising that mutations in these regulatory proteins can influence hematopoietic outcomes. Indeed, many central hematopoietic regulators, when dysfunctional, contribute to leukemia pathogenesis [88, 79, 89, 68, 12, 87].

Leukemia pathogenesis is intimately associated with genetic alterations involving the heterodimeric transcriptional regulator RUNX1/CBFβ. *RUNX1* is a promiscuous target for leukemogenic mutations, participating in not less than 18 translocations. Particularly common among these are t(8;21) in M2 acute myeloid leukemia (AML) and t(12;21) in precursor B-cell acute lymphoblastic leukemia. RUNX1 point mutations in sporadic and familial leukemias have also been described [64]. The CBFβ locus is affected by two known proleukemic alterations; inv(16)(p13;q22) and t(16;16)(p13;q22) that characterize an eosinophil predominant subtype of AML (M4eo-AML) [84]. Through study of these mutations and their consequences common themes are emerging regarding the pathogenesis of acute leukemia.

The RUNX1/CBFβ Transcription Factor

RUNX1/CBFβ is a heterodimeric transcriptional regulator whose components were first identified through studies of chromosomal translocations and inversions observed in acute leukemia [2]. Each subunit contributes distinct and complimentary functions to the complex [64, 84, 50, 65]. RUNX1, RUNX2, and RUNX3 are expressed in mammals and interact with CBFβ to form a DNA-binding complex [95, 9, 74, 75, 47]. The RUNX proteins are evolutionarily conserved across diverse species, localize to the nucleus, and display largely distinct patterns of expression during development [52, 42]. In contrast, CBFβ expression is ubiquitous both in the developing embryo and in adult tissues [74].

RUNX1 and CBFβ in Hematopoiesis

Genetic experiments in mice have established a central role for RUNX1/CBFβ in definitive hematopoiesis. Mice homozygous for deletion of *Runx1* are morphologically normal, but display midgestation lethality and central nervous hemorrhage secondary to failure of definitive hematopoiesis [77, 105]. Furthermore, embryonic stem cells from *Runx1*-null mice fail to contribute to definitive hematopoiesis in chimeric mice and those heterozygous for *Runx1* deletion show a reduced number of myeloid and erythroid progenitors. Similarly, *Cbfβ*-null mice manifest the same spectrum of abnormalities, which likely reflects the interdependence between RUNX1 and CBFβ in hematopoiesis. These findings confirm the critical role of RUNX1/CBFβ in normal hematopoiesis [106].

RUNX1/CBFβ Structure

RUNX proteins, represented by the prototypical family member RUNX1 (Fig. 1A), are evolutionarily related to the *Drosophila* pair rule protein, Runt [29]. This evolutionary relationship is most evident within a 128 amino acid region of RUNX proteins, the Runt homology domain (RHD), that binds to DNA and mediates interactions with other transcription factors. The RHD resembles an S-type immunoglobulin (Ig)-fold motif that characterizes the DNA-binding domains of many other transcriptional regulators including p53, NF-κB, NFAT, STAT, and T-box proteins [71, 10, 109]. RUNX proteins were initially purified from nuclear extracts using *cis*-regulatory elements from either polyomavirus or Moloney murine leukemia virus enhancers [42, 41, 83]. RUNX proteins bind directly to the DNA sequence 5′-TG(T/c)GGT-3′ found not only in these viral enhancers but also in the promoter/enhancer regions of many genes that govern cell proliferation and hematopoiesis [83, 103, 98, 120, 97, 73, 108, 111, 82, 107, 23, 49, 54, 115, 57].

CBFβ (Fig. 1A) indirectly interacts with DNA by binding the RHD of RUNX proteins. Otherwise, CBFβ is localized to the cytoplasm bound to filamin-A [117]. The interaction between RUNX proteins and CBFβ stabilizes the association of the complex with DNA at the RUNX response element [64, 45, 53, 18]. Furthermore, CBFβ protects RUNX1 from ubiquitin-mediated degradation by the 26S proteasome to augment RUNX1/CBFβ functions in transcriptional control [37].

RUNX Proteins in Transcriptional Activation

RUNX binding sites are necessary, but not sufficient for robust transcriptional activation of target genes. Instead, RUNX proteins cooperate with lineage-specific transcriptional regulators whose response elements are often adjacent to RUNX binding sites in DNA (Fig. 2A). In hematopoiesis these

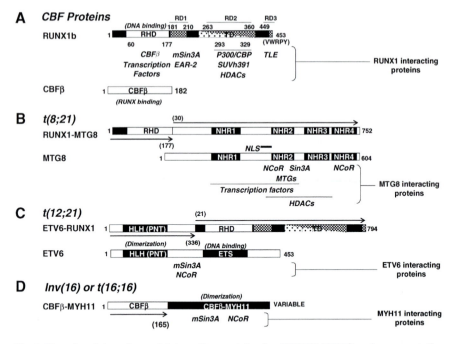

Fig. 1 Functional domains and interacting proteins for RUNX1/CBFβ and representative translocation partners. (**A**) RUNX1b is shown to represent the α-subunit group. The Runt homology domain (RHD) (60–177), transactivation domain (TD), and repression domains RD1 (181–220), RD2 (263–360), and RD3 (449–453) are shown. Dual shading of the TD reflects its role in both transcriptional activation and repression. Interacting proteins are shown below the domains they bind. CBFβ is shown. (**B**) MTG8 and the RUNX1-MTG8 fusion protein are shown. For MTG8, nervy homology regions (NHR) 1–4 are indicated, along with a nuclear localization sequence (NLS). Interacting proteins are shown below domains/regions with which they interact. RUNX1 and MTG8 contributions to the fusion protein are represented by arrows and amino acid numbers in parentheses, below and above the fusion protein, respectively. (**C**) ETV6 and the ETV6-RUNX1 fusion proteins are shown. Helix-loop-helix (HLH)/pointed (PNT) dimerization domain and the ETS DNA-binding domain are indicated. Corepressor-binding regions in ETV6 are shown. ETV6 and RUNX1 contributions to the fusion protein are indicated by *arrows* and amino acid numbers in *parentheses* below and above the fusion protein, respectively. (**D**) CBFβ-MYH11 is shown. Binding regions in MYH11 for NCoR and mSin3A are indicated. The VARIABLE designation reflects that differing lengths of MYH11 have been described in distinct fusion proteins. The most commonly observed CBFβ contribution (165 amino acids) is shown

transcriptional partners include C/EBP-α, TCF/LEF-1, PU.1, c-Myb, ETS, and GATA-1. RUNX1 can interact directly with many of these transcription factors and stabilize their binding to target promoters [111, 32, 26, 119, 96, 80, 58]. RUNX proteins also recruit coactivators of transcription, including p300/CBP, MOZ/ZNF220, and ALY to target promoters and synergize with them to potentiate target gene expression. Thus, RUNX1 acts as a transcriptional organizer. Through coincident contacts with DNA, adjacent, site-specific transcription factors, cofactors and the basal transcription machinery, RUNX

Fig. 2 RUNX1 as a transcriptional organizer for activation and repression. (**A**) RUNX1 cooperates with multiple transcription factors and cofactors to recruit the basal transcription machinery and to promote target gene expression. (**B**) Through assembly of a repressor complex that includes corepressors and effector proteins such as HDACs and RUNX1 coordinates site-specific transcriptional repression. The consensus RUNX1 binding site is shown. See text for details

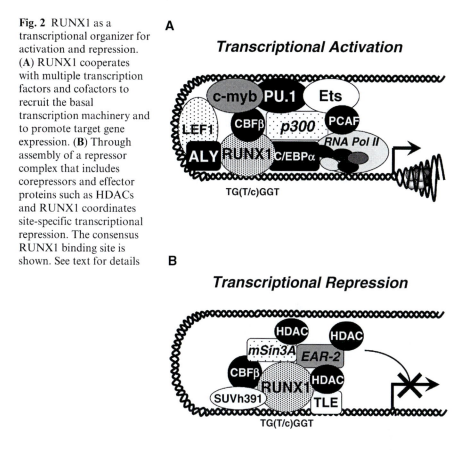

A

Transcriptional Activation

TG(T/c)GGT

B

Transcriptional Repression

TG(T/c)GGT

proteins facilitate expression of target genes while simultaneously enabling regulatory events that make gene expression context dependent.

RUNX Proteins in Transcriptional Repression

In the proper context, RUNX1 also represses gene expression (Fig. 2B). RUNX1 contains three discrete transcriptional repression domains; RD1, RD2, and RD3 [54, 7]. The first, immediately C-terminal to the RHD, interacts with both erbA-related gene (EAR)-2, an orphan nuclear hormone receptor, and mSin3A, a transcriptional corepressor [54, 3]. RD2 is found downstream of RD1, associates with histone deacetylases (HDACs) and the histone methyltransferase Suv39H1, and contributes to transcriptional repression and more stable gene silencing [54, 100]. At its most C-terminal boundary is the RD3 repressor region. This motif, a 5-amino acid stretch whose sequence is VWRPY, is required for interaction with corepressors transducin-like enhancer of split (TLE)-1 through 4, mammalian homologues of the *Drosophila* protein,

Groucho [54, 7, 48, 72]. These corepressors bind HDACs, suggesting that recruitment of epigenetic effectors is a central feature of RUNX-mediated transcriptional repression.

RUNX1 and CBFβ Mutations in Leukemia Pathogenesis

RUNX1 and CBFβ are preferred targets of proleukemic mutations in humans. Chromosomal translocations are most commonly observed, but inversions, amplifications and point mutations have been described as well. Collectively, alterations in RUNX1 or CBFβ are found in a significant percentage of leukemia cases, and studies of these mutations have revealed much about leukemia pathogenesis.

AML Associated with the t(8;21) Translocation

The best characterized translocation involving *RUNX1* is that which occurs between chromosomes 8q22 and 21q22 and is observed in nearly 50% of M2 and 15% of all cases of AML [79, 65]. This translocation fuses exons 1–5 of *RUNX1* with nearly the entire coding region of *Myeloid Translocation Gene (MTG)-8* (also known as *ETO, RUNX1T1,* or *CBFA2T1*). The most commonly observed chimeric protein juxtaposes amino acids 1–177 of RUNX1 with residues 30–604 of MTG8. Consequently, the fusion protein combines those functions attributed to the RHD in RUNX1 with those of MTG8.

MTG8 is the prototypical member of a gene family that includes *MTG16* (also known as *ETO2* or *CBFA2T3*) and *MTGR1* (also known as *CBFA2T2*), all homologues of the *Drosophila* protein, Nervy [20, 39]. MTG16 is a target of a proleukemic translocation with RUNX1 in therapy-related MDS/AML [27]. *MTGR1* resides on chromosome 20q within a region that is frequently deleted in MDS [8]. MTG family members display regulated patterns of expression.

Biochemical and transcriptional analysis suggests that MTG proteins are essential for epigenetic regulation of gene expression. MTG proteins coordinate the association of transcriptional corepressors and HDACs with site-specific DNA-binding proteins to form higher-order repression complexes at target promoters [39]. The resulting complexes favor repression of the target locus by altering acetylation of histones and other regulatory proteins, in turn causing changes in chromatin structure that control access of the basal transcription machinery to target genes.

MTG proteins share four conserved domains with Nervy, known as nervy homology regions (NHR) 1–4 [20]. These domains contribute to interactions governing repressor complex composition and assembly. The NHR1 domain binds multiple, site-specific transcriptional regulators representing distinct structural classes. Included among these are Runx1, Gfi-1 [59] (Moore, 2005, personal communication) and Gfi-1b, PLZF [66], TAL1/SCL [61], TCF4 [91], Ldb1 [31], and BCL-6 [17]. Each of these factors has been implicated in normal hematopoiesis, and in several cases (e.g., PLZF, TAL1, BCL-6), have definitive roles in leukemogenesis or lymphomagenesis. Interactions such as these support the privileged status of MTG proteins in normal hematopoiesis.

NHR domains 2 through 4, with intervening sequences and a short C-terminal tail, comprise the C-terminal half of MTG proteins. This region is largely responsible for recruiting "effectors" of transcriptional repression, including HDACs, other MTG proteins and corepressors. Specifically, NHR2 contains a hydrophobic heptad repeat (HHR) domain necessary for both homo- and hetero-oligomerization of MTG proteins [20]. In MTG8, the NHR2 domain contributes to binding of the corepressor mSin3A [4] and lies near a functional nuclear localization motif [36]. Interactions with HDACs and possibly other transcriptional regulators are mediated by the NHR3 domain and adjacent structures. The zinc-finger-containing NHR4 domain is one of two domains that contact the nuclear hormone corepressors NCoR and SMRT [55]. Additionally, it contributes binding sites for multiple HDACs. Between NHR domains, MTG proteins contain three proline/serine/threonine (PST) rich regions [20]. These regions show the greatest divergence among MTG proteins, and in conjunction with divergent N- and C-termini, are likely to specify family member-specific functions and regulation.

Despite absolute conservation of the zinc finger motif in the NHR4 domain, MTG proteins fail to bind DNA directly. Rather, they are tethered to target promoters by interactions with site-specific DNA-binding proteins. Their modular organization enables a diverse collection of DNA-binding transcription factors to access common "effector" mechanisms of transcriptional repression. Conceptually, MTG proteins provide a bridge between these "effectors" and carefully selected sites where repression is to occur. As such, they are uniquely poised to reconcile inputs from multiple afferent signal transduction cascades to regulate gene expression programs.

MTG8 is predominantly expressed in the nervous system, preadipocytes, smooth muscle, and gut, and homozygous deletion of *Mtg8* in mice results in an incompletely penetrant, midgut deletion phenotype. MTG16 and MTGR1 are more widely expressed [20, 19]. *Mtgr1*-null mice [5] display abnormal development of secretory lineage cells and increased proliferation in the gut. Within hematopoietic lineages, MTG8 is expressed in CD34[+] hematopoietic progenitors but is lost in more differentiated cells. MTG16 is most highly expressed in the CD34[+] compartment, and its expression declines as differentiation proceeds. Nevertheless, MTG16 expression is retained in more mature hematopoietic cells. MTGR1 is also widely expressed in hematopoietic lineages [20]. These patterns of expression intimate a fundamental role for the family in normal hematopoiesis, and when combined with the RUNX1 RHD can contribute significantly to leukemia pathogenesis.

The RUNX1-MTG8 fusion protein juxtaposes the RHD with oligomerization, corepressor, and HDAC-binding domains of MTG8. However, the binding sites for transcriptional coactivators p300/CBP and histone acetyltransferases, as well as sites for post-translational modifications that increase transactivation by RUNX1 are absent in the fusion protein. By coupling the RHD of RUNX1 with most of MTG8, the fusion protein acts as a constitutive, transdominant transcriptional repressor at RUNX1-regulated promoters (Fig. 3A, B). This view is consistent with the observation that RUNX1-MTG8 directly represses the TCRβ enhancer [62], the GM-CSF promoter [26], the p14^{Arf} promoter [49], and the neurofibromatosis (NF)-1 promoter [115]. In these instances, transcriptional repression depends upon recruitment of corepressors and HDACs, and HDAC inhibitors, such as trichostatin-A, block these effects. Moreover, by more effectively competing for CBFβ, RUNX1-MTG8 may further negate the effects of wild-type RUNX1 [99, 78]. As such, those promoters normally regulated by RUNX1 would be inappropriately repressed through RUNX1-MTG8 binding and repressor complex assembly.

Because RUNX1-MTG8 retains structural motifs responsible for the full spectrum of MTG8 interactions, it may also interfere with gene expression by

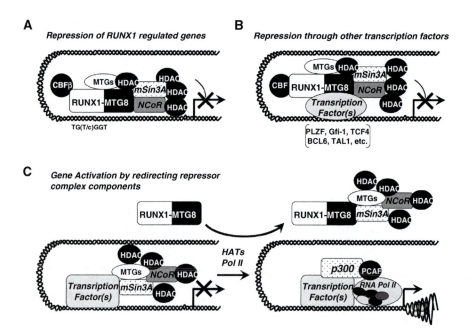

Fig. 3 Models for altered gene expression by the RUNX1-MTG8 fusion protein. (**A**) Because RUNX1-MTG8 retains the ability to bind DNA and CBFβ, it can inappropriately repress RUNX1-regulated promoters. (**B**) MTG8 interacts with multiple site-specific DNA-binding proteins (see text). Because the RUNX1-MTG8 protein retains the MTG8 domains responsible for these interactions, the fusion protein may alter transcription of MTG8-regulated genes. (**C**) The MTG8 domains of RUNX1-MTG8 interact with other MTG proteins, corepressors, and HDACs. In so doing, the fusion protein may relieve repression of MTG-regulated genes by redirecting these factors

redirecting MTG8-binding proteins away from the promoters normally regulated by them. By redirecting corepressors and HDACs, gene expression may be paradoxically activated by RUNX1-MTG8 (Fig. 3C). For example, the M-CSF receptor promoter is activated by RUNX1-MTG8 both in vitro and in vivo [86]. Likewise, RUNX1-MTG8 activates the TCF4/β-catenin-responsive reporter TOPFLASH in vitro independent of DNA binding and instead dependent upon corepressor binding [67]. Moreover, unbiased screens for genes whose expression is regulated by RUNX1-MTG8 identified multiple candidates not normally regulated by RUNX1 [92, 93]. Expression of a subset of these genes required an intact NHR2 domain. These data are consistent with fusion protein-regulated gene expression that involves diverting and sequestering transcriptional regulators away from their normal sites of action.

The effects of RUNX1-MTG8 on gene expression are extensive and the underlying mechanisms complex. In an effort to transcend this complexity, mouse models of t(8;21) AML have been developed to definitively address the role of RUNX1-MTG8 in leukemia pathogenesis. A "gene knock-in" strategy was used to mimic t(8;21) by inserting a RUNX1-MTG8 expression cassette into the *RUNX1* locus [78]. Embryos heterozygous for the fusion gene manifested impaired fetal liver definitive hematopoiesis and suffered catastrophic central nervous system hemorrhage at embryonic day 13.5. This phenotype closely matches that seen for homozygous deletion of either the *RUNX1* or the *CBFβ* loci, but with important differences. Heterozygous embryos harbored dysplastic, multilineage hematopoietic progenitors with abnormally high self-renewal capacity in vitro. The numbers of hematopoietic progenitors and mature progeny were dramatically reduced relative to wild-type controls.

To overcome the embryonic lethality of RUNX1-MTG8 expression, a tetracycline inducible *RUNX1-MTG8* transgene was employed [85]. Induction of RUNX1-MTG8 expression in adult hematopoietic progenitors again resulted in elevated self-renewal capacity. These and related data in retrovirally transduced mouse and human hematopoietic progenitors suggest that RUNX1-MTG8 expression enhances self-renewal capacity and impairs hematopoietic differentiation [70].

The t(8;21) translocation seems insufficient to cause leukemia in isolation [85, 118, 35]. Transgenic mice expressing RUNX1-MTG8 under control of the myeloid-specific hMRP8 promoter have only mild hematopoietic abnormalities, but develop a myeloproliferative disorder following treatment with *N*-ethyl-*N*-nitrosurea (ENU). However, a naturally occurring splice variant of RUNX1-MTG8 whose encoded protein lacks the C-terminus, including the NHR3 and NHR4 domains from MTG8, rapidly induces an immature myeloid leukemia in mice [114]. Also, the NHR2 domain contributes to RUNX1-MTG8-mediated immortalization of hematopoietic progenitors [51]. These findings suggest additional events influencing repressor complex integrity are required for RUNX1-MTG8 expressing cells to progress to a fully malignant phenotype. Identifying the mediators of these cooperating events is an active area of investigation.

The cumulative data support a model of RUNX1-MTG8-mediated leukemogenesis characterized by alterations in the expression of RUNX1- and MTG8-regulated genes creating an imbalance between hematopoietic stem cell self-renewal and lineage specification (Fig. 4). This imbalance creates a permissive environment for acquiring secondary mutations that contribute to malignant progression. Three target genes appear to be critical for

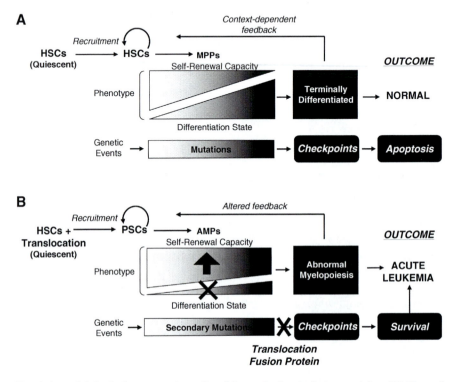

Fig. 4 A model for leukemogenesis mediated by proleukemic fusion proteins. (**A**) Normal hematopoiesis. Quiescent hematopoietic stem cells (HSCs) are recruited into the active pool dictated by the needs of the individual. These cells undergo symmetric divisions to maintain the HSC compartment, and asymmetric divisions giving rise to multipotent progenitors (MPPs). MPPs differentiate directionally toward terminally differentiated formed elements, progressively losing self-renewal capacity. HSCs and MPPs are highly regulated by growth and differentiation factors and context-dependent feedback. Replicative errors that occur trigger cell cycle checkpoints and apoptosis. This produces an appropriate balance of circulating formed elements for diverse circumstances. (**B**) Malignant hematopoiesis in setting of proleukemic fusion proteins. Latent HSCs with leukemic translocations are recruited to the active pool, giving rise to preleukemic stem cells (PSCs). PSCs undergo symmetric divisions that theoretically increase the population of PSCs in the marrow. Asymmetric divisions give rise to abnormal myeloid progenitors (AMPs) with altered/restricted differentiation potential and enhanced self-renewal capacity. This supports abnormal myelopoiesis and presumably altered feedback controls. Proleukemic fusion proteins may also alter checkpoint functions. Combined, this state promotes and propagates cooperating secondary mutations, enhancing survival and precipitating acute leukemia

leukemogenesis: $C/EBP\alpha$, $PU.1$, and $p14^{ARF}$. The suppression of $C/EBP\alpha$ and $PU.1$ is likely to affect cell cycle control, promoting self-renewal of immature progenitor cells. The consequences of these cell cycle effects may be compounded by fusion protein-mediated repression of $p14^{ARF}$, allowing affected cells to escape the p53 oncogenic checkpoint. The resultant survival advantage would further accelerate progression toward the fully transformed phenotype.

This model predicts that t(8;21) would be an early event in myeloid leukemogenesis. Indeed, retrospective studies of neonatal blood spots reveal the presence of the t(8;21) translocation in patients that went on to develop myeloid leukemia, and cord blood samples identify the t(8;21) translocation at an incidence that is approximately 100-fold greater than the incidence of myeloid leukemia in the general population [69, 112]. These findings suggest that the translocation fusion protein acts as an initiator in myeloid leukemia pathogenesis.

The t(12;21) Translocation in Acute Leukemia

The t(12;21)(p13;q22) translocation is frequently observed in pediatric B-lineage ALL [94, 30]. This translocation fuses the ETS family transcriptional regulator ETV6 (also known as TEL) with RUNX1 to generate ETV6-RUNX1. ETV6 is characterized by an N-terminal, helix-loop-helix region known as the pointed (PNT) domain and a highly conserved, 85-amino acid, C-terminal ETS domain [76, 11]. The PNT domain mediates oligomerization of ETV6, while the ETS domain coordinates DNA binding at a GGAA/T motif. Because of its ability to bind NCoR, SMRT, and mSin3A, ETV6 is thought to function as a transcriptional repressor, and analysis of *Etv6*-null mice reveals its role in homing of hematopoietic cells to the bone marrow [104, 24, 16, 33]. The ETV6– RUNX1 fusion protein excludes the ETS domain, but otherwise contains all of ETV6 fused in frame with nearly the entire RUNX1 protein. ETV6–RUNX1 binds DNA and associates with CBFβ through the RHD. The ETV6 portion of the fusion protein contributes oligomerization and corepressor recruitment functions, and thus, like RUNX1-MTG8, can act as a transdominant repressor toward RUNX1-regulated promoters [102, 34].

Like the t(8;21), cooperating secondary events are believed to be necessary for leukemogenesis related to the t(12;21). For example, transgenic mice expressing ETV6-RUNX1 in lymphoid cells fail to develop leukemia after 24 months and show no phenotypic abnormalities in their hematopoietic cells [6]. This finding is consistent with the characteristically indolent nature and generally good prognosis of B-lineage ALL cases harboring t(12;21). They also predict that t(12;21) leukemias would have a long latency between acquiring the translocation and developing a fully transformed phenotype. Monozygotic twin studies support this notion [25, 113].

Inv(16) and t(16;16) in Acute Myeloid Leukemia

By either an inversion or more rarely a translocation mechanism, *CBFβ* is disrupted in M4 eosinophil predominant (M4eo) AML [44, 21, 101, 90, 110]. Both alterations juxtapose a portion of CBFβ with varying lengths of the tail domain of the myosin heavy chain gene, MYH11. The most common variant of inv(16)(p13;q22) contains 165 residues of CBFβ coupled with 446 residues of MYH11 [84]. The CBFβ portion retains RUNX1 binding, while the MYH11 region promotes oligomerization and may also contribute to RUNX1 binding.

Two models for CBFβ-MYH11-mediated leukemogenesis, a "cytoplasmic sequestration" model and a "dominant repressor" model, have been proposed [38]. The dominant inhibitory function of CBFβ-MYH11 toward CBFβ is best illustrated in transgenic mice whose CBFβ locus was replaced by a CBFβ-MYH11 expression cassette. Just as was seen for *Cbfβ* null and *Runx1* null mice, CBFβ-MYH11 heterozygous mice lack definitive hematopoiesis and experience fatal central nervous system hemorrhages at E12.5 [13]. Since mice heterozygous for deletion of either *CBFβ* or *RUNX1* locus display phenotypically normal hematopoietic differentiation, this result strongly suggests that the CBFβ-MYH11 fusion protein dominantly interferes with hematopoiesis directed by RUNX1/CBFβ.

The "cytoplasmic sequestration" model is based on the observation that CBFβ-MYH11 preferentially co-localizes with RUNX1 in the cytoplasm through interactions between the RUNX1 and the CBFβ domain and concurrent interactions between the MYH11 domain and the actin cytoskeleton [1]. This model predicts that CBFβ–MYH11 effectively negates the functions of RUNX1 in hematopoietic differentiation by controlling its bioavailability. However, cytoplasmic sequestration of RUNX1 is not seen in RUNX1-MTG8 "knock-in" mice (see above), whose phenotype is similar. Moreover, CBFβ-MYH11 expression does not phenocopy *Runx1* deletion in adult hematopoietic progenitors. As such, the importance of RUNX1 cytoplasmic sequestration to leukemia pathogenesis is uncertain.

The "dominant repressor" model is based on the observation that CBFβ-MYH11 cooperates with RUNX1 to enhance repression of RUNX1-regulated genes and can interact with both mSin3A and HDAC8 through the MYH11 domain. This model predicts that through concurrent interactions with RUNX1, mSin3A, and HDAC8 (and perhaps other repressor complex components), CBFβ-MYH11 becomes a transdominant repressor of RUNX1-regulated genes [56, 22]. Thus, CBFβ-MYH11 would act in a manner similar to that of RUNX1-MTG8, which is consistent with the murine "knock-in" phenotypes and with studies where integration of retroviral enhancers near RUNX loci cooperates with CBFβ-MYH11 to induce AML.

Accumulating evidence suggests that CBFβ-MYH11 cannot transform hematopoietic stem and progenitor cells alone, but instead requires cooperating secondary events. Chimeric CBFβ-MYH11 mice fail to develop leukemia even after a long latency [13], yet readily do so after either treatment with a single dose of ENU [15], retroviral integration [14], coexpression of the E7 oncoprotein from human papillomavirus, or expression on a p16^{INK4a}/p14^{ARF}-deficient background [116]. The need for cooperating secondary events in order to progress to leukemia is consistent with the detection of the inv(16) mutation in a neonatal blood spot from a patient who went on to develop inv(16) AML [60]. The clonality of the neonatal and diagnostic specimens was established by sequencing of the corresponding chromosomal breakpoints. Of note, the latency period for the patient in question was 9.4 years. The long latency is thought to reflect postnatal persistence of inv(16)-positive, quiescent multipotent cells that when recruited into the myelopoietic pool acquire secondary changes needed for malignant transformation. A CBFβ-MYH11 conditional "knock-in" approach was recently used to demonstrate this point in vivo, providing evidence for recruitment of a latent "leukemic stem cell" into a proliferative pool [43].

Alternative Mechanisms of RUNX Dysfunction in Leukemogenesis

The pivotal contributions of RUNX1/CBFβ to normal and malignant hematopoiesis have prompted searches for more subtle mutations within *RUNX1* in sporadic and familial leukemia syndromes [28]. *RUNX1* nonsense and missense mutations, as well as monoallelic *RUNX1* deletion contribute to the autosomal dominant FPD/AML syndrome. The latter is consistent with haploinsufficiency as a contributing factor in FPD/AML. Nonsense codons introduced by frameshift mutations have been identified in two FPD/AML patient families. In vitro, the truncated proteins fail to bind DNA and are unable to transactivate a RUNX1-responsive promoter [63]. However, the likelihood that they are expressed in vivo is low in the setting of nonsense-mediated RNA decay. Missense mutations, concentrated at DNA binding surfaces within the RHD, variably impair the interaction of RUNX1 with target promoters. However, these mutants typically spare interactions with CBFβ and transcriptional regulators that bind RUNX1 beyond the C-terminal boundary of the RHD. Collectively, these findings suggest that haploinsufficiency through deletion or epigenetic gene silencing, or redirection of limiting transcriptional regulators, may contribute to the primary manifestations and leukemic predisposition in FPD/AML syndrome.

RUNX1 point mutations have also been described in approximately 10% of sporadic cases of AML, often of the undifferentiated M0 subclass [28]. These mutants cluster in the RHD and impair DNA-binding and transactivating functions. Interestingly, leukemias characterized by sporadic RUNX1 point mutations typically lack wild-type RUNX1 expression. This may be due to epigenetic silencing, somatic mutation, or deletion of the wild-type allele with duplication of the germ-line mutation [81]. Of note, while RUNX1 point mutations are commonly observed, point mutations within *CBFβ* have not been described [46].

The overexpression of RUNX family members has also been implicated in leukemogenesis. For example, *RUNX1* amplification, through either tandem duplication or polysomy, has been observed in pediatric ALL, whereas haploinsufficiency and point mutations of RUNX1 have been observed in AML and preleukemic, myelodysplastic syndromes [64]. These findings imply that strict controls must be maintained over the abundance and activity of RUNX1 in normal hematopoiesis.

Conclusions

Emerging from the study of translocation fusion proteins involving RUNX1 and CBFβ are common themes regarding the pathogenesis of leukemia. Mutations in these proteins appear to act as initiators in leukemia pathogenesis, altering the expression of genes necessary for coordinate control of growth, and differentiation in hematopoiesis. The resulting changes in stem cell self-renewal and cell fate specification reflect inappropriate repression and activation of RUNX1/CBFβ target gene transcription, creating a fertile ground for cooperating secondary events that precipitate the fully transformed phenotype. The prolonged latency that follows primary alterations in RUNX1/CBFβ functions represents an opportunity for chemoprevention strategies that target these cooperating secondary events. Defining these changes and the genetic and environmental influences that promote them is a major challenge for us all. Likewise, understanding those signaling events that regulate RUNX1, CBFβ, and the proleukemic fusion proteins to which they contribute will provide us with new targets for active chemoprevention and therapy.

Acknowledgment The authors greatly appreciate the thoughtful comments and support from members of the Hiebert Laboratory during the preparation of this monograph.

References

1. Adya N, Stacy T, Speck NA, Liu PP. The leukemic protein core binding factor beta (CBFbeta)-smooth-muscle myosin heavy chain sequesters CBFalpha2 into cytoskeletal filaments and aggregates. *Mol Cell Biol.* 1998;18(12):7432–7443.
2. Adya N, Castilla LH, Liu PP. Function of CBFbeta/Bro proteins. *Semin Cell Dev Biol.* 2000;11(5):361–368.
3. Ahn MY, Huang G, Bae SC, Wee HJ, Kim WY, Ito Y. Negative regulation of granulocytic differentiation in the myeloid precursor cell line 32Dcl3 by ear-2, a mammalian homolog of Drosophila seven-up, and a chimeric leukemogenic gene, AML1/ETO. *Proc Natl Acad Sci USA.* 17 1998;95(4):1812–1817.

4. Amann JM, Nip J, Strom DK, et al. ETO, a target of t(8;21) in acute leukemia, makes distinct contacts with multiple histone deacetylases and binds mSin3A through its oligomerization domain. *Mol Cell Biol.* 2001;21(19):6470–6483.
5. Amann JM, Chyla BJ, Ellis TC, et al. Mtgr1 is a transcriptional corepressor that is required for maintenance of the secretory cell lineage in the small intestine. *Mol Cell Biol.* 2005;25(21):9576–9585.
6. Andreasson P, Schwaller J, Anastasiadou E, Aster J, Gilliland DG. The expression of ETV6/CBFA2 (TEL/AML1) is not sufficient for the transformation of hematopoietic cell lines in vitro or the induction of hematologic disease in vivo. *Cancer Genet Cytogenet.* 2001;130(2):93–104.
7. Aronson BD, Fisher AL, Blechman K, Caudy M, Gergen JP. Groucho-dependent and -independent repression activities of Runt domain proteins. *Mol Cell Biol.* 1997;17(9): 5581–5587.
8. Asimakopoulos FA, Green AR. Deletions of chromosome 20q and the pathogenesis of myeloproliferative disorders. *Br J Haematol.* 1996;95(2):219–226.
9. Bae SC, Yamaguchi-Iwai Y, Ogawa E, et al. Isolation of PEBP2 alpha B cDNA representing the mouse homolog of human acute myeloid leukemia gene, AML1. *Oncogene.* 1993;8(3):809–814.
10. Berardi MJ, Sun C, Zehr M, et al. The Ig fold of the core binding factor alpha Runt domain is a member of a family of structurally and functionally related Ig-fold DNA-binding domains. *Structure.* 1999;7(10):1247–1256.
11. Bohlander SK. ETV6: a versatile player in leukemogenesis. *Semin Cancer Biol.* 2005;15(3):162–174.
12. Bresnick EH, Chu J, Christensen HM, Lin B, Norton J. Linking Notch signaling, chromatin remodeling, and T-cell leukemogenesis. *J Cell Biochem Suppl.* 2000;(suppl 35):46–53.
13. Castilla LH, Wijmenga C, Wang Q, et al. Failure of embryonic hematopoiesis and lethal hemorrhages in mouse embryos heterozygous for a knocked-in leukemia gene CBFB-MYH11. *Cell.* 1996;87(4):687–696.
14. Castilla LH, Perrat P, Martinez NJ, et al. Identification of genes that synergize with Cbfb-MYH11 in the pathogenesis of acute myeloid leukemia. *Proc Natl Acad Sci USA.* 2004;101(14):4924–4929.
15. Castilla LH, Garrett L, Adya N, et al. The fusion gene Cbfb-MYH11 blocks myeloid differentiation and predisposes mice to acute myelomonocytic leukaemia. *Nat Genet.* 1999;23(2):144–146.
16. Chakrabarti SR, Nucifora G. The leukemia-associated gene TEL encodes a transcription repressor which associates with SMRT and mSin3A. *Biochem Biophys Res Commun.* 1999;264(3):871–877.
17. Chevallier N, Corcoran CM, Lennon C, et al. ETO protein of t(8;21) AML is a corepressor for Bcl-6 B-cell lymphoma oncoprotein. *Blood.* 2004;103(4):1454–1463.
18. Coffman JA. Runx transcription factors and the developmental balance between cell proliferation and differentiation. *Cell Biol Int.* 2003;27(4):315–324.
19. Davis JN, Williams BJ, Herron JT, Galiano FJ, Meyers S. ETO-2, a new member of the ETO-family of nuclear proteins. *Oncogene.* 1999;18(6):1375–1383.
20. Davis JN, McGhee L, Meyers S. The ETO (MTG8) gene family. *Gene.* 2003;303:1–10.
21. de la Chapelle A, Lahtinen R. Chromosome 16 and bone-marrow eosinophilia. *N Engl J Med.* 1983;309(22):1394.
22. Durst KL, Lutterbach B, Kummalue T, Friedman AD, Hiebert SW. The inv(16) fusion protein associates with corepressors via a smooth muscle myosin heavy-chain domain. *Mol Cell Biol.* 2003;23(2):607–619.
23. Erman B, Cortes M, Nikolajczyk BS, Speck NA, Sen R. ETS-core binding factor: a common composite motif in antigen receptor gene enhancers. *Mol Cell Biol.* 1998;18(3): 1322–1330.

24. Fenrick R, Amann JM, Lutterbach B, et al. Both TEL and AML-1 contribute repression domains to the t(12;21) fusion protein. *Mol Cell Biol.* 1999;19(10):6566–6574.

25. Ford AM, Bennett CA, Price CM, Bruin MC, Van Wering ER, Greaves M. Fetal origins of the TEL-AML1 fusion gene in identical twins with leukemia. *Proc Natl Acad Sci USA.* 1998;95(8):4584–4588.

26. Frank R, Zhang J, Uchida H, Meyers S, Hiebert SW, Nimer SD. The AML1/ETO fusion protein blocks transactivation of the GM-CSF promoter by AML1B. *Oncogene.* 1995;11(12):2667–2674.

27. Gamou T, Kitamura E, Hosoda F, et al. The partner gene of AML1 in t(16;21) myeloid malignancies is a novel member of the MTG8(ETO) family. *Blood.* 1998;91(11): 4028–4037.

28. Ganly P, Walker LC, Morris CM. Familial mutations of the transcription factor RUNX1 (AML1, CBFA2) predispose to acute myeloid leukemia. *Leuk Lymphoma.* 2004;45(1):1–10.

29. Gergen JP, Butler BA. Isolation of the Drosophila segmentation gene runt and analysis of its expression during embryogenesis. *Genes Dev.* 1988;2(9):1179–1193.

30. Golub TR, Barker GF, Bohlander SK, et al. Fusion of the TEL gene on 12p13 to the AML1 gene on 21q22 in acute lymphoblastic leukemia. *Proc Natl Acad Sci USA.* 1995;92(11):4917–4921.

31. Grosveld F, Rodriguez P, Meier N, et al. Isolation and characterization of hematopoietic transcription factor complexes by in vivo biotinylation tagging and mass spectrometry. *Ann N Y Acad Sci.* 2005;1054:55–67.

32. Hernandez-Munain C, Krangel MS. Regulation of the T-cell receptor delta enhancer by functional cooperation between c-Myb and core-binding factors. *Mol Cell Biol.* 1994;14(1):473–483.

33. Hiebert SW, Lutterbach B, Amann J. Role of co-repressors in transcriptional repression mediated by the t(8;1), t(16;21), t(12;21), and inv(16) fusion proteins. *Curr Opin Hematol.* 2001;8(4):197–200.

34. Hiebert SW, Sun W, Davis JN, et al. The t(12;21) translocation converts AML-1B from an activator to a repressor of transcription. *Mol Cell Biol.* 1996;16(4):1349–1355.

35. Higuchi M, O'Brien D, Kumaravelu P, Lenny N, Yeoh EJ, Downing JR. Expression of a conditional AML1-ETO oncogene bypasses embryonic lethality and establishes a murine model of human t(8;21) acute myeloid leukemia. *Cancer Cell.* 2002;1(1):63–74.

36. Hoogeveen AT, Rossetti S, Stoyanova V, et al. The transcriptional corepressor MTG16a contains a novel nucleolar targeting sequence deranged in t (16; 21)-positive myeloid malignancies. *Oncogene.* 2002;21(43):6703–6712.

37. Huang G, Shigesada K, Ito K, Wee HJ, Yokomizo T, Ito Y. Dimerization with PEBP2-beta protects RUNX1/AML1 from ubiquitin-proteasome-mediated degradation. *Embo J.* 2001;20(4):723–733.

38. Huang G, Shigesada K, Wee HJ, Liu PP, Osato M, Ito Y. Molecular basis for a dominant inactivation of RUNX1/AML1 by the leukemogenic inversion 16 chimera. *Blood.* 2004;103(8):3200–3207.

39. Hug BA, Lazar MA. ETO interacting proteins. *Oncogene.* 2004;23(24):4270–4274.

40. Javed A, Guo B, Hiebert S, et al. Groucho/TLE/R-esp proteins associate with the nuclear matrix and repress RUNX (CBF(alpha)/AML/PEBP2(alpha)) dependent activation of tissue-specific gene transcription. *J Cell Sci.* 2000;113 (Pt 12):2221–2231.

41. Kamachi Y, Ogawa E, Asano M, et al. Purification of a mouse nuclear factor that binds to both the A and B cores of the polyomavirus enhancer. *J Virol.* 1990;64(10): 4808–4819.

42. Kania MA, Bonner AS, Duffy JB, Gergen JP. The Drosophila segmentation gene runt encodes a novel nuclear regulatory protein that is also expressed in the developing nervous system. *Genes Dev.* 1990;4(10):1701–1713.

43. Kuo YH, Landrette SF, Heilman SA, et al. Cbf beta-SMMHC induces distinct abnormal myeloid progenitors able to develop acute myeloid leukemia. *Cancer Cell.* 2006;9(1):57–68.

44. Le Beau MM, Larson RA, Bitter MA, Vardiman JW, Golomb HM, Rowley JD. Association of an inversion of chromosome 16 with abnormal marrow eosinophils in acute myelomonocytic leukemia. A unique cytogenetic-clinicopathological association. *N Engl J Med.* 1983;309(11):630–636.

45. Lenny N, Westendorf JJ, Hiebert SW. Transcriptional regulation during myelopoiesis. *Mol Biol Rep.* 1997;24(3):157–168.

46. Leroy H, Roumier C, Grardel-Duflos N, et al. Unlike AML1, CBFbeta gene is not deregulated by point mutations in acute myeloid leukemia and in myelodysplastic syndromes. *Blood.* 2002;99(10):3848–3850.

47. Levanon D, Negreanu V, Bernstein Y, Bar-Am I, Avivi L, Groner Y. AML1, AML2, and AML3, the human members of the runt domain gene-family: cDNA structure, expression, and chromosomal localization. *Genomics.* 1994;23(2):425–432.

48. Levanon D, Goldstein RE, Bernstein Y, et al. Transcriptional repression by AML1 and LEF-1 is mediated by the TLE/Groucho corepressors. *Proc Natl Acad Sci USA.* 1998;95(20):11590–11595.

49. Linggi B, Muller-Tidow C, van de Locht L, et al. The t(8;21) fusion protein, AML1 ETO, specifically represses the transcription of the p14(ARF) tumor suppressor in acute myeloid leukemia. *Nat Med.* 2002;8(7):743–750.

50. Liu P, Tarle SA, Hajra A, et al. Fusion between transcription factor CBF beta/PEBP2 beta and a myosin heavy chain in acute myeloid leukemia. *Science.* 1993;261(5124):1041–1044.

51. Liu Y, Cheney MD, Gaudet JJ, et al. The tetramer structure of the Nervy homology two domain, NHR2, is critical for AML1/ETO's activity. *Cancer Cell.* 2006;9(4):249–260.

52. Lu J, Maruyama M, Satake M, et al. Subcellular localization of the alpha and beta subunits of the acute myeloid leukemia-linked transcription factor PEBP2/CBF. *Mol Cell Biol.* 1995;15(3):1651–1661.

53. Lutterbach B, Hiebert SW. Role of the transcription factor AML-1 in acute leukemia and hematopoietic differentiation. *Gene.* 21 2000;245(2):223–235.

54. Lutterbach B, Westendorf JJ, Linggi B, Isaac S, Seto E, Hiebert SW. A mechanism of repression by acute myeloid leukemia-1, the target of multiple chromosomal translocations in acute leukemia. *J Biol Chem.* 2000;275(1):651–656.

55. Lutterbach B, Westendorf JJ, Linggi B, et al. ETO, a target of t(8;21) in acute leukemia, interacts with the N-CoR and mSin3 corepressors. *Mol Cell Biol.* 1998;18(12):7176–7184.

56. Lutterbach B, Hou Y, Durst KL, Hiebert SW. The inv(16) encodes an acute myeloid leukemia 1 transcriptional corepressor. *Proc Natl Acad Sci USA.* 1999;96(22):12822–12827.

57. Lutterbach B, Sun D, Schuetz J, Hiebert SW. The MYND motif is required for repression of basal transcription from the multidrug resistance 1 promoter by the t(8;21) fusion protein. *Mol Cell Biol.* 1998;18(6):3604–3611.

58. Mao S, Frank RC, Zhang J, Miyazaki Y, Nimer SD. Functional and physical interactions between AML1 proteins and an ETS protein, MEF: implications for the pathogenesis of t(8;21)-positive leukemias. *Mol Cell Biol.* 1999;19(5):3635–3644.

59. McGhee L, Bryan J, Elliott L, et al. Gfi-1 attaches to the nuclear matrix, associates with ETO (MTG8) and histone deacetylase proteins, and represses transcription using a TSA-sensitive mechanism. *J Cell Biochem.* 2003;89(5):1005–1018.

60. McHale CM, Wiemels JL, Zhang L, et al. Prenatal origin of childhood acute myeloid leukemias harboring chromosomal rearrangements t(15;17) and inv(16). *Blood.* 2003;101(11):4640–4641.

61. Schuh AH, Tipping AJ, Clark AJ, et al. ETO-2 associates with SCL in erythroid cells and megakaryocytes and provides repressor functions in erythropoiesis. *Mol Cell Biol.* 2005;25(23):10235–10250.

62. Meyers S, Lenny N, Hiebert SW. The t(8;21) fusion protein interferes with AML-1B-dependent transcriptional activation. *Mol Cell Biol.* 1995;15(4):1974–1982.
63. Michaud J, Wu F, Osato M, et al. In vitro analyses of known and novel RUNX1/AML1 mutations in dominant familial platelet disorder with predisposition to acute myelogenous leukemia: implications for mechanisms of pathogenesis. *Blood.* 2002;99(4): 1364–1372.
64. Mikhail FM, Sinha KK, Saunthararajah Y, Nucifora G. Normal and transforming functions of RUNX1: a perspective. *J Cell Physiol.* 2006;207(3):582–593.
65. Miyoshi H, Shimizu K, Kozu T, Maseki N, Kaneko Y, Ohki M. t(8;21) breakpoints on chromosome 21 in acute myeloid leukemia are clustered within a limited region of a single gene, AML1. *Proc Natl Acad Sci USA.* 1991;88(23):10431–10434.
66. Melnick AM, Westendorf JJ, Polinger A, et al. The ETO protein disrupted in t(8;21)-associated acute myeloid leukemia is a corepressor for the promyelocytic leukemia zinc finger protein. *Mol Cell Biol.* 2000;20(6):2075–2086.
67. Moore A. Activation of TCF-dependent gene expression by the RUNX1-MTG8 fusion protein. Paper presented at: FASEB Conference on Hematologic Malignancies. 2005. Saxon's River, Vermont.
68. Moore MA. Converging pathways in leukemogenesis and stem cell self-renewal. *Exp Hematol.* 2005;33(7):719–737.
69. Mori H, Colman SM, Xiao Z, et al. Chromosome translocations and covert leukemic clones are generated during normal fetal development. *Proc Natl Acad Sci USA.* 2002;99(12):8242–8427.
70. Mulloy JC, Cammenga J, MacKenzie KL, Berguido FJ, Moore MA, Nimer SD. The AML1-ETO fusion protein promotes the expansion of human hematopoietic stem cells. *Blood.* 2002;99(1):15–23.
71. Nagata T, Gupta V, Sorce D, et al. Immunoglobulin motif DNA recognition and heterodimerization of the PEBP2/CBF Runt domain. *Nat Struct Biol.* 1999;6(7):615–619.
72. Nishimura M, Fukushima-Nakase Y, Fujita Y, et al. VWRPY motif-dependent and -independent roles of AML1/Runx1 transcription factor in murine hematopoietic development. *Blood.* 2004;103(2):562–570.
73. Nuchprayoon I, Meyers S, Scott LM, Suzow J, Hiebert S, Friedman AD. PEBP2/CBF, the murine homolog of the human myeloid AML1 and PEBP2 beta/CBF beta proto-oncoproteins, regulates the murine myeloperoxidase and neutrophil elastase genes in immature myeloid cells. *Mol Cell Biol.* 994;14(8):5558–5568.
74. Ogawa E, Inuzuka M, Maruyama M, et al. Molecular cloning and characterization of PEBP2 beta, the heterodimeric partner of a novel Drosophila runt-related DNA binding protein PEBP2 alpha. *Virology.* 1993;194(1):314–331.
75. Ogawa E, Maruyama M, Kagoshima H, et al. PEBP2/PEA2 represents a family of transcription factors homologous to the products of the Drosophila runt gene and the human AML1 gene. *Proc Natl Acad Sci USA.* 1993;90(14):6859–6863.
76. Oikawa T. ETS transcription factors: possible targets for cancer therapy. *Cancer Sci.* 2004;95(8):626–633.
77. Okuda T, van Deursen J, Hiebert SW, Grosveld G, Downing JR. AML1, the target of multiple chromosomal translocations in human leukemia, is essential for normal fetal liver hematopoiesis. *Cell.* 1996;84(2):321–330.
78. Okuda T, Cai Z, Yang S, et al. Expression of a knocked-in AML1-ETO leukemia gene inhibits the establishment of normal definitive hematopoiesis and directly generates dysplastic hematopoietic progenitors. *Blood.* 1998;91(9):3134–3143.
79. Peterson LF, Zhang DE. The 8;21 translocation in leukemogenesis. *Oncogene.* 2004;23(24):4255–4262.
80. Petrovick MS, Hiebert SW, Friedman AD, Hetherington CJ, Tenen DG, Zhang DE. Multiple functional domains of AML1: PU.1 and C/EBPalpha synergize with different regions of AML1. *Mol Cell Biol.* 1998;18(7):3915–3925.

81. Preudhomme C, Warot-Loze D, Roumier C, et al. High incidence of biallelic point mutations in the Runt domain of the AML1/PEBP2 alpha B gene in Mo acute myeloid leukemia and in myeloid malignancies with acquired trisomy 21. *Blood.* 2000;96(8):2862–2869.
82. Prosser HM, Wotton D, Gegonne A, et al. A phorbol ester response element within the human T-cell receptor beta-chain enhancer. *Proc Natl Acad Sci USA.* 1992;89(20):9934–9938.
83. Redondo JM, Pfohl JL, Hernandez-Munain C, Wang S, Speck NA, Krangel MS. Indistinguishable nuclear factor binding to functional core sites of the T-cell receptor delta and murine leukemia virus enhancers. *Mol Cell Biol.* 1992;12(11):4817–4823.
84. Reilly JT. Pathogenesis of acute myeloid leukaemia and inv(16)(p13;q22): a paradigm for understanding leukaemogenesis. *Br J Haematol.* 2004;128:18–34.
85. Rhoades KL, Hetherington CJ, Harakawa N, et al. Analysis of the role of AML1-ETO in leukemogenesis, using an inducible transgenic mouse model. *Blood.* 2000;96(6): 2108–2115.
86. Rhoades KL, Hetherington CJ, Rowley JD, et al. Synergistic up-regulation of the myeloid-specific promoter for the macrophage colony-stimulating factor receptor by AML1 and the t(8;21) fusion protein may contribute to leukemogenesis. *Proc Natl Acad Sci USA.* 1996;93(21):11895–11900.
87. Rosmarin AG, Yang Z, Resendes KK. Transcriptional regulation in myelopoiesis: hematopoietic fate choice, myeloid differentiation, and leukemogenesis. *Exp Hematol.* 2005;33(2):131–143.
88. Rowley JD. The role of chromosome translocations in leukemogenesis. *Semin Hematol.* 1999;36(4 Suppl 7):59–72.
89. Rubnitz JE, Look AT. Molecular basis of leukemogenesis. *Curr Opin Hematol.* 1998;5(4):264–270.
90. Schmitz N, Godde-Salz E, Gassmann W, Loffler H. Acute myelomonocytic leukemia with involvement of eosinophils and inversion of chromosome 16. *Blut.* 1984;48(5):263–267.
91. Moore AC, Amann JM, Williams CS, et al. Myeloid translocation gene family members associate with T-cell factors (TCFs) and influence TCF-dependent transcription. *Mol Cell Biol.* 2008;28(3):977–987.
92. Shimada H, Ichikawa H, Nakamura S, et al. Analysis of genes under the downstream control of the t(8;1) fusion protein AML1-MTG8: overexpression of the TIS11b (ERF-1, cMG1) gene induces myeloid cell proliferation in response to G-CSF. *Blood.* 2000;96(2):655–663.
93. Shimada H, Ichikawa H, Ohki M. Potential involvement of the AML1-MTG8 fusion protein in the granulocytic maturation characteristic of the t(8;21) acute myelogenous leukemia revealed by microarray analysis. *Leukemia.* 2002;16(5):874–885.
94. Shurtleff SA, Buijs A, Behm FG, et al. TEL/AML1 fusion resulting from a cryptic t(12;21) is the most common genetic lesion in pediatric ALL and defines a subgroup of patients with an excellent prognosis. *Leukemia.* 1995;9(12):1985–1989.
95. Speck NA, Terryl S. A new transcription factor family associated with human leukemias. *Crit Rev Eukaryot Gene Expr.* 1995;5(3–4):337–364.
96. Sun W, Graves BJ, Speck NA. Transactivation of the Moloney murine leukemia virus and T-cell receptor beta-chain enhancers by cbf and ets requires intact binding sites for both proteins. *J Virol.* 1995;69(8):4941–4949.
97. Suzow J, Friedman AD. The murine myeloperoxidase promoter contains several functional elements, one of which binds a cell type-restricted transcription factor, myeloid nuclear factor 1 (MyNF1). *Mol Cell Biol.* 1993;13(4):2141–2151.
98. Takahashi A, Satake M, Yamaguchi-Iwai Y, et al. Positive and negative regulation of granulocyte-macrophage colony-stimulating factor promoter activity by AML1-related transcription factor, PEBP2. *Blood.* 1995;86(2):607–616.
99. Tanaka K, Tanaka T, Kurokawa M, et al. The AML1/ETO(MTG8) and AML1/Evi-1 leukemia-associated chimeric oncoproteins accumulate PEBP2beta(CBFbeta) in the nucleus more efficiently than wild-type AML1. *Blood.* 1998;91(5):1688–1699.

100. Telfer JC, Hedblom EE, Anderson MK, Laurent MN, Rothenberg EV. Localization of the domains in Runx transcription factors required for the repression of CD4 in thymocytes. *J Immunol.* 2004;172(7):4359–4370.

101. Testa JR, Hogge DE, Misawa S, Zandparsa N. Chromosome 16 rearrangements in acute myelomonocytic leukemia with abnormal eosinophils. *N Engl J Med.* 1984;310(7):468–469.

102. Uchida H, Downing JR, Miyazaki Y, Frank R, Zhang J, Nimer SD. Three distinct domains in TEL-AML1 are required for transcriptional repression of the IL-3 promoter. *Oncogene.* 1999;18(4):1015–1022.

103. Uchida H, Zhang J, Nimer SD. AML1A and AML1B can transactivate the human IL-3 promoter. *J Immunol.* 1997;158(5):2251–2258.

104. Wang LC, Kuo F, Fujiwara Y, Gilliland DG, Golub TR, Orkin SH. Yolk sac angiogenic defect and intra-embryonic apoptosis in mice lacking the Ets-related factor TEL. *Embo J.* 1997;16(14):4374–4383.

105. Wang Q, Stacy T, Binder M, Marin-Padilla M, Sharpe AH, Speck NA. Disruption of the Cbfa2 gene causes necrosis and hemorrhaging in the central nervous system and blocks definitive hematopoiesis. *Proc Natl Acad Sci USA.* 1996;93(8):3444–3449.

106. Wang Q, Stacy T, Miller JD, et al. The CBFbeta subunit is essential for CBFalpha2 (AML1) function in vivo. *Cell.* 1996;87(4):697–708.

107. Wang S, Wang Q, Crute BE, Melnikova IN, Keller SR, Speck NA. Cloning and characterization of subunits of the T-cell receptor and murine leukemia virus enhancer core-binding factor. *Mol Cell Biol.* 1993;13(6):3324–3339.

108. Wargnier A, Legros-Maida S, Bosselut R, et al. Identification of human granzyme B promoter regulatory elements interacting with activated T-cell-specific proteins: implication of Ikaros and CBF binding sites in promoter activation. *Proc Natl Acad Sci USA.* 1995;92(15):6930–6394.

109. Warren AJ, Bravo J, Williams RL, Rabbitts TH. Structural basis for the heterodimeric interaction between the acute leukaemia-associated transcription factors AML1 and CBFbeta. *Embo J.* 2000;19(12):3004–3015.

110. Wessels HW, Dauwerse HG, Breuning MH, Beverstock GC. Inversion 16 and translocation (16;16) in ANLL M4eo break in the same subregion of the short arm of chromosome 16. *Cancer Genet Cytogenet.* 1991;57(2):225–228.

111. Westendorf JJ, Yamamoto CM, Lenny N, Downing JR, Selsted ME, Hiebert SW. The t(8;21) fusion product, AML-1-ETO, associates with C/EBP-alpha, inhibits C/EBP-alpha-dependent transcription, and blocks granulocytic differentiation. *Mol Cell Biol.* 1998;18(1):322–333.

112. Wiemels JL, Xiao Z, Buffler PA, et al. In utero origin of t(8;21) AML1-ETO translocations in childhood acute myeloid leukemia. *Blood.* 2002;99(10):3801–3805.

113. Wiemels JL, Ford AM, Van Wering ER, Postma A, Greaves M. Protracted and variable latency of acute lymphoblastic leukemia after TEL-AML1 gene fusion in utero. *Blood.* 1999;94(3):1057–1062.

114. Yan M, Kanbe E, Peterson LF, et al. A previously unidentified alternatively spliced isoform of t(8;21) transcript promotes leukemogenesis. *Nat Med.* 2006;12(8):945–949.

115. Yang G, Khalaf W, van de Locht L, et al. Transcriptional repression of the Neurofibromatosis-1 tumor suppressor by the t(8;21) fusion protein. *Mol Cell Biol.* 2005;25(14):5869–5879.

116. Yang Y, Wang W, Cleaves R, et al. Acceleration of G(1) cooperates with core binding factor beta-smooth muscle myosin heavy chain to induce acute leukemia in mice. *Cancer Res.* 2002;62(8):2232–2235.

117. Yoshida N, Ogata T, Tanabe K, et al. Filamin A-bound PEBP2beta/CBFbeta is retained in the cytoplasm and prevented from functioning as a partner of the Runx1 transcription factor. *Mol Cell Biol.* 2005;25(3):1003–1012.

118. Yuan Y, Zhou L, Miyamoto T, et al. AML1-ETO expression is directly involved in the development of acute myeloid leukemia in the presence of additional mutations. *Proc Natl Acad Sci USA*. 2001;98(18):10398–10403.

119. Zhang DE, Hetherington CJ, Meyers S, et al. CCAAT enhancer-binding protein (C/EBP) and AML1 (CBF alpha2) synergistically activate the macrophage colony-stimulating factor receptor promoter. *Mol Cell Biol*. 1996;16(3):1231–1240.

120. Zhang DE, Fujioka K, Hetherington CJ, et al. Identification of a region which directs the monocytic activity of the colony-stimulating factor 1 (macrophage colony-stimulating factor) receptor promoter and binds PEBP2/CBF (AML1). *Mol Cell Biol*. 1994;14(12):8085–8095.

Acute Myeloid Leukemia with Mutated Nucleophosmin *(NPM1)*: Molecular, Pathological, and Clinical Features

Brunangelo Falini

Abstract The *NPM1* gene encodes for nucleophosmin, a nucleolus-located shuttling protein that is involved in multiple cell functions, including regulation of ribosome biogenesis, control of centrosome duplication and preservation of ARF tumor suppressor integrity. The *NPM1* gene is specifically mutated in about 30% acute myeloid leukemia (AML) but not in other human neoplasms. Mutations cause crucial changes at the C-terminus of the NPM1 protein that are responsible for the aberrant nuclear export and accumulation of NPM1 mutants in the cytoplasm of leukemic cells. Diagnosis of AML with mutated *NPM1* can be done using molecular techniques, immunohistochemistry (looking at cytoplasmic dislocation of nucleophosmin that is predictive of *NPM1* mutations) and Western blotting with antibodies specifically directed against NPM1 mutants. Because of its distinctive molecular, pathological, immunophenotypic and prognostic features, AML with mutated *NPM1* (synonym: NPMc+ AML) has been included, as a new provisional entity, in the 2008 World Health Organization (WHO) classification of myeloid neoplasms.

Introduction

The human nucleophosmin *(NPM1)* gene [17] contains 12 exons ranging in size from 58 to 358 bp and maps to chromosome 5q35. Knockout mice models showed that *NPM1* is an essential gene, since its inactivation caused death at mid-gestation [48], mainly due to ineffective hemopoiesis and developmental defects in the brain.

The *NPM1* gene encodes a ubiquitously expressed nucleolar phosphoprotein (also known as B23, NO38, and numatrin) [91] which constantly shuttles between nucleus and cytoplasm [11]. Nucleophosmin belongs to the nucleophosmin/nucleophosmin family of nuclear chaperones [29, 92] and is involved

B. Falini (✉)
From the Institute of Hematology, University of Perugia, Perugia, Italy
e-mail: faliniem@unipg.it

L. Nagarajan (ed.), *Acute Myelogenous Leukemia*,
Cancer Treatment and Research 145, DOI 10.1007/978-0-387-69259-3_9,
© Springer Science+Business Media, LLC 2010

in multiple cell functions, including regulation of ribosome biogenesis, prevention of unscheduled centrosome duplication during the cell cycle, and control of integrity of the ARF tumor suppressor protein [49].

Much evidence has implicated the *NPM1* gene in the pathogenesis of several human malignancies [49]. *NPM1* is over-expressed in various solid tumors [75] or is involved in tumor progression [6, 50, 106]. In lymphomas and leukemias, the *NPM1* gene fuses with other gene partners to generate chimeric proteins (NPM-ALK, NPM-RARα, NPM-MLF1) which are thought to play a critical role in tumorigenesis [39]. About one-third of adult acute myeloid leukemia (AML) harbor *NPM1* gene mutations, usually at exon-12, that aberrantly dislocate the nucleophosmin protein into leukemic cell cytoplasm, hence the term NPMc+ (cytoplasmic-positive) AML [38].

Here, I review the structure and functions of the wild-type nucleophosmin protein, as well as the distinctive molecular, pathological, and clinical features of AML with mutated *NPM1*.

Structure of Nucleophosmin Protein

Alternative splicing from a single gene generates at least two nucleophosmin isoforms: NPM1 (or B23.1), a 294-amino acid protein which is abundant in all tissues [110] and resides mainly in the nucleolus; and NPM1.2 (or B23.2), a truncated 259-amino acid protein, which lacks the last 35 C-terminal amino acids of NPM1, is expressed at very low levels in tissues and is localized mainly in nucleoplasm [24].

Several domains [53] in NPM1 structure dictate its multiple biochemical functions (Fig. 1A). Like NPM2 and NPM3, other members of the nucleoplasmin protein family to which nucleophosmin belongs [29, 92], NPM1 shares a conserved N-terminal core structure (the nucleoplasmin domain) which is required for oligomerization [51] and for histone chaperone activities (Fig. 1A). The vast majority of cellular NPM1 exists in an oligomeric form [15] (mostly a hexamer) [116], and the ratio of oligomeric to monomeric forms may play a role in regulating the chaperone activity and the nucleic acid binding capability of NPM1 during ribosome biogenesis [19, 51]. NPM1 chaperone activity serves to prevent protein aggregation in the crowded environment of the nucleolus, to favor histone and nucleosome assembly [72, 80, 101], and to increase acetylation-dependent transcriptional activity [100]. The N-terminal region of NPM1 also harbors two leucine-rich nuclear export signal (NES) motifs [109, 114] that are targeted by the nuclear export receptor Crm1/exportin1 and play a critical role in the nucleo-cytoplasmic traffic of nucleophosmin.

The mid-portion of NPM1 contains two highly acidic regions which consist of stretches of aspartic and glutamic acids [53] (Fig. 1A). They are likely to serve as binding sites for the positive charge of basic proteins [80] such as histones and

A

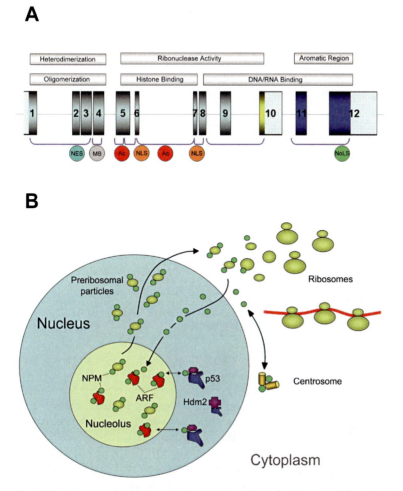

B

Fig. 1 The *NPM1* gene encodes for a protein involved in multiple functions. (**A**) The *NPM1* gene contains 12 exons. NPM1 is translated from exons 1 to 9 and 11 to 12. The portion encoding isoform B23.2 contains exons 1 to 10. In the protein, the N-terminus is characterized by a nonpolar domain responsible for oligomerization and heterodimerization. Two functional Nuclear Export Signals (NES) and a metal binding domain are present in this region. The central portion of the protein contains two acidic stretches (Ac) that are important for binding to histones, and a bipartite Nuclear Localization Signal (NLS); this region confers ribonuclease activity. The C-terminus of the protein contains basic regions involved in nucleic-acid binding, followed by an aromatic stretch unique to NPM isoform 1, which contains two tryptophan residues (288 and 290) that are required for nucleolar localization of the protein (NoLS). (**B**) Nucleophosmin is a nucleolar phosphoprotein that shuttles between the nucleus and the cytoplasm. Shuttling plays a fundamental role in ribosome biogenesis, since NPM1 transports preribosomal particles. In cytoplasm, NPM1 binds to the unduplicated centrosome and regulates its duplication during cell division. Furthermore, NPM1 interacts with p53 and its regulatory molecules (ARF, Hdm2/Mdm2) influencing the ARF-Hdm2/Mdm2-p53 oncosuppressive pathway. (This figure was originally published in Blood, Falini B. et al. Acute myeloid leukemia carrying cytoplasmic/mutated nucleophosmin (NPMc + AML): biologic and clinical features. *Blood.* 2007;109: 874–885, copyright 2007 by the American Society of Hematology)

ribosomal proteins, so as to minimize non-specific interactions within the congested nucleolar environment. The NPM1 region between the two acidic stretches underlies NPM1 ribonuclease activity (critical for ribosome biogenesis) and also contains one bipartite nuclear localization signal (NLS) [27] which is responsible for NPM1 nuclear import.

Unique to NPM1 is an RNA-binding domain at C-terminal end (Fig. 1A). The ability of NPM1 to bind through this domain to target RNA appears to be strictly dependent upon the oligomerization and phosphorylation status of the protein [81]. This is supported by the evidence that molecules which interact with the N-terminal core domain of NPM1 (e.g., the protein NPM3) [55] may inhibit the RNA-binding activity of NPM1 by destabilizing nucleophosmin oligomer formation. The two tryptophans residues at positions 288 and 290 within the RNA-binding domain are also critical for NPM1 localization to the nucleolus [74]. These residues, together with the two N-terminus NES motifs and the nuclear localization signal (NLS) of NPM1, play a key role in regulating the nucleophosmin shuttling across different cell compartments (nucleolus, nucleoplasm, and cytoplasm).

Phosphorylation of NPM1 Protein

NPM1 exists in cells as a phosphoprotein and several sites of phosphorylation by casein kinase II (CKII), by nuclear kinase II (N-II kinase), and by cd2 type kinase have been identified in its highly acid central region [16, 56, 84]. Therefore, phosphorylation plays a critical role in the functional regulation of NPM1. Phosphorylation at the cdc2 sites appears to be closely related to nucleolus breakdown during mitosis. On the other hand, CKII phosphorylation of a serine residue during the G2 phase increases NPM1 binding affinity to nucleolar components and modulated its chaperone activity. In the C-terminal region of the protein, several potential cyclin-dependent kinase 1 (CDK1) phosphorylation sites appear to influence binding of NPM1 to the nucleolus and other subnuclear compartments [73, 102] and, therefore, its mobility and location. This effect may be important in tuning NPM1 control of ribosome biogenesis. Finally, phosphorylation of NPM1 at different sites appears to be a critical step in regulating centrosome duplication during mitosis [79, 109] (see below).

Expression of NPM1 Protein

NPM1 protein expression is higher in proliferating than in resting cells [26] and increased in response to mitogenic stimuli [43]. Conversely, NPM1 levels drop after retinoic acid-induced differentiation of leukemic cells lines [54].

The NPM1 protein is found in the granular component of the nucleolus [60] which contains maturing pre-ribosomal particles [98]. Analysis of NPM1

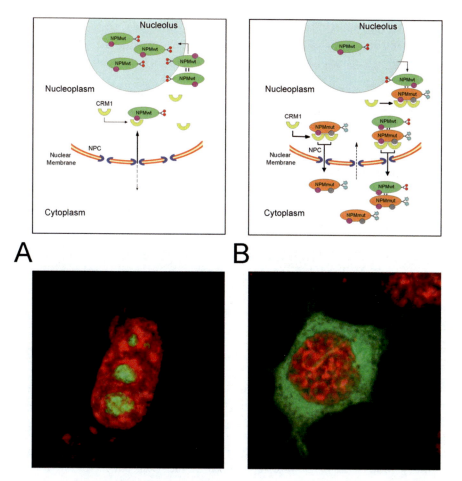

Fig. 2 Mechanism of altered traffic of wild-type and mutant NPM1 proteins. (A) Higher panel. Export from the nucleus to the cytoplasm of wild-type NPM1 is minimum since it contains only weak physiological NES motifs. Moreover, the two tryptophans at the C-terminus drive wild-type NPM1 to the nucleolus, where the bulk of protein accumulates. Immunofluorescence analysis (lower panel) clearly shows nucleolar localization of NPM1 in NIH3T3 mouse fibroblasts transfected with eGFP/wild-type NPM1. **(B)** Higher panel. NPM1 mutants accumulate in cytoplasm since the mutation creates an additional NES motif at C-terminus which is available for Crm1-mediated nuclear export. Moreover, mutation of the two tryptophans impairs nucleolar localization. Dimerization between mutated and wild-type (encoded by the normal allele) NPM1 partially delocalize the latter into cytoplasm. The image in the lower panel is a confocal micrograph of NIH3T3 cells transfected with eGFP-NPM mutant A showing complete dislocation of the mutant in the cytoplasm. Nuclei are counterstained red with propidium iodide. Cells were analysed with a Zeiss LSM 510 confocal microscope and reconstructed with Imaris software (Bitplane, Zurich, CH). (This figure was originally published in Falini B. et al. Translocations and mutations involving the nucleophosmin (NPM1) gene in lymphomas and leukemias. *Haematologica.* 2007;92:519–532)

functional domains explains why the protein localizes selectively in the nucleolus [60, 98], despite its ability to shuttle from nucleus to cytoplasm [11, 115]. The following scenario can be envisaged. The NLS signal drives NPM1 from cytoplasm to nucleoplasm, where it is translocated to the nucleolus by the C-terminus tryptophans 288 and 290 [74]. Since the two NES motifs within residues 94–102 [109] and 42–49 [114] have low nuclear export strength [10], little NPM1 is exported from nucleus to cytoplasm (Fig. 2A). Therefore, as nuclear import of wild-type NPM1 predominates over export, most NPM1 resides within the nucleolus (Fig. 2A), where it represents one of the most abundant proteins [4].

Functions of the NPM1 Protein

The ability of the NPM1 protein to act as a key regulator for several cellular activities mainly depends upon its chaperone [53, 101] and shuttling activities [109, 114]. Here, we focus on the main NPM1 functions that have been deciphered to date (Fig. 1B).

Regulation of Ribosome Biogenesis

Given its predominant expression in the nucleolus (an intranuclear organelle which is primarily dedicated to forming ribosome particles), NPM1 is assumed to play a pivotal role in ribosome biogenesis. This function is mediated through NPM1 nucleo-cytoplasmic shuttling properties [109, 114] and intrinsic RNAse activity [52], as well as through its ability to bind nucleic acids [108], to process pre-RNA molecules [90] and to act as a chaperone [101]. The main functions of NPM1 are to impede protein aggregation in the nucleolus during ribosome assembly [116] and to facilitate transport of large (60S) and small (40S) ribosomal subunits into cytoplasm where, together with mRNAs, they form the polyribosomes that are required for protein synthesis. NPM1 was recently shown to mediate 5S rRNA nuclear export through interaction with the ribosomal protein L5 [114]. Interaction between NPM1 and its binding partner ARF also appears crucial in regulating ribosome biogenesis [5].

Control of Centrosome Duplication

The NPM1 protein plays a fundamental role in maintaining genomic stability [20, 48] through a tight control of DNA repair mechanisms [62, 112] and centrosome duplication [77]. Regulation of centrosome duplication depends closely on the shuttling properties of nucleophosmin. NPM1 traffic is, in turn, controlled by the Ran/Cmr1 complex which, through interaction with the nucleophosmin NES motifs, dictates NPM1 transportation to the centrosome [12]. Ran-Cmr1-mediated transport of NPM1 also seems to be influenced by phosphorylation on Ser-4 of nucleophosmin by Plk1 and Nek2A [113]. The following events involving NPM1, Ran/Cmr1, and centrosomes are likely to take place during the different phases of the cell cycle. In resting cells, NPM1 associates with unduplicated centrosomes. During the late G1 phase, CDK2/ cyclinE-mediated phosphorylation on Thr199 causes NPM1 to dissociate from centrosomes [78], thereby enabling proper chromosome duplication. ROCK II kinase serves as the effector of the CDK2/cyclinE-nucleophosmin pathway [66]. During cell cycle progression from the S to the G2 phase, most NPM1 protein is located within the nucleolus where it regulates ribosome biogenesis. At the beginning of mitosis, when the nuclear membrane breaks down, nucleophosmin relocalizes to centrosome [117], thus impeding centrosome re-duplication. This regulation network ensures that centrosomes duplicate only once per cell cycle which is a major pre-requisite for proper cell division and prevention of aneu-ploidy, an important factor in neoplastic transformation [63]. The critical role played by NPM1 in protecting cells from aneuploidy is supported by evidence that, in $NPM1^{-/+}$ mice models, the unrestricted centrosome duplication caused by $NPM1$ haploinsufficiency leads to development of a myelodysplastic-like syndrome [48].

Control of p53 Activity and Stability

The NPM1 protein is involved in the apoptotic response to stress and oncogenic stimuli [87, 88] and is believed to modulate p53 protein activity and stability [21, 58, 59, 61]. Disruption of nucleolar integrity (e.g., following stimuli which induce a cellular stress) causes NPM1 to re-locate from nucleolus to nucleo-plasm where NPM1 stabilizes p53 by binding to and inhibiting Mdm2 [59]. These events in turn activate p53 [58, 67] with consequent arrest of cell growth.

Interaction of NPM1 with the Oncosuppressor ARF

In the nucleolus, the NPM1 protein co-localizes with the oncosuppressor ARF, forming molecular complexes of high molecular weight [7]. Since ARF assumes a stable structure only when bound to NPM1 [96], the NPM1–ARF interaction appears to protect ARF from degradation. ARF stabilization by NPM1 contributes to maintain the basal levels of ARF, ensuring it escapes the rapid proteasome-mediated destruction [20, 57] which many misfolded proteins are subject to [47]. NPM1 is responsible not only for protecting ARF from degra-dation but also for dictating its nucleolar location [96]. Since stability and subcellular distribution of ARF are essential for its functional activity [46], NPM1 appears to play an important role in modulating growth-suppressive pathways.

NPM1 Acts as an Oncogene and Oncosuppressor

The above features explain why NPM1 may act as oncogene and oncosuppressor [49]. When NPM1 expression aberrantly increases, the protein acts as oncogene by promoting abnormal cell growth through enhancement of ribosome machinery [89] and, possibly, by inhibiting programmed cell death [49]. Conversely, NPM1 may paradoxically behave as oncosuppressor since its reduced expression due to haploinsufficiency and/or altered subcellular distribution (as in AML carrying a mutated *NPM1* gene) may facilitate leukemogenesis, possibly through destabilization and functional impairment of the ARF-tumor suppressor pathway and increased genomic instability [49]. *NPM1* gene translocations or mutations may contribute to malignant transformation through alteration of NPM1 subcellular distribution and functions.

AML with Mutated NPM1

In 2005, we reported that leukemic cells in about 35% of adult AML showed aberrant cytoplasmic expression of nucleophosmin [38] caused by mutations occurring at exon-12 of the *NPM1* gene [38]. This discovery derived from our immunohistochemical studies on anaplastic large cell lymphoma carrying the t(2;5)/NPM-ALK fusion gene [69], a lymphoma entity with distinct clinical and pathological features [36]. Using specific anti-NPM monoclonal antibodies [23], we had observed that NPM1, instead of showing the expected nucleolar-restricted location, was ectopically expressed in the cytoplasm of ALCL cells with t(2;5) [41]. We linked this finding to the presence of the NPM-ALK fusion protein in the cytoplasm of lymphoma cells and reasoned that, by looking at aberrant NPM1 expression in cytoplasm, immunohistochemistry would serve as a simple, rapid screening test for putative *NPM1* gene alterations in a wide range of human neoplasms. This led us to discover the first AML cases which we named as NPMc+ (cytoplasmic positive) to distinguish them from the NPMc⁻ cases which carried the expected nucleus-restricted expression of NPM1 [38]. Subsequent studies clearly demonstrated that NPMc+ AML had distinct biological and clinical features, including mutual exclusion of recurrent genetic abnormalities, close association with normal karyotype, and CD34 negativity [38]. All this information reinforced our hypothesis that aberrant NPM cytoplasmic dislocation in AML reflected an underlying, as yet unidentified, genetic lesion. Absence of the known *NPM1* fusion gene/transcripts in the leukemic cells of NPMc+ AML prompted us to sequence the entire coding region of the *NPM1* gene, where we discovered mutations at exon-12 [38].

General Characteristics of NPM1 Mutations

One of the most remarkable features of *NPM1* mutations is their specificity for AML [64]. The rare occurrence of *NPM1* mutations in patients with chronic myelo monocytic leukemia [13, 76] should be interpreted cautiously, since these cases usually progressed rapidly to AML and, according to our experience, might have been examples of M4 or M5 AML with marked monocytic differentiation. The finding of *NPM1* mutations in about 5% of myelodysplastic syndromes [118] remains also controversial, since no *NPM1* gene mutations were detected in a recently reported series of myelodysplastic syndromes carrying various alterations in chromosome 5 [30, 97]. Moreover, a significant fraction of AML with mutated *NPM1* shows myelodysplasia-related changes which can lead to a misdiagnosis of myelodysplastic syndrome.

NPM1 gene mutations are consistently heterozygous [40] and usually occur at exon-12 [8, 28, 38, 93, 99, 105, 107]. Extremely rare mutations involving exons other than exon-12 have been reported [1, 31, 68]. Analysis of thousands of AML patients has identified to date about 50 molecular variants of *NPM1* mutations. Mutation A (a TCTG tetranucleotide duplication at positions 956 to 959) is the most common variant, which accounts for 75–80% of all *NPM1*-mutated adult AML [38]. Mutations B and D have been detected in about 10 and 5% of cases, while other *NPM1* mutations are very rare. Despite their heterogeneity, all *NPM1* mutations lead to common alterations at the C-terminus of the NPM1 which are responsible for the aberrant nuclear export and accumulation in cytoplasm of the protein (see below). *NPM1* mutation is very stable in AML [18, 34, 94], suggesting it is a founder genetic alteration.

Frequency of *NPM1* mutations among human myeloid leukemia cell lines is interesting. Although AML with mutated *NPM1* accounts for about one-third of adult AML, the *NPM1* mutation was detected only in 1/79 myeloid cell lines (i.e., the OCI-AML3 cell line [85]), suggesting that microenviromental factors may favor the growth of leukemic cells. Notably, the OCI-AML3 cell line exhibits the distinctive features of AML with mutated *NPM1*, i.e., mutation A and expression of cytoplasmic nucleophosmin [85], and it may, therefore, serve as in vitro model for AML with mutated *NPM1*.

AML with cytoplasmic-mutated NPM1 closely associates with normal karyotype and is mutually exclusive of major recurrent genetic abnormalities [37, 38]. About 14% NPMc+ AML carries secondary chromosomal aberrations [38]. Notably, about 40% of AML with mutated *NPM1* harbor the *FLT3*-ITD mutation [38]; *FLT3*-ITD is thought to be a secondary event [105] that may even disappear during the course of the disease. Partial tandem duplications within the *MLL* gene (*MLL*-PTD) and cytoplasmic mutated *NPM1* are mutually exclusive [38, 93, 105]. In two large studies, *CEBPA*, *KIT*, and *NRAS* mutations did not appear to differ in *NPM1*-mutated versus *NPM1*-unmutated AML with normal karyotype (AML-NK) [28, 93]. Mutations of the *p53* gene have been detected only in 3% of AML with mutated *NPM1*[99]. This finding is in keeping with previous reports showing that *p53* mutations, unlike *NPM1* mutations, usually occur in AML carrying karyotype abnormalities [42, 111].

Aberrant Cytoplasmic Location of NPM1 Mutants

Aberrant accumulation of nucleophosmin in the cytoplasm of the leukemic cells [38] (Fig. 2B) is the consequence of two changes that all *NPM1* mutations induce at the C-terminus of NPM1 protein: i) generation of a new nuclear export signal (NES) motif [71] and ii) loss of tryptophans at positions 288 and 290 (or tryptophan 290 alone) [38] which are critical for binding of NPM1 to nucleoli [74]. These events are both essential in dislocating NPM1 to cytoplasm [32] (Fig. 2B). Interestingly, the type of mutation-induced C-terminus NES motif appears to vary according to whether both 288 and 290 or 290 alone were mutated [32]. Mutants lacking tryptophans 288 and 290 carry the weak NES motif L-xxx-V-xx-V-x-L, while the rare mutants retaining tryptophan 288 harbor C-terminus NES variants where Valine at the second position is replaced by another hydrophobic amino acid (leucine, phenylalanine, cysteine, or methionine) [32]. To unravel the nature of this close correlation, we used a Rev(1.4)-based shuttling assay that measured the nuclear export efficiency of different C-terminal NES motifs in leukemic mutants [10]. Notably we found that, when both tryptophans were mutated, NPM1 mutants lost their nucleolus-binding capability and only needed a weak NES motif (**L-xxx-V-xx-V-x-L**) to be exported from the nucleus into cytoplasm [10]. In contrast, the rare NPM1 mutants carrying only mutated tryptophan 290 had to harbor a stronger NES motif to counterbalance tryptophan 288 which drives mutants to the nucleolus [10]. Thus, the opposing balance of forces (tryptophans and NES) seems to determine the subcellular localization of NPM1. The fact that the NPM1 mutants retaining tryptophan 288 always combine with the strongest NES motif reveals a mutational selection pressure toward efficient export of the leukemic mutants into cytoplasm, pointing to this event as critical for AML development [10].

Since mutants retain their ability to bind wild-type NPM1 (through formation of heterodimers) and the oncosuppressor ARF, they might recruit these molecules into ectopic sites (nucleoplasm and cytoplasm) [22, 25, 32] (Fig. 2B) and interfere with their functions. Contribution of these events to leukemogenesis remains to be clarified.

Diagnosis of AML with Mutated NPM1

AML with mutated *NPM1* is easily diagnosed using molecular biology techniques, immunohistochemistry, or western blotting. Several molecular assays have been described for detecting and characterizing *NPM1* mutations [3, 86, 95, 103]. Immunohistochemical detection of cytoplasmic NPM1 in B5-fixed/EDTA decalcified bone marrow biopsies is predictive of *NPM1* mutations [35] and may be used as a simple, rapid, and specific first-line screening to rationalize cytogenetic and molecular studies in AML. Moreover, immunohistochemistry is particularly useful in diagnosing AML presenting with "dry tap" or myeloid sarcoma [9, 33]. Smears or cytospins are not suitable for NPM1 immunostaining [38], probably because the protein diffuse out of the cells.

Three different types of antibodies directed against fixative-resistant epitopes of nucleophosmin are available for immunohistochemistry: those recognizing wild-type and mutant NPM1 proteins and those specifically directed against either the mutant or the wild-type NPM1 protein. Monoclonal antibodies that recognize both wild-type and mutated NPM1 [23, 32, 38] are the most reliable reagents for immunohistochemical diagnosis of AML with mutated *NPM1* [35]. They label leukemic cells in cytoplasm (which contains mutant and wild-type NPM1) and nucleus (which contains only wild-type NPM1) (Fig. 3, top). Polyclonal antibodies which recognize mutant but not wild-type NPM1 [32, 83, 85] label only the cytoplasm of leukemic cells (Fig. 3, middle), providing evidence that mutants are completely dislocated in the cytoplasm. These reagents are very useful for research purposes but are not suitable for routine diagnosis of AML with mutated *NPM1* in paraffin sections, since the epitopes they recognize are frequently denatured by fixation and embedding procedures [83]. Finally, a monoclonal antibody recognizing only wild-type NPM1 [32, 83] stains the leukemic cells in nucleus and cytoplasm, indicating that, in AML with mutated *NPM1*, the mutant recruits wild-type NPM1 into the cytoplasm. Although useful for research purposes, it is of little diagnostic relevance. The best control for specificity of aberrant cytoplasmic expression of nucleophosmin is immunostaining with an antibody against C23/nucleolin, another abundant shuttling nucleolar protein [11] which, in AML with mutated *NPM1*, is located only in the nucleus [38] (Fig. 3, bottom).

Morphological Features and Immunophenotype of AML with Mutated NPM1

AML with mutated *NPM1* frequently exhibits the morphological features of myelomonocytic (M4) or acute monocytic (M5) leukemia [28, 38, 93]. *NPM1* mutations can be detected at a lower frequency in other FAB subtypes (M1, M2, and M6). A significant fraction of AML with mutated *NPM1* shows multilineage dysplasia.

Notably, at phenotypic analysis, over 95% of AML with mutated *NPM1* is CD34-negative [38]. Paradoxically, CD34 negativity associates with overexpression of most *HOX* and *TALE* genes [2, 107] which are implicated in stem cell maintenance and leukemogenesis [65, 82]. *HOX* gene dysregulation in AML with mutated *NPM1* resembles, but clearly differs from what is observed in AML with *MLL* rearrangements, since upregulation of *HOXB2* and *B6* is typical of *NPM1* mutated but not of *MLL*-rearranged AML [70].

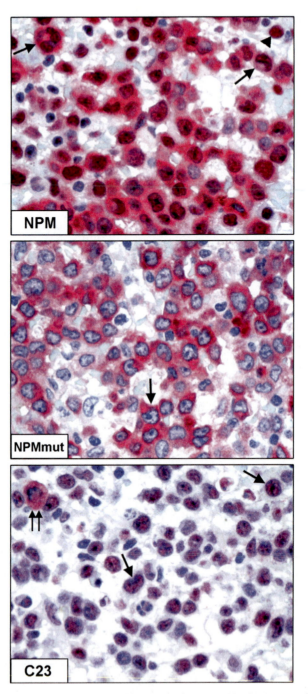

Fig. 3 Aberrant cytoplasmic expression of nucleophosmin in AML with mutated *NPM1*. (*Top*) Nuclear plus cytoplasmic expression of nucleophosmin in leukemic blasts (*arrows*). Normal residual haemopoietic cells show nucleus-restricted NPM1 (*arrowhead*). Bone marrow

Clinical Features of AML with Mutated NPM1

The frequency of AML with mutated *NPM1* varies with age [38, 93, 99, 105]. It accounts for 2.1–6.5% of AML in children [14, 18, 70] (9–26.9% of pediatric AML-NK [14, 18, 70]) and 25–35% [8, 18, 28, 38, 93, 99, 105, 107] of AML in adults (45.7–63.8 of adult AML-NK [8, 18, 28, 38, 93, 99, 105, 107]). Adult and children with AML also appear to differ in the frequency of *NPM1* mutation type. Mutation A, the most frequent in adults, appears to be rare in children [104]. AML with mutated *NPM1* tend to occur more frequently in women [28, 93, 105].

AML with mutated *NPM1* is characteristically a de novo leukemia [38]. Patients with this form of leukemia have higher white blood cell and platelet counts than AML-NK without *NPM1* mutations [28, 105] and usually show a

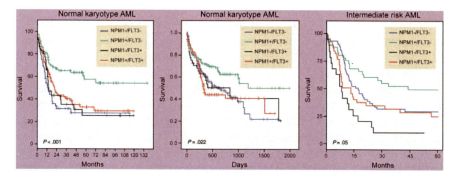

Fig. 4 Prognostic value of *NPM1* mutations in AML. Kaplan–Meier analysis of overall survival in normal karyotype or intermediate-risk AML in subgroups defined by the presence or absence of frameshift mutations in *NPM1* and/or internal tandem duplication mutations of *FLT3*. The panels were modified from Fig. 3B of Döhner et al (*left*), Fig. 5A of Schnittger et al. (*middle*), and Fig. 4A, B of Verhaak et al. (*right*). The *vertical hash marks* in the left and middle panels indicate censored patients. Illustration by Frank Forney. (This figure was originally published in Blood, Gallagher R. Dueling mutations in normal karyotype AML. *Blood.* 2005;106:3681–3682, copyright 2005 by the American Society of Hematology)

◄────────────────────────────────────

Fig. 3 (continued) biopsy paraffin section stained with monoclonal antibody 376 that recognizes wild-type and mutated NPM1 (APAAP technique; hematoxylin counterstaining; × 1,000). (*Middle*) Cytoplasmic-restricted expression of mutant NPM1 protein in NPMc + AML (*arrow*). Bone marrow biopsy paraffin section stained with a polyclonal antibody (Sil-A) that recognizes mutated but not wild-type NPM1 (APAAP technique; hematoxylin counterstaining; × 1,000). (*Bottom*) Nucleus-restricted expression of C23/nucleolin in NPMc + AML. Bone marrow biopsy paraffin section stained with a specific anti-C23 monoclonal antibody (APAAP technique; hematoxylin counterstaining; × 1,000). (This figure was originally published in Falini B. et al. Translocations and mutations involving the nucleophosmin (NPM1) gene in lymphomas and leukemias. *Haematologica.* 2007;92:519–532)

good response to induction therapy [38, 93, 99, 105]. AML with mutated *NPM1* without concomitant *FLT3*-ITD in patients under 60 years of age has a favorable prognosis [28, 45, 93, 105, 107] (Fig. 4) which is comparable to AML carrying favorable genetic abnormalities, such as t(8;21) or inv(16). These patients might possibly be exempted from allogeneic stem cell transplantation [28] as first-line treatment. Interestingly, in the Medical Research Council series [44], *NPM1* mutations emerged as an independent prognostic value, since *NPM1*-mutated/*FLT3*-ITD-positive AML had better prognosis than *NPM1*-unmutated/*FLT3*-ITD-negative AML.

Unsolved Issues and Future Directions

Future efforts should focus on further elucidating the role *NPM1* mutations in leukemogenesis and its significance in the management of AML patients. As wild-type NPM1 protein performs multiple functions, unraveling the molecular mechanism underlying AML with mutated *NPM1* is anticipated to be a difficult task. However, several factors that might contribute to AML development can be envisaged. The perturbed cellular traffic of the *NPM1* mutants is likely to interfere with the activities of wild-type NPM1 and ARF proteins, and possibly other yet unidentified NPM1 interacting partners, by recruiting them into ectopic sites and interfering with their functions. Given the close association of *FLT3*-ITD with *NPM1* mutations which is observed in clinical practice [28, 38, 93, 105, 107], the cooperative role of *FLT3*-ITD in AML with mutated *NPM1* should be explored in animal models. Moreover, it remains to be clarified whether the *HOX* gene upregulation typical of AML with mutated *NPM1* reflects its derivation from a normal hematopoietic progenitor which physiologically harbors this molecular signature or whether it is causally related to the NPM1 mutant that could genetically re-program a common myeloid progenitor to confer to this cell self-renewal capabilities.

Several major diagnostic and clinical issues are also on the agenda. Because of its distinctive biological and clinical features, AML with mutated *NPM1* has been included as provisional entity in the 2008 World Health Organization (WHO) classification of hemopoietic neoplasms. The good favorable prognostic impact of *NPM1*-mutated/*FLT3*-ITD-negative genotype in AML patients older than 60 years needs to be established. As *NPM1* mutations are stable in the course of the disease, they represent an ideal target for quantitative monitoring of minimal residual disease in about one-third of adult AML. The value of this approach in predicting early relapse and long-term survival in AML with mutated *NPM1* should be assessed in prospective clinical trials. Finally, better understanding of *NPM1*-mediated leukemogenesis may eventually lead to the development of targeted therapies. In this regard, the OCI-AML3 cell line which exhibits the distinctive features of AML with mutated *NPM1*[85] might serve as in vitro model for testing the activity of new drugs.

Acknowledgments Supported by the Associazione Italiana per la Ricerca sul Cancro (AIRC). We would like to thank Mrs. Claudia Tibidò for her excellent secretarial assistance and Dr. Geraldine Anne Boyd for her help in editing this paper. B. Falini applied for a patent on clinical use of NPM1 mutants.

References

1. Albiero E, Madeo D, Bolli N, et al. Identification and functional characterization of a cytoplasmic nucleophosmin leukaemic mutant generated by a novel exon-11 NPM1 mutation. *Leukemia.* 2007;21:1099–1103.
2. Alcalay M, Tiacci E, Bergomas R, et al. Acute myeloid leukemia bearing cytoplasmic nucleophosmin (NPMc+ AML) shows a distinct gene expression profile characterized by up-regulation of genes involved in stem-cell maintenance. *Blood.* 2005;106:899–902.
3. Ammatuna E, Noguera NI, Zangrilli D, et al. Rapid detection of nucleophosmin (NPM1) mutations in acute myeloid leukemia by denaturing HPLC. *Clin Chem.* 2005;51:2165–2167.
4. Andersen JS, Lam YW, Leung AK, et al. Nucleolar proteome dynamics. *Nature.* 2005;433:77–83.
5. Apicelli AJ, Maggi LB, Jr, Hirbe AC, et al. A non-tumor suppressor role for basal p19ARF in maintaining nucleolar structure and function. *Mol Cell Biol.* 2008;28(3):1068–1080.
6. Bernard K, Litman E, Fitzpatrick JL, et al. Functional proteomic analysis of melanoma progression. *Cancer Res.* 2007;63:6716–6725.
7. Bertwistle D, Sugimoto M, Sherr CJ. Physical and functional interactions of the Arf tumor suppressor protein with nucleophosmin/B23. *Mol Cell Biol.* 2004;24:985–996.
8. Boissel N, Renneville A, Biggio V, et al. Prevalence, clinical profile, and prognosis of NPM mutations in AML with normal karyotype. *Blood.* 2005;106:3618–3620.
9. Bolli N, Galimberti S, Martelli MP, et al. Cytoplasmic nucleophosmin in myeloid sarcoma occurring 20 years after diagnosis of acute myeloid leukaemia. *Lancet Oncol.* 2006;7:350–352.
10. Bolli N, Nicoletti I, De Marco MF, et al. Born to be exported: COOH-terminal nuclear export signals of different strength ensure cytoplasmic accumulation of nucleophosmin leukemic mutants. *Cancer Res.* 2007;67:6230–6237.
11. Borer RA, Lehner CF, Eppenberger HM, et al. Major nucleolar proteins shuttle between nucleus and cytoplasm. *Cell.* 1989;56:379–390.
12. Budhu AS, Wang XW. Loading and unloading: orchestrating centrosome duplication and spindle assembly by Ran/Crm1. *Cell Cycle.* 2005;4:1510–1514.
13. Caudill JS, Sternberg AJ, Li CY, et al. C-terminal nucleophosmin mutations are uncommon in chronic myeloid disorders. *Br J Haematol.* 2006;133:638–641.
14. Cazzaniga G, Dell'Oro MG, Mecucci C, et al. Nucleophosmin mutations in childhood acute myelogenous leukemia with normal karyotype. *Blood.* 2005;106:1419–1422.
15. Chan PK, Chan FY. Nucleophosmin/B23 (NPM) oligomer is a major and stable entity in HeLa cells. *Biochim Biophys Acta.* 1995;1262:37–42.
16. Chan PK, Liu QR, Durban E. The major phosphorylation site of nucleophosmin (B23) is phosphorylated by a nuclear kinase II. *Biochem J.* 1990;270:549–552.
17. Chan WY, Liu QR, Borjigin J, et al. Characterization of the cDNA encoding human nucleophosmin and studies of its role in normal and abnormal growth. *Biochemistry.* 1989;28:1033–1039.
18. Chou WC, Tang JL, Lin LI, et al. Nucleophosmin mutations in de novo acute myeloid leukemia: the age-dependent incidences and the stability during disease evolution. *Cancer Res.* 2006;66:3310–3316.

19. Chou YH, Yung BY. Cell cycle phase-dependent changes of localization and oligomerization states of nucleophosmin/B23. *Biochem Biophys Res Commun.* 1995;217:313–325.
20. Colombo E, Bonetti P, Lazzerini Denchi E, et al. Nucleophosmin is required for DNA integrity and p19Arf protein stability. *Mol Cell Biol.* 2005;25:8874–8886.
21. Colombo E, Marine JC, Danovi D, et al. Nucleophosmin regulates the stability and transcriptional activity of p53. *Nat Cell Biol.* 2002;4:529–533.
22. Colombo E, Martinelli P, Zamponi R, et al. Delocalization and destabilization of the Arf tumor suppressor by the leukemia-associated NPM mutant. *Cancer Res.* 2006;66:3044–3050.
23. Cordell JL, Pulford KA, Bigerna B, et al. Detection of normal and chimeric nucleophosmin in human cells. *Blood.* 1999;93:632–642.
24. Dalenc F, Drouet J, Ader I, et al. Increased expression of a COOH-truncated nucleophosmin resulting from alternative splicing is associated with cellular resistance to ionizing radiation in HeLa cells. *Int J Cancer.* 2002;100:662–668.
25. den Besten W, Kuo ML, Williams RT, et al. Myeloid leukemia-associated nucleophosmin mutants perturb p53-dependent and independent activities of the Arf tumor suppressor protein. *Cell Cycle.* 2005;4:1593–1598.
26. Dergunova NN, Bulycheva TI, Artemenko EG, et al. A major nucleolar protein B23 as a marker of proliferation activity of human peripheral lymphocytes. *Immunol Lett.* 2002;83:67–72.
27. Dingwall C, Dilworth SM, Black SJ, et al. Nucleoplasmin cDNA sequence reveals polyglutamic acid tracts and a cluster of sequences homologous to putative nuclear localization signals. *EMBO J.* 1987;6:69–74.
28. Dohner K, Schlenk RF, Habdank M, et al. Mutant nucleophosmin (NPM1) predicts favorable prognosis in younger adults with acute myeloid leukemia and normal cytogenetics: interaction with other gene mutations. *Blood.* 2005;106:3740–3746.
29. Eirin-Lopez JM, Frehlick LJ, Ausio J. Long-term evolution and functional diversification in the members of the nucleophosmin/nucleoplasmin family of nuclear chaperones. *Genetics.* 2006;173:1835–1850.
30. Falini B. Any role for the nucleophosmin (NPM1) gene in myelodysplastic syndromes and acute myeloid leukemia with chromosome 5 abnormalities? *Leuk Lymphoma.* 2007;48:2093–2095.
31. Falini B, Albiero E, Bolli N, et al. Aberrant cytoplasmic expression of C-terminal-truncated NPM leukaemic mutant is dictated by tryptophans loss and a new NES motif. *Leukemia.* 2007;21:2052–2054; author reply 2054; discussion 2055–2056.
32. Falini B, Bolli N, Shan J, et al. Both carboxy-terminus NES motif and mutated tryptophan(s) are crucial for aberrant nuclear export of nucleophosmin leukemic mutants in NPMc+ AML. *Blood.* 2006;107:4514–4523.
33. Falini B, Lenze D, Hasserjian R, et al. Cytoplasmic mutated nucleophosmin (NPM) defines the molecular status of a significant fraction of myeloid sarcomas. *Leukemia.* 2007;21:1566–1570.
34. Falini B, Martelli M, Mecucci C, et al. Cytoplasmic mutated nucleophosmin is stable in primary leukemic cells and in a xenotransplant model of NPMc+ AML in SCID mice. *Haematologica.* 2008;93(5):775–779.
35. Falini B, Martelli MP, Bolli N, et al. Immunohistochemistry predicts nucleophosmin (NPM) mutations in acute myeloid leukemia. *Blood.* 2006;108:1999–2005.
36. Falini B, Mason DY. Proteins encoded by genes involved in chromosomal alterations in lymphoma and leukemia: clinical value of their detection by immunocytochemistry. *Blood.* 2002;99:409–426.
37. Falini B, Mecucci C, Saglio G, et al. NPM1 mutations and cytoplasmic nucleophosmin are mutually exclusive of recurrent genetic abnormalities: a comparative analysis of 2562 AML patients. *Haematologica.* 2008;93(3):439–442.
38. Falini B, Mecucci C, Tiacci E, et al. Cytoplasmic nucleophosmin in acute myelogenous leukemia with a normal karyotype. *N Engl J Med.* 2005;352:254–266.

39. Falini B, Nicoletti I, Bolli N, et al. Translocations and mutations involving the nucleophosmin (NPM1) gene in lymphomas and leukemias. *Haematologica.* 2007;92:519–532.
40. Falini B, Nicoletti I, Martelli MF, et al. Acute myeloid leukemia carrying cytoplasmic/mutated nucleophosmin (NPMc+ AML): biologic and clinical features. *Blood.* 2007;109:874–885.
41. Falini B, Pulford K, Pucciarini A, et al. Lymphomas expressing ALK fusion protein(s) other than NPM-ALK. *Blood.* 1999;94:3509–3515.
42. Fenaux P, Jonveaux P, Quiquandon I, et al. P53 gene mutations in acute myeloid leukemia with 17p monosomy. *Blood.* 1991;78:1652–1657.
43. Feuerstein N, Chan PK, Mond JJ. Identification of numatrin, the nuclear matrix protein associated with induction of mitogenesis, as the nucleolar protein B23. Implication for the role of the nucleolus in early transduction of mitogenic signals. *J Biol Chem.* 1988;263:10608–10612.
44. Gale RE, Green C, Allen C, et al. The impact of FLT3 internal tandem duplication mutant level, number, size and interaction with NPM1 mutations in a large cohort of young adult patients with acute myeloid leukemia. *Blood.* 2008;111(5):2776–2784.
45. Gallagher R. Dueling mutations in normal karyotype AML. *Blood.* 2005;106:3681–3682.
46. Gjerset RA, Bandyopadhyay K. Regulation of p14ARF through subnuclear compartmentalization. *Cell Cycle.* 2006;5:686–690.
47. Goldberg AL. Protein degradation and protection against misfolded or damaged proteins. *Nature.* 2003;426:895–899.
48. Grisendi S, Bernardi R, Rossi M, et al. Role of nucleophosmin in embryonic development and tumorigenesis. *Nature.* 2005;437:147–153.
49. Grisendi S, Mecucci C, Falini B, et al. Nucleophosmin and cancer. *Nat Rev Cancer.* 2006;6:493–505.
50. Hayami R, Sato K, Wu W, et al. Down-regulation of BRCA1-BARD1 ubiquitin ligase by CDK2. *Cancer Res.* 2005;65:6–10.
51. Herrera JE, Correia JJ, Jones AE, et al. Sedimentation analyses of the salt- and divalent metal ion-induced oligomerization of nucleolar protein B23. *Biochemistry.* 1996;35:2668–2673.
52. Herrera JE, Savkur R, Olson MO. The ribonuclease activity of nucleolar protein B23. *Nucleic Acids Res.* 1995;23:3974–3979.
53. Hingorani K, Szebeni A, Olson MO. Mapping the functional domains of nucleolar protein B23. *J Biol Chem.* 2000;275:24451–24457.
54. Hsu CY, Yung BY. Down-regulation of nucleophosmin/B23 during retinoic acid-induced differentiation of human promyelocytic leukemia HL-60 cells. *Oncogene.* 1998;16:915–923.
55. Huang N, Negi S, Szebeni A, et al. Protein NPM3 interacts with the multifunctional nucleolar protein B23/nucleophosmin and inhibits ribosome biogenesis. *J Biol Chem.* 2005;280:5496–5502.
56. Jones CE, Busch H, Olson MO. Sequence of a phosphorylation site in nucleolar protein B23. *Biochim Biophys Acta.* 1981;667:209–212.
57. Kuo ML, den Besten W, Bertwistle D, et al. N-terminal polyubiquitination and degradation of the Arf tumor suppressor. *Genes Dev.* 2004;18:1862–1874.
58. Kurki S, Peltonen K, Laiho M. Nucleophosmin, HDM2 and p53: players in UV damage incited nucleolar stress response. *Cell Cycle.* 2004;3:976–979.
59. Kurki S, Peltonen K, Latonen L, et al. Nucleolar protein NPM interacts with HDM2 and protects tumor suppressor protein p53 from HDM2-mediated degradation. *Cancer Cell.* 2004;5:465–475.
60. Lam YW, Trinkle-Mulcahy L, Lamond AI. The nucleolus. *J Cell Sci.* 2005;118:1335–1337.
61. Lambert B, Buckle M. Characterisation of the interface between nucleophosmin (NPM) and p53: potential role in p53 stabilisation. *FEBS Lett.* 2006;580:345–350.

62. Lee SY, Park JH, Kim S, et al. A proteomics approach for the identification of nucleo-phosmin and heterogeneous nuclear ribonucleoprotein C1/C2 as chromatin-binding proteins in response to DNA double-strand breaks. *Biochem J*. 2005;388:7–15.
63. Lingle WL, Salisbury JL. The role of the centrosome in the development of malignant tumors. *Curr Top Dev Biol*. 2000;49:313–329.
64. Liso A, Bogliolo A, Freschi V, et al. In human genome, generation of a nuclear export signal through duplication appears unique to nucleophosmin (NPM1) mutations and is restricted to AML. *Leukemia*. 2008;22(6):1285–1289.
65. Look AT. Oncogenic transcription factors in the human acute leukemias. *Science*. 1997;278:1059–1064.
66. Ma Z, Kanai M, Kawamura K, et al. Interaction between ROCK II and Nucleophosmin/B23 in the regulation of centrosome duplication. *Mol Cell Biol*. 2006;26:9016–9034.
67. Maiguel DA, Jones L, Chakravarty D, et al. Nucleophosmin sets a threshold for p53 response to UV radiation. *Mol Cell Biol*. 2004;24:3703–3711.
68. Mariano AR, Colombo E, Luzi L, et al. Cytoplasmic localization of NPM in myeloid leukemias is dictated by gain-of-function mutations that create a functional nuclear export signal. *Oncogene*. 2006;25:4376–4380.
69. Morris SW, Kirstein MN, Valentine MB, et al. Fusion of a kinase gene, ALK, to a nucleolar protein gene, NPM, in non-Hodgkin's lymphoma. *Science*. 1994;263: 1281–1284.
70. Mullighan CG, Kennedy A, Zhou X, et al. Pediatric acute myeloid leukemia with NPM1 mutations is characterized by a gene expression profile with dysregulated HOX gene expression distinct from MLL-rearranged leukemias. *Leukemia*. 2007;21:2000–2009.
71. Nakagawa M, Kameoka Y, R. S. Nucleophosmin in acute myelogenous leukemia (letter). *N Engl J Med*. 2005;352:1819–1820.
72. Namboodiri VM, Akey IV, Schmidt-Zachmann MS, et al. The structure and function of Xenopus NO38-core, a histone chaperone in the nucleolus. *Structure*. 2004;12:2149–2160.
73. Negi SS, Olson MO. Effects of interphase and mitotic phosphorylation on the mobility and location of nucleolar protein B23. *J Cell Sci*. 2006;119:3676–3685.
74. Nishimura Y, Ohkubo T, Furuichi Y, et al. Tryptophans 286 and 288 in the C-terminal region of protein B23.1 are important for its nucleolar localization. *Biosci Biotechnol Biochem*. 2002;66:2239–2242.
75. Nozawa Y, Van Belzen N, Van der Made AC, et al. Expression of nucleophosmin/B23 in normal and neoplastic colorectal mucosa. *J Pathol*. 1996;178:48–52.
76. Oki Y, Jelinek J, Beran M, et al. Mutations and promoter methylation status of NPM1 in myeloproliferative disorders. *Haematologica*. 2006;91:1147–1148.
77. Okuda M. The role of nucleophosmin in centrosome duplication. *Oncogene*. 2002;21: 6170–6174.
78. Okuda M, Horn HF, Tarapore P, et al. Nucleophosmin/B23 is a target of CDK2/cyclin E in centrosome duplication. *Cell*. 2000;103:127–140.
79. Okuwaki M. The structure and functions of NPM1/Nucleophsmin/B23, a multifunc-tional nucleolar acidic protein. *J Biochem*. 2008;143(4):441–448.
80. Okuwaki M, Matsumoto K, Tsujimoto M, et al. Function of nucleophosmin/B23, a nucleolar acidic protein, as a histone chaperone. *FEBS Lett*. 2001;506:272–276.
81. Okuwaki M, Tsujimoto M, Nagata K. The RNA binding activity of a ribosome biogen-esis factor, nucleophosmin/B23, is modulated by phosphorylation with a cell cycle-dependent kinase and by association with its subtype. *Mol Biol Cell*. 2002;13:2016–2030.
82. Owens BM, Hawley RG. HOX and non-HOX homeobox genes in leukemic hematopoi-esis. *Stem Cells*. 2002;20:364–379.
83. Pasqualucci L, Liso A, Martelli MP, et al. Mutated nucleophosmin detects clonal multi-lineage involvement in acute myeloid leukemia: impact on WHO classification. *Blood*. 2006;108:4146–4155.
84. Peter M, Nakagawa J, Doree M, et al. Identification of major nucleolar proteins as candidate mitotic substrates of cdc2 kinase. *Cell*. 1990;60:791–801.

85. Quentmeier H, Martelli MP, Dirks WG, et al. Cell line OCI/AML3 bears exon-12 NPM gene mutation-A and cytoplasmic expression of nucleophosmin. *Leukemia.* 2005;19: 1760–1767.

86. Roti G, Rosati R, Bonasso R, et al. Denaturing high-performance liquid chromatography: a valid approach for identifying NPM1 mutations in acute myeloid leukemia. *J Mol Diagn.* 2006;8:254–259.

87. Rubbi CP, Milner J. Disruption of the nucleolus mediates stabilization of p53 in response to DNA damage and other stresses. *EMBO J.* 2003;22:6068–6077.

88. Rubbi CP, Milner J. p53–guardian of a genome's guardian? *Cell Cycle.* 2003;2:20–21.

89. Ruggero D, Pandolfi PP. Does the ribosome translate cancer? *Nat Rev Cancer.* 2003;3:179–192.

90. Savkur RS, Olson MO. Preferential cleavage in pre-ribosomal RNA by protein B23 endoribonuclease. *Nucl Acids Res.* 1998;26:4508–4515.

91. Schmidt-Zachmann MS, Franke WW. DNA cloning and amino acid sequence determination of a major constituent protein of mammalian nucleoli. Correspondence of the nucleoplasmin-related protein NO38 to mammalian protein B23. *Chromosoma.* 1988;96:417–426.

92. Schmidt-Zachmann MS, Hugle-Dorr B, Franke WW. A constitutive nucleolar protein identified as a member of the nucleoplasmin family. *EMBO J.* 1987;6:1881–1890.

93. Schnittger S, Schoch C, Kern W, et al. Nucleophosmin gene mutations are predictors of favorable prognosis in acute myelogenous leukemia with a normal karyotype. *Blood.* 2005;106:3733–3739.

94. Scholl S, Luftner J, Mugge LO, et al. Sustained expression of nucleophosmin (NPM1) mutation at late relapse presenting as isolated myeloid sarcoma in a patient with acute myeloid leukemia. *Ann Hematol.* 2007;86:763–765.

95. Scholl S, Mugge LO, Landt O, et al. Rapid screening and sensitive detection of NPM1 (nucleophosmin) exon 12 mutations in acute myeloid leukaemia. *Leuk Res.* 2007;31:1205–1211.

96. Sherr CJ. Divorcing ARF and p53: an unsettled case. *Nat Rev Cancer.* 2006;6:663–673.

97. Shiseki M, Kitagawa Y, Wang YH, et al. Lack of nucleophosmin mutation in patients with myelodysplastic syndrome and acute myeloid leukemia with chromosome 5 abnormalities. *Leuk Lymphoma.* 2007;48:2141–2144.

98. Spector DL, Ochs RL, Busch H. Silver staining, immunofluorescence, and immunoelectron microscopic localization of nucleolar phosphoproteins B23 and C23. *Chromosoma.* 1984;90:139–148.

99. Suzuki T, Kiyoi H, Ozeki K, et al. Clinical characteristics and prognostic implications of NPM1 mutations in acute myeloid leukemia. *Blood.* 2005;106:2854–2861.

100. Swaminathan V, Kishore AH, Febitha KK, et al. Human histone chaperone nucleophosmin enhances acetylation-dependent chromatin transcription. *Mol Cell Biol.* 2005;25:7534–7545.

101. Szebeni A, Olson MO. Nucleolar protein B23 has molecular chaperone activities. *Protein Sci.* 1999;8:905–912.

102. Tarapore P, Shinmura K, Suzuki H, et al. Thr199 phosphorylation targets nucleophosmin to nuclear speckles and represses pre-mRNA processing. *FEBS Lett.* 2006;580: 399–409.

103. Thiede C, Creutzig E, Illmer T, et al. Rapid and sensitive typing of NPM1 mutations using LNA-mediated PCR clamping. *Leukemia.* 2006;20:1897–1899.

104. Thiede C, Creutzig E, Reinhardt D, et al. Different types of NPM1 mutations in children and adults: evidence for an effect of patient age on the prevalence of the TCTG-tandem duplication in NPM1-exon 12. *Leukemia.* 2007;21:366–367.

105. Thiede C, Koch S, Creutzig E, et al. Prevalence and prognostic impact of NPM1 mutations in 1485 adult patients with acute myeloid leukemia (AML). *Blood.* 2006;107:4011–4020.

106. Tsui KH, Cheng AJ, Chang PL, et al. Association of nucleophosmin/B23 mRNA expression with clinical outcome in patients with bladder carcinoma. *Urology.* 2004;64:839–844.
107. Verhaak RG, Goudswaard CS, van Putten W, et al. Mutations in nucleophosmin (NPM1) in acute myeloid leukemia (AML): association with other gene abnormalities and previously established gene expression signatures and their favorable prognostic significance. *Blood.* 2005;106:3747–3754.
108. Wang D, Baumann A, Szebeni A, et al. The nucleic acid binding activity of nucleolar protein B23.1 resides in its carboxyl-terminal end. *J Biol Chem.* 1994;269:30994–30998.
109. Wang W, Budhu A, Forgues M, et al. Temporal and spatial control of nucleophosmin by the Ran-Crm1 complex in centrosome duplication. *Nat Cell Biol.* 2005;7:823–830.
110 Wang D, Umekawa H, Olson MO. Expression and subcellular locations of two forms of nucleolar protein B23 in rat tissues and cells. *Cell Mol Biol Res.* 1993;39:33–42.
111. Wattel E, Preudhomme C, Hecquet B, et al. p53 mutations are associated with resistance to chemotherapy and short survival in hematologic malignancies. *Blood.* 1994;84: 3148–3157.
112. Wu MH, Chang JH, Yung BY. Resistance to UV-induced cell-killing in nucleophosmin/ B23 over-expressed NIH 3T3 fibroblasts: enhancement of DNA repair and up-regulation of PCNA in association with nucleophosmin/B23 over-expression. *Carcinogenesis.* 2002;23:93–100.
113. Yao J, Fu C, Ding X, et al. Nek2A kinase regulates the localization of numatrin to centrosome in mitosis. *FEBS Lett.* 2004;575:112–118.
114. Yu Y, Maggi LB, Jr., Brady SN, et al. Nucleophosmin is essential for ribosomal protein L5 nuclear export. *Mol Cell Biol.* 2006;26:3798–3809.
115. Yun JP, Chew EC, Liew CT, et al. Nucleophosmin/B23 is a proliferate shuttle protein associated with nuclear matrix. *J Cell Biochem.* 2003;90:1140–1148.
116. Yung BY, Chan PK. Identification and characterization of a hexameric form of nucleolar phosphoprotein B23. *Biochim Biophys Acta.* 1987;925:74–82.
117. Zhang H, Shi X, Paddon H, et al. B23/nucleophosmin serine 4 phosphorylation mediates mitotic functions of polo-like kinase 1. *J Biol Chem.* 2004;279:35726–35734.
118. Zhang Y, Zhang M, Yang L, et al. NPM1 mutations in myelodysplastic syndromes and acute myeloid leukemia with normal karyotype. *Leuk Res.* 2006;31:109–111.

MicroRNAs: New Players in AML Pathogenesis

Milena S. Nicoloso, Bharti Jasra, and George A. Calin

MicroRNAs: Biogenesis and Function

MicroRNAs (miRNAs) constitute a class of regulatory non-coding RNAs involved in gene silencing pathways. They are single-stranded RNA molecules ranging from 19 to 25 nucleotides and may be considered the natural counterparts of small interfering RNAs (siRNAs). Because of their fine regulation of gene expression at the post-transcriptional level, miRNAs are crucial players in several regulatory pathways, including developmental timing, organogenesis, differentiation, hematopoiesis, apoptosis, cell proliferation, and tumorigenesis [2]. The regulatory network of miRNA and messenger RNA (mRNA) is combinatorial and unique to each cell type; in fact, a single miRNA molecule can bind and repress several different mRNAs and multiple miRNAs cooperate to control a single mRNA target, in a cell-customized manner [36]. Because of their pleiotropic gene expression regulation potential, miRNAs are also considered cellular micromanagers [2, 36].

miRNAs are transcribed from endogenous genes by RNA polymerase II [5, 35] and the primary transcripts (pri-miRNAs), which can be up to 1-kb long, contain cap structures as well as poly(A) tails, like all class II gene transcripts. To generate mature miRNAs, primary transcripts undergo two sequential digestion steps (Fig. 1). The first step occurs in the nucleus through the action of DROSHA, a double-stranded RNA-specific ribonuclease III [33] that is responsible for cropping the hairpin-shaped secondary structures of pri-miRNAs into precursor miRNAs (pre-miRNAs), which are 60–110 nucleotides long. After the nuclear export of the pre-miRNAs via EXPORTIN-5/RAN-GTP [37, 49], the second step occurs in the cytoplasm through DICER, another class III ribonuclease, initially generating a double-stranded product 19–25 nucleotides long [3, 27]. Only one strand of the miRNA duplex will be

G.A. Calin (✉)
Department of Experimental Therapeutics and Department of Cancer Genetics,
The University of Texas MD Anderson Cancer Center, Houston, TX 77030, USA
e-mail: gcalin@mdanderson.org

L. Nagarajan (ed.), *Acute Myelogenous Leukemia*,
Cancer Treatment and Research 145, DOI 10.1007/978-0-387-69259-3_10,
© Springer Science+Business Media, LLC 2010

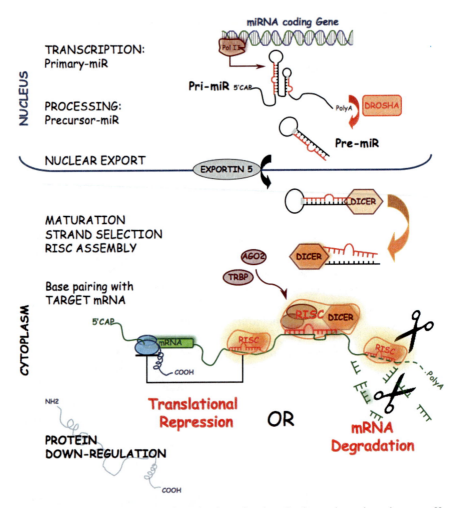

Fig. 1 MicroRNA biogenesis and mechanism of action. In the nucleus, the polymerase II-dependent primary microRNA transcript is processed into the hair pin-shaped precursor miRNA. The maturation of the precursor miRNA occurs in the cytoplasm, where the active RISC–miRNA complex is assembled and recognizes its multiple substrates. According to the degree of the miRNA–mRNA complementarity, the silencing mechanism is based on the inhibition of protein translation or mRNA degradation. In either case, the final effect is a reduction in total protein expression levels

incorporated into an RNA–protein effector complex, the RNA-induced silencing complex (RISC) [17, 46]. Dicer plays an active part in RISC assembly, along with the double-stranded RNA-binding protein TARBP2 and proteins of the Argonaute family, which are the core components of RISC [23, 47]. RISC distinguishes the guide strand of the duplex miRNA from the passenger strand and specifically incorporates the single-stranded mature miRNA, coupling miRNA biogenesis and target-RNA interference [12]. Mature miRNAs, loaded onto RISC, mediate target gene regulation by base-pairing with the 3′ untranslated regions (3′ UTR) of mRNAs [1]. Finally, miRNA-bound mRNAs remain untranslated, with reduction in corresponding protein levels, or are degraded, with reduction in both the mRNA transcript and the protein levels.

miRNAs are commonly associated with complex transcriptional loci and can be classified into exonic or intronic subsets according to their genomic locations [45]. Since most miRNAs are found in intergenic regions (>1 kb away from annotated or predicted genes) or are oriented antisense-wise to neighboring genes it has been postulated that miRNAs can be transcribed as autonomous units [47, 48]. When localized close to each other, miRNAs give rise to a single polycistronic transcription unit [34] with its own promoter. Additionally, it is very likely that the tissue-specific expression profiles of intronic miRNAs, which are oriented sense-wise to the host gene, correlate with those of the corresponding mRNAs.

MiRNA in Hematopoiesis

Cell differentiation during hematopoiesis requires a precise gene reprogramming that has recently been shown to rely on the fine modulation of gene expression by miRNA ([8]; Table 1). miRNA expression patterns display large differences between hematopoietic and non-hematopoietic cells and specific hematopoietic lineages have unique characteristics. Fully differentiated effector cells (Th1 and Th2 (T helper) lymphocytes and mast cells) and precursors at comparable stages of differentiation (double-negative thymocytes and pro-B cells) are characterized by similar miRNA motifs. The distinctive miRNA milieu of a cell affects not only the process of cellular lineage commitment but also the process of cell differentiation and the maintenance of cell identity [38].

Table 1 MiRNAs involved in hematopoiesis

MicroRNA	Lineage	Function in hematopoiesis	Reference
miR-142	Myelopoiesis	Highly expressed in myeloid lineages/ induces T-cell lineage commitment when overexpressed.	[8]
miR-223	Myelopoiesis/ granulopoiesis Megakaryocytopoiesis	Highly expressed in myeloid lineages/ induces T-cell lineage commitment when overexpressed; increased during RA-induced granulocytic differentiation; reduced at first and then increased during megakaryocytic differentiation.	[8, 15, 19–21]
miR-15a, miR-15b	Megakaryocytopoiesis Granulocytic differentiation	Reduced at first and then increased during	[19, 20]

Table 1 (continued)

MicroRNA	Lineage	Function in hematopoiesis	Reference
		megakaryocytic differentiation; increased by ATRA treatment.	
miR-16	Megakaryocytopoiesis granulopoiesis, monocytopoiesis, B-lymphopoiesis	Reduced at first and then increased during megakaryocytic differentiation; increased by ATRA treatment.	[19–21]
Let-7a, Let-7c, Let-7d, miR-342,	Granulopoiesis	Increased by ATRA treatment.	[20]
miR-181	B-Lymphopoiesis and all branches of hematopoiesis	Induces B-cell lineage commitment when overexpressed; blocked in early differentiation.	[8, 21]
miR-128	All branches of hematopoiesis	Blocked in early differentiation.	[21]
miR-221 and miR-222	Erythropoiesis	Induce growth arrest and differentiation.	[16, 21]
miR-155, miR-24, miR-17	Myelopoiesis and B-lymphopoiesis		[21]
miR-103,	Granulopoiesis, monocytopoiesis, B-lymphopoiesis		[21]
miR-150	Lymphopoiesis	Highly expressed during B and T development and maturation; reduced in Th1 and Th2 differentiated state	[38]
miR-146	Lymphopoiesis	Upregulated in Th2 differentiated state	[21, 38]
miR-17-5p, miR-20a, miR-106a	Monocytopoiesis	Downregulated during initial phases of monocytic differentiation and maturation	[18, 38]
miR17-92 cluster	Erythropoiesis	Overexpression in erythroblastic cell lines triggers EPO-induced proliferation.	[10]
miR-10a, miR-10b, miR-30c, miR-106, miR-126, miR-130a, miR-132	Megakaryocytopoiesis	Downregulated during differentiation.	[19]
miR-143	Megakariocytopoiesis	Downregulated during differentiation.	[19]
miR-107	Granulopoiesis, monocytopoiesis, B-lymphopoiesis	Increased by ATRA treatment.	[20, 21]

Chen et al. performed the first study confirming the importance of miRNAs in hematopoiesis by dissecting miRNA expression in murine hematopoietic lineages [8]. This work demonstrated how certain miRNAs are specifically expressed in hematopoietic cells and how their expression is dynamically regulated during early lineage commitment. According to this expression study, murine B cell progenitors express abundant *miR-181a*, while myeloid lineages highly express *miR-223* and *miR-142*. In fact, ectopic expression of *miR-181* in hematopoietic stem and progenitor cells enhanced the number of B-lineage cells, whereas *miR-223* and *miR-142* expression unexpectedly induced a T-cell lineage commitment [8].

Monticelli's group's systematic miRNA gene profiling of murine hematopoietic cell types at immature, mature, and effector stages also confirmed that miRNA expression patterns were modulated along each differentiation and maturation step [38]. In particular, they observed a high expression of *miR-150* during the developmental stages of B- and T-cell maturation, followed by its downmodulation during the differentiation of naïve T cells into effector Th1 and Th2 cells, while *miR-146* was upregulated solely in Th1-committed cells.

Felli et al. [16] showed that *miR-221* and miR-*222* expression decreases during early erythropoiesis, with a concomitant accumulation of c-KIT protein, which is in turn responsible for erythroblast expansion. Indeed, when they overexpressed *miR-221* and *miR-222* in CD34$^+$ cell lines or in erythroid leukemic cell lines, they observed growth arrest and accelerated differentiation accompanied by a loss of c-KIT protein. Furthermore, transplantation experiments in NOD–SCID (non-obese diabetic severe combined immunedeficient) mice revealed that the expression of *miR-221* and *miR-222* in CD34$^+$ cells impairs their engraftment capacity and stem cell activity, suggesting a more generalized regulation of differentiation [16].

In another study, Fazi et al. [15] explored how human granulopoiesis is controlled by the mini-circuitry comprising *miR-223* and transcription factors NF1-A and C/EBPα. NF1-A and C/EBPα act as *miR-223* transactivators, with NF1-A constitutively interacting with and repressing the *miR-223* promoter and C/EBPα activating *miR-223* expression in a differentiation-specific way. During retinoic acid (RA)-induced granulocytic differentiation, C/EBPα displaces the repressive NF1-A from the *miR-223* promoter, upregulating *miR-223* expression; *miR-223* in turn acts as part of an autoregulatory loop, targeting *NF1-A* mRNA and blocking its translation. Therefore, *miR-223* reinforces its own sustained expression, so as to mediate lineage-specific reprogramming. Indeed, in acute promyelocytic (APL) cell lines, both *NF1-A* RNA interference and *miR-223* overexpression enhanced differentiation, whereas *miR-223* knockdown inhibited the differentiation response of granulocytes to RA [15].

The miRNA expression profile of CD34$^+$ hematopoietic stem–progenitor cells (HSPC) correlates with the mRNA expression profile of CD34$^+$ cells based on predicted miRNA interactions This finding, supported by experimental evidence (e.g., results from luciferase and differentiation assays), allowed Georgantas et al. to identify miRNAs participating in each step of human hematopoiesis, based on the differentiation potential of the predicted targets

[21]. Specifically, *miR-181* and *miR-128* were found to inhibit differentiation in all hematopoietic lineages, whereas *miR-146* was specifically found to participate in lymphoid differentiation; *miR-155*, *miR-24*, and *miR-17* in granulocytic, monocytic, erythroid, megakaryocytic, and B-lymphoid differentiation; and *miR-16*, *miR-103*, and *miR-107* in granulocytic, monocytic, and B-lymphoid differentiation [21]. Georgantas' study also pinpointed miRNAs specific to a single lineage; for example, *miR-221* and *miR-222* were found to be important in erythroid development and *miR-223* was found to be important in granulocytic development, both in agreement with results from other investigators [15, 16]. On the contrary, the miRNA profiling in hematopoiesis from Georgantas et al. does not resemble the miRNA behavior observed for mouse *miR-142*, *miR-181*, and *miR-223* [42], suggesting that miRNAs may act differently in human hematopoiesis and mouse hematopoiesis.

Fontana et al. identified three miRNAs (*miR-17-5p*, *miR-20a*, and *miR-106a*) from the *miRNA 17~92* and *miR-106a~92* clusters to have regulatory roles during monocytic differentiation. Their report proposes the existence of a cascade-controlling monocytopoiesis associated with a sequential decrease in *miR-17-5p*, *miR-20a*, and *miR-106a* levels and a concomitant increase of *AML1* and *M-CSFR* (monocytic-colony-stimulating factor receptor) expression. Blast cell proliferation and the inhibition of monocytic differentiation and maturation were enhanced by high levels of *miR-17-5p*, *miR-20a*, and *miR-106a*, which are found highly expressed in CD34$^+$ hematopoietic progenitor cells (HPCs). These miRNAs were shown to interact directly with *AML1* 3' UTR and to block *AML1* protein synthesis. The downregulation of *AML1* leads to the downregulation of *M-CSFR* expression, which in turn promotes the maintenance of an undifferentiated state [4]. These miRNAs were also shown to be downregulated during the initial stages of monocytopoiesis to enhance *AML1* translation. Furthermore, *AML1* has been shown to form a negative feedback loop by inhibition of the *miRNA 17~92* and *106a~92* transcription through direct promoter binding; as a consequence, *AML1* reinforces the monocytic differentiation and maturation program [18].

The expression patterns of *miR-17-5p*, *miR-20a*, and *miR-106a* are different in monocytic and erythroid lineages [16]. In monocytic differentiation, the levels of these miRNAs are inversely correlated with that of *AML1* protein, whereas in erythroid differentiation no inverse correlation was found. Therefore, the processing of these miRNAs is controlled in part by lineage-specific factors [18].

miRNAs have also been demonstrated to be important in megakaryocytopoiesis. By studying miRNA expression in megakaryocytes differentiated in vitro from CD34$^+$ hematopoietic progenitors, Garzon et al. identified a distinctive miRNA signature characterized by the downregulation of *miR-10a*, *miR-10b*, *miR-30c*, *miR-106*, *miR-126*, *miR-130a*, and *miR-132* and by the upregulation of *miR-143* [19]. Interestingly, *miR-223*, *miR-15*, and *miR-16* showed biphasic expression with downregulation during the initial phases of megakaryocytic differentiation and gradual upregulation in later phases. The significance of this modulation was confirmed by the finding that mRNA targets for these miRNAs are well-known regulators of megakaryocytopoiesis. Additionally, *miR-130a* was shown to regulate the transcription factor *MAFB*, which is involved in the activation of the *GPIIB* promoter, a key protein in platelet function [19].

MicroRNAs in Acute Myelogenous Leukemia

A large body of evidence has accumulated since the original discovery of the loss of function of *miR-15a/16-1* in chronic lymphocytic leukemia (CLL) [6] that miRNAs are now widely accepted as active players in tumorigenesis (Table 2). miRNA genes are frequently located at fragile sites, minimal regions of loss of heterozygosity, regions of amplification, and common chromosomal breakpoints [7]. Many miRNAs can act either as tumor suppressor genes or as oncogenes and have been proven to accelerate the oncogenic process, depending on the mRNA they target [14]. For example, the first oncomir, i.e., the cluster *miR-17~92*, accelerates *c-Myc*-induced lymphomagenesis in mice [25]. Nonetheless, very few studies have been published on the expression and role of miRNAs in acute myelogenous leukemia (AML).

In one study, Ramkissoon et al. [42] compared the expression of hematopoiesis-specific miRNAs (*miR-142*, *miR-155*, *miR-181*, and *miR-223*) from 17 commercially available malignant hematopoietic cell lines with miRNAs from normal human B, T, monocytic, and granulocytic lineages. The group demonstrated that, although expression patterns were similar to those in normal human hematopoietic lineages, malignant cells had altered miRNA expression levels, indicating a role for miRNA deregulation in human hematopoietic malignancies as well.

Table 2 Demonstrated miRNA targets relevant in myeloid malignancies

MicroRNA	Target	Target gene function	Reference
miR-223, miR-107	*NFI-A*	Transcription factor known to regulate genes involved in cell proliferation.	[15, 20]
miR-221 and miR-222	*c-KIT*	Transmembrane tyrosine-kinase receptor for stem cell factor, required for normal hematopoiesis.	[16]
miR-17-5p, miR-20a, miR-106a	*AML1*	DNA binding subunit of the hematopoietic transcription factor CBF, controlling multiple genes involved in myeloid differentiation.	[18, 35]
miR-130a	*MAFB*	Transcription factor involved in the activation of GPII8, a key protein for platelet physiology.	[19]
miR-10a	*HOXA1*	Homeodomain transcription factor with a role in regulating definitive hematopoiesis.	[19]
miR-15a, miR-16-1	*BCL2*	Anti-apoptotic protein.	[9, 19]
Let-7a	*RAS*	Small GTPase involved in the regulation of cell proliferation, survival, and apoptosis.	[20]

Recently, it was proven that miRNA expression levels could be used to sub-classify AML with normal karyotypes and a possible function has since been proposed for *miR-181a* in lineage development and differentiation state of AML [11]. In their study, Debernardi et al. used quantitative real-time PCR to compare five specific miRNA expression levels with the transcriptome profile of 30 diploid karyotype AML samples. The five miRNAs included *miR-10a*, *miR-10b*, *miR-196a-1*, located within the HOX clusters and known to target homeobox genes, and *miR-181a* and *miR-223*, which are important in hemato-poietic development. Of these miRNAs, *miR-181a* had the highest number of correlations with its predicted target genes, indicating that it may directly regulate the transcript level of the negatively correlated target genes or indir-ectly regulate positively correlated genes. Furthermore, investigators were able to use *miR-181a* to classify the AML samples into FAB (French-American-British classification of leukemia) subsets, as it was expressed at distinctively higher levels in M1 and M2 AML samples than in M4 or M5 AML samples. This observation strongly supports the possibility of using in the near future miRNA expression profiles to stratify AML patients according to diagnosis, and eventually also to prognosis and response to treatment, leading overall to better patient management.

In Debernardi's study it was also reported that the three miRNAs located in the intergenic regions of the HOX clusters, *miR-10a*, *miR-10b*, and *miR-196a-1*, had high positive correlations with almost all the genes of the *HOXA* and *HOXB* clusters, except for *HOXA1* and *HOXA13* in cluster A and *HOXB1* and *HOXB13* in cluster B [11]. This finding was in agreement with previous work by Garzon et al. showing that *miR-10a* and *HOXA1* interact directly and that their levels are inversely correlated during megakaryocytic differentiation [19]. Additionally, Garzon et al. described a common regulatory transcription mechanism for the HOX cluster genes and the miRNAs located within them that are positively correlated (i.e., *HOXB4* and *HOXB5* and *miR10a*) [19]. Notably, the HOX family of homeodomain transcription factors are critical regulators of definitive hematopoiesis, and the deregulation of HOX gene expression is a characteristic of poor AML prognosis [13].

Altogether these findings suggest that miRNA alteration in AML is a part of the deregulated HOX code relevant to the pathogenesis of myeloid malignan-cies [48]. In fact, in AML samples with normal karyotypes, 30% of the genes correlating with *miR-10a*, *miR-10b*, and *miR-196a-1* expression are known for their oncogenic potential, and many are transcription factors, further support-ing a HOX-driven oncogene activation in leukemogenesis. Although prelimin-ary and merely descriptive, the study by Debernardi et al. [11] clearly opened the field of miRNA involvement in AML pathogenesis and suggested that a more extensive analysis of a larger number of miRNAs could provide valuable insights into leukemogenesis.

AML1/RUNX1 is a gene frequently deregulated in AML by t(8;21) translocations, which are associated with 12% of de novo AML cases and up to 40% of AML subtypes with M2 morphology. These translocations have also been reported in a small portion of M0, M1, and M4 AML samples [41]. The fusion protein AML1–ETO that results from the translocation process interferes with AML1-dependent gene transactivation [26]. The overexpression of *miR-17-5p, miR-18a, miR-20a,* and *miR-106a* downregulates AML1 and can mimic the role of the AML1–ETO fusion protein, which promotes the proliferation of immature cells and blocks monocytopoiesis by inhibiting the *AML1*-dependent transcription of the *M-CSFR* gene [39, 43]. Specifically, the effects of miRNA overexpression are reminiscent of the phenotype of M2/M4 AML subtypes that are associated with AML1–ETO fusion protein expression (i.e., blockade of myeloid and monocytic differentiation coupled with extensive self-renewal capacity). However, *miR 17-5p, miR-18a, miR-20a,* and *miR-106a* overexpression is not leukemogenic per se and requires additional genetic alterations, as seen in AML1–ETO knock-in mice [30].

This evidence that miRNAs are relevant in leukemia has opened a new area of targeted drug development. For example, antagomirs—chemically modified anti-miRNA oligonucleotides conjugated with cholesterol—inhibit miRNA activity in mice and might be useful as therapeutic agents [29]. By using antagomirs, Krutzfeldt et al. were able to inhibit endogenous *miRNA-17-5p* and *miR–20a* in HPCs and to induce more rapid monocytic differentiation and inhibition of proliferation. In AML cases in which these miRNAs are overexpressed, the administration of antagomirs may become a therapeutic option. Additionally, the experiments of Felli et al. in c-KIT+ erythroleukemic cells, which differentiated in response to *miR-221* and *miR-222* overexpression, suggest a potential role for *miR-221* and *miR-222* genes in AML therapy [16].

A novel oncogenic role in erythroid leukemia has been attributed to the miR-17~92 cluster [10]. In particular, by studying the murine model of F-MuLV-induced erythroleukemia, Cui et al. identified a novel gene, *FLI-3,* that was activated or amplified by insertional mutagenesis and highly conserved in the human *C13orf25* gene. Interestingly, *C13orf25* is also frequently amplified in several types of human tumors, including lymphoma and lung cancer [24, 44], and it has been demonstrated to encode the *miR-17~92* cluster. In the *FLI-3*-driven erythroleukemic cell lines, the *miR-17~92* cluster was upregulated, suggesting an oncogenic role in erythroleukemia, analogous to its role in the *c-MYC*-driven B-cell lymphoma mouse model [24]. According to Cui et al., the overexpression of the cluster in erythroblastic cell lines switched Erythropoietin (EPO)-induced differentiation to EPO-induced proliferation and activated the RAS and PI3K pathways. The direct target genes of the *miR-17~92* cluster that are involved in the erythroleukemic transformation process have yet to be identified, although the cellular program the miR-17~92 cluster activates overlaps with the one induced by *Fli-1* expression in erythroblastic cell lines and is characterized by the increase of SPI-1 and the downregulation of the cell cycle inhibitor p27 [10].

Recent developments now indicate that miRNA may have a role in acute promyelocytc leukemia (APL), a form of AML characterized by maturation arrest at the promyelocytic stage, chromosomal translocation t(15;17), and consequent expression of the fusion gene PML-RARα. It is well known that the PML-RARα gene product inhibits myeloid differentiation by the transcriptional repression of retinoic acid (RA)-responsive genes and that pharmacological levels of all-trans retinoic acid (ATRA) can induce terminal differentiation by reverting the PML-RARα dominant-negative effect [22]. In RA-responsive myeloid cell lines, Fazi et al. found that *miR-223* expression is strongly increased during ATRA treatment and that the differentiation effect of RA treatment is mediated by *miR-223*. They report that the upregulation of *miR-223* may be due to the direct involvement of RA binding to either a functional PML-RARα or a endogenous wild-type RARα [15]. As shown by the expression profiling study of Garzon et al., ATRA treatment of APL cells upregulated a number of miRNAs (*miR-15a, miR-15b, miR-16-1, Let-7a-3, Let-7c, Let-7d, miR-223, miR-107*, and *miR-342*), whereas only miR-181b was downregulated [20]. In agreement with the conclusions of Fazi et al., Garzon et al. found *miR-223* to be upregulated in ATRA-treated APL cell lines. Additionally, Garzon et al. showed that the upregulated *miR-107* targets *NF1-A* and therefore mimics the regulatory circuitry of *miR-223* and *C/EBPα* during granulocytic differentiation. In APL, the expression levels of *miR-15a* and *miR-16-1* in ATRA-treated samples displayed a negative correlation with the anti-apoptotic protein BCL2, as previously shown for CLL [9], suggesting that a similar pathway is involved in APL pathogenesis. Similarly, also *Let-7a* expression induction by ATRA was followed by a reduction of its known targets, such as *RAS* [28], resembling a tumorigenic pathway common in solid tumors. Additionally, Garzon et al. found that *NF-kB*, an ATRA-regulated transcription factor, binds the Let-7a-3/let-7b cluster upstream and is essential for its transactivation [20].

Garzon et al. have further proven the widespread deregulation of miRNAs in AML through the miRNA profiling of acute megakaryoblastic leukemia (AMKL) cell lines [19]. In this study, five miRNAs important in megakaryocytic differentiation were found to be upregulated during leukemogenesis: *miR-101, miR-126, miR-99a, miR-135*, and *miR-20*. Whether this profile merely represents a state of AMKL cell differentiation or miRNAs play a pathogenic role in AMKL remains to be elucidated. So far, one report [40] supports the second hypothesis, specifically naming *miR-106, miR-135*, and *miR-20* as the miRNAs predicted to target AML1, a transcription factor frequently affected by translocations in AML.

Concluding Remarks

Deciphering the miRNA expression profiles of tumors with greater accuracy is a novel and informative tool that can classify AML cases; furthermore, the identification of alterations in miRNA expression patterns may disclose pathogenic pathways offering alternative strategies to target AMLs. We are only beginning to understand the contribution of miRNAs to AML, but promising and exciting possibilities have already emerged.

Acknowledgments Dr. Calin is supported by the CLL Global Research Foundation, and in part as a University of Texas System Regents Research Scholar and a Fellow of The University of Texas MD Anderson Research Trust. Dr. Jasra participated in the preparation of this chapter during her observership in Dr. Calin's laboratory.

References

1. Ambros V. MicroRNA pathways in flies and worms: growth, death, fat, stress, and timing. *Cell.* 2003;113:673–676.
2. Bartel DP, Chen CZ. Micromanagers of gene expression: the potentially widespread influence of metazoan microRNAs. *Nat Revi.* 2004;5:396–400.
3. Bernstein E, Caudy AA, Hammond SM, Hannon GJ. Role for a bidentate ribonuclease in the initiation step of RNA interference. *Nature.* 2001;409:363–366.
4. Bourette RP, Rohrschneider LR. Early events in M-CSF receptor signaling. *Growth Factors (Chur, Switzerland).* 2000;17:155–166.
5. Cai X, Hagedorn CH, Cullen BR. Human microRNAs are processed from capped, polyadenylated transcripts that can also function as mRNAs. *RNA.* 2004;10:1957–1966.
6. Calin GA, Dumitru CD, Shimizu M, et al. Frequent deletions and down-regulation of micro-RNA genes miR15 and miR16 at 13q14 in chronic lymphocytic leukemia. *Proc Natl Acad Sci USA.* 2002;99:15524–15529.
7. Calin GA, Sevignani C, Dumitru CD, et al. Human microRNA genes are frequently located at fragile sites and genomic regions involved in cancers. *Proc Natl Acad Sci USA.* 2004;101:2999–3004.
8. Chen CZ, Li L, Lodish HF, Bartel DP. MicroRNAs modulate hematopoietic lineage differentiation. *Science (New York, NY).* 2004;303:83–86.
9. Cimmino A, Calin GA, Fabbri M, et al. miR-15 and miR-16 induce apoptosis by targeting BCL2. *Proc Natl Acad Sci USA.* 2005;102:13944–13949.
10. Cui JW, Li YJ, Sarkar A, et al. Retroviral insertional activation of the Fli-3 locus in erythroleukemias encoding a cluster of microRNAs that convert Epo-induced differentiation to proliferation. *Blood.* 2007;110(7):2631–2640.
11. Debernardi S, Skoulakis S, Molloy G, Chaplin T, Dixon-McIver A, Young BD. Micro-RNA miR-181a correlates with morphological sub-class of acute myeloid leukaemia and the expression of its target genes in global genome-wide analysis. *Leukemia.* 2007;21:912–916.
12. Doi N, Zenno S, Ueda R, Ohki-Hamazaki H, Ui-Tei K, Saigo K. Short-interfering-RNA-mediated gene silencing in mammalian cells requires Dicer and eIF2C translation initiation factors. *Curr Biol.* 2003;13:41–46.
13. Eklund EA. The role of HOX genes in malignant myeloid disease. *Curr Opin Hematol.* 2007;14:85–89.
14. Esquela-Kerscher A, Slack FJ. Oncomirs – microRNAs with a role in cancer. *Nat Rev Cancer.* 2006;6:259–269.
15. Fazi F, Rosa A, Fatica A, et al. A minicircuitry comprised of microRNA-223 and transcription factors NFI-A and C/EBPalpha regulates human granulopoiesis. *Cell.* 2005;123:819–831.
16. Felli N, Fontana L, Pelosi E, et al. MicroRNAs 221 and 222 inhibit normal erythropoiesis and erythroleukemic cell growth via kit receptor down-modulation. *Proc Natl Acad Sci USA.* 2005;102:18081–18086.
17. Filipowicz W. RNAi: the nuts and bolts of the RISC machine. *Cell.* 2005;122:17–20.

18. Fontana L, Pelosi E, Greco P, et al. MicroRNAs 17-5p-20a-106a control monocytopoiesis through AML1 targeting and M-CSF receptor upregulation. *Nat Cell Biol.* 2007;9:775–787.
19. Garzon R, Pichiorri F, Palumbo T, et al. MicroRNA fingerprints during human megakaryocytopoiesis. *Proc Natl Acad Sci USA.* 2006;103:5078–5083.
20. Garzon R, Pichiorri F, Palumbo T, et al. MicroRNA gene expression during retinoic acid-induced differentiation of human acute promyelocytic leukemia. *Oncogene.* 2007;26:4148–4157.
21. Georgantas RW, 3rd, Hildreth R, Morisot S, et al. CD34+ hematopoietic stem-progenitor cell microRNA expression and function: a circuit diagram of differentiation control. *Proc Natl Acad Sci USA.* 2007;104:2750–2755.
22. Glass CK, Rosenfeld MG. The coregulator exchange in transcriptional functions of nuclear receptors. *Genes Dev.* 2000;14:121–141.
23. Hammond SM, Boettcher S, Caudy AA, Kobayashi R, Hannon GJ. Argonaute2, a link between genetic and biochemical analyses of RNAi. *Science (New York, NY).* 2001;293:1146–1150.
24. Hayashita Y, Osada H, Tatematsu Y, et al. A polycistronic microRNA cluster, miR-17-92, is overexpressed in human lung cancers and enhances cell proliferation. *Cancer Res.* 2005;65:9628–9632.
25. He L, Thomson JM, Hemann MT, et al. A microRNA polycistron as a potential human oncogene. *Nature.* 2005;435:828–833.
26. Hiebert SW, Sun W, Davis JN, et al. The t(12;21) translocation converts AML-1B from an activator to a repressor of transcription. *Mol Cell Biol.* 1996;16:1349–1355.
27. Hutvagner G, McLachlan J, Pasquinelli AE, Balint E, Tuschl T, Zamore, PD. A cellular function for the RNA-interference enzyme Dicer in the maturation of the let-7 small temporal RNA. *Science (New York, NY).* 2001;293:834–838.
28. Johnson SM, Grosshans H, Shingara J, et al. RAS is regulated by the let-7 microRNA family. *Cell.* 2005;120:635–647.
29. Krutzfeldt J, Rajewsky N, Braich R, et al. Silencing of microRNAs in vivo with 'antagomirs'. *Nature.* 2005;438:685–689.
30. Kuchenbauer F, Feuring-Buske M, Buske C. AML1-ETO needs a partner: new insights into the pathogenesis of t(8;21) leukemia. *Cell Cycle (Georgetown, Tex).* 2005;4:1716–1718.
31. Lagos-Quintana M, Rauhut R, Lendeckel W, Tuschl T. Identification of novel genes coding for small expressed RNAs. *Science (New York, NY).* 2001;294:853–858.
32. Lau NC, Lim LP, Weinstein EG, Bartel DP. An abundant class of tiny RNAs with probable regulatory roles in Caenorhabditis elegans. *Science (New York, NY).* 2001;294:858–862.
33. Lee Y, Ahn C, Han J, et al. The nuclear RNase III Drosha initiates microRNA processing. *Nature.* 2003;425:415–419.
34. Lee Y, Jeon K, Lee JT, Kim S, Kim VN. MicroRNA maturation: stepwise processing and subcellular localization. *EMBO J.* 2002;21:4663–4670.
35. Lee Y, Kim M, Han J, et al. MicroRNA genes are transcribed by RNA polymerase II. *EMBO J.* 2004;23:4051–4060.
36. Lewis BP, Shih IH, Jones-Rhoades MW, Bartel DP, Burge CB. Prediction of mammalian microRNA targets. *Cell.* 2003;115:787–798.
37. Lund E, Guttinger S, Calado A, Dahlberg JE, Kutay U. Nuclear export of microRNA precursors. *Science (New York, NY).* 2004;303:95–98.
38. Monticelli S, Ansel KM, Xiao C, et al. 2005. MicroRNA profiling of the murine hematopoietic system. *Genome Biol.* 2004;6:R71.
39. Mulloy JC, Cammenga J, MacKenzie KL, Berguido FJ, Moore MA, Nimer SD. The AML1-ETO fusion protein promotes the expansion of human hematopoietic stem cells. *Blood.* 2002;99:15–23.

40. Nakao M, Horiike S, Fukushima-Nakase Y, et al. Novel loss-of-function mutations of the haematopoiesis-related transcription factor, acute myeloid leukaemia 1/runt-related transcription factor 1, detected in acute myeloblastic leukaemia and myelodysplastic syndrome. *Br J Haematol.* 2004;125:709–719.

41. Peterson LF, Zhang DE. The 8;21 translocation in leukemogenesis. *Oncogene.* 2004;23:4255–4262.

42. Ramkissoon SH, Mainwaring LA, Ogasawara Y, et al. Hematopoietic-specific micro-RNA expression in human cells. *Leuk Res.* 2006;30:643–647.

43. Rhoades KL, Hetherington CJ, Harakawa N, et al. Analysis of the role of AML1-ETO in leukemogenesis, using an inducible transgenic mouse model. *Blood.* 2000;96:2108–2115.

44. Rinaldi A, Poretti G, Kwee I, Zucca E, Catapano CV, Tibiletti MG, Bertoni F. Concomitant MYC and microRNA cluster miR-17-92 (C13orf25) amplification in human mantle cell lymphoma. *Leuk Lymphoma.* 2007;48:410–412.

45. Rodriguez A, Griffiths-Jones S, Ashurst JL, Bradley A. Identification of mammalian microRNA host genes and transcription units. *Genome Res.* 2004;14:1902–1910.

46. Sontheimer EJ, Carthew RW. Molecular biology. Argonaute journeys into the heart of RISC. *Science (New York, NY).* 2004;305:1409–1410.

47. Sontheimer EJ, Carthew RW. Silence from within: endogenous siRNAs and miRNAs. *Cell.* 2005;122:9–12.

48. Thorsteinsdottir U, Mamo A, Kroon E, Jerome L, Bijl J, Lawrence HJ, Humphries K, Sauvageau G. Overexpression of the myeloid leukemia-associated Hoxa9 gene in bone marrow cells induces stem cell expansion. *Blood.* 2002;99:121–129.

49. Yi R, Qin Y, Macara IG, Cullen BR. Exportin-5 mediates the nuclear export of pre-microRNAs and short hairpin RNAs. *Genes Dev.* 2003;17:3011–3016.

Murine Models of Human Acute Myeloid Leukemia

Julie M. Fortier and Timothy A. Graubert

Abstract Primary human AML cells can be isolated and studied in vitro, but many experimental questions can only be addressed using in vivo models. In particular, tractable animal models are needed to test novel therapies. The genetic complexity of human AML poses significant challenges for the generation of reliable animal models.

The hematopoietic systems of both zebrafish (*Danio rerio*) and Drosophila have been well characterized (reviewed in [5, 31]). Both organisms are well suited to forward genetics mutagenesis screens. Although this approach has been useful for identification of mutants with hematopoietic phenotypes (e.g., *cloche*), the impact on cancer biology and hematopoietic malignancies in particular has been limited. A zebrafish model of acute lymphoblastic leukemia has been generated [37] and Drosophila models have shed light on the biology of epithelial tumors (reviewed in [60]). Nonetheless, in vivo modeling of human AML relies most heavily on mice. Most cellular, molecular, and developmental features of the hematopoietic system are well conserved across mammalian species. The availability of the human and mouse genome sequences and the capability of manipulating the mouse genome make mice the most valuable model organism for AML research. Mice have additional practical value because they have a short reproductive cycle and are relatively inexpensive to house.

Introduction: Strategies for Modeling Human AML in Mice

An ideal mouse model of AML would be robust, reproducible, and should recapitulate the key phenotypic features of human AML (e.g., leukocytosis, organ infiltration, lethality). Short latency, high penetrance models are preferred because of cost savings.

T.A. Graubert (✉)

Division of Oncology, Stem Cell Biology Section, Washington University School of Medicine, Campus Box 8007, 660 South Euclid Avenue, St. Louis, MO 63110, USA
e-mail: tgrauber@dom.wustl.edu

L. Nagarajan (ed.), *Acute Myelogenous Leukemia*,
Cancer Treatment and Research 145, DOI 10.1007/978-0-387-69259-3_11,
© Springer Science+Business Media, LLC 2010

Despite intensive effort from many laboratories around the world and substantial progress, none of the existing models is ideal. A key obstacle is that AML is not caused by a single genetic event. Using a variety of complementary approaches, considerable progress has been made. This review will emphasize the principle advantages and drawbacks of the most commonly employed systems for modeling human AML in mice (see Table 1).

Table 1 Comparison of strategies used for mouse models of human AML

	Xenotransplantation	Retroviral transduction	Transgenesis
Advantages	Primary human cells	Relatively high throughput	Regulated expression
Disadvantages	Incomplete phenotype	Unregulated expression	Resource intensive
	Sample variability	Insertional mutagenesis	Integration site effects
	Short life span		

Xenotransplantation

Incremental technical advances over the past two decades have made it possible to adoptively transfer primary human AML cells into immunodeficient mice with good efficiency. These murine xenograft models provide a powerful system for observation and characterization of human AML without significant a priori knowledge of the genetic and/or epigenetic changes present in these cells.

The murine xenotransplantation models for human AML rest on a foundation of prior success engrafting normal human hematopoietic cells in immunodeficient mice [39]. B6.CB17-$Prkdc^{scid}$/SzJ severe combined immunodeficiency (SCID) mice lack functional mature B and T cells due to a spontaneous mutation in DNA-dependent protein kinase catalytic subunit ($Prkdc$), a protein kinase required for antigen receptor rearrangement [7]. Primary human AML cells engraft variably and at low levels in SCID mice and require exogenous human hematopoietic cytokines (e.g., IL-3, SCF, GM-CSF) for survival [38, 9]. Non-obese diabetic/severe combined immune deficiency (NOD/SCID) mice are more permissive hosts, allowing higher levels of engraftment and circumventing the need for cytokine supplementation in most cases [41, 47]. These mice lack hemolytic complement and have additional defects in innate immunity due to the NOD background, which causes deficiencies in macrophage and natural killer cell function. Third-generation immunodeficient hosts combining additional mutations (e.g., beta-2 microglobulin $-/-$, beige, IL-2 receptor γ $-/-$) on the NOD/SCID background are more efficient recipients for normal human HSC engraftment [47, 35, 28]. Newborn NOD/SCID/Il2rg $-/-$ ("NOG") hosts support the highest level of engraftment reported to date for primary, unmanipulated human AML cells [28].

In a typical transplant, 5–40×10^6 donor bone marrow or peripheral blood mononuclear cells from AML patients are injected via the lateral tail vein or retro-orbital venous plexus of sublethally irradiated (350–400 cGy) hosts [9, 1]. The intrafemoral route has been used by several groups and appears to result in more rapid engraftment of normal cells and higher levels of chimerism [43]. Pretreatment of NOD/SCID hosts with an anti-CD122 antibody also increases the level of human cell engraftment [55]. This antibody is directed against the interleukin-2 receptor beta chain, targeting both NK cells and macrophages.

Adoptive transfer of human AML cells infrequently causes leukemia in these immunodeficient hosts. Some of the clinical features of AML are reproduced in the chimeric mice (e.g., organ infiltration), but others are generally not (e.g., leukocytosis, lethal dissemination) [38]. Instead, the readout in these experiments is measurable engraftment of human cells and recapitulation of some phenotypic features of the primary disease. Engraftment is measured by Southern blot or flow cytometric analysis of the peripheral blood or bone marrow of recipient mice. Chimerism is often low ($<0.5\%$) and rarely exceeds 50% [47, 53]. AML chimerism can be overestimated if the assay does not account for cotransplanted normal hematopoietic cells. Multiparameter flow cytometry can be used to distinguish engraftment of normal cells (CD45 + CD19 + CD20 + B cells predominate) from AML engraftment (typically, CD45 + CD34 + CD33 + CD14 +) [9], although there can be significant drift in the immunophenotype as the cells are passaged in NOD/SCID mice [53]. RNA profiling has been applied as a rigorous measure of genetic stability in the transplanted leukemic cells. In two small studies in which expression profiling was performed pre- and post-transplant, fewer than 5% of genes had more than a 2-fold change in expression level [41, 47].

The xenograft model has proven most useful in helping to characterize the leukemic stem cell, operationally defined as the "SCID leukemia-initiating cell" (SL-IC) [9]. These cells are rare, ranging from 1 to 100 per 10^6 cells in the peripheral blood of AML patients [38]. Analysis of SL-ICs has shown them to be enriched in the Lin-CD34 + CD38− compartment and depleted in the Lin-CD34 + CD38 + fraction. This supports the model that human AML is arranged in a hierarchy that resembles normal hematopoiesis [9]. Although normal human HSC are characterized by high ALDH activity [25], this property is not consistently observed in SL-ICs [48]. Side population (SP) cells, identified using the DNA-binding dye Hoechst 33342, are enriched for HSC [20] and are also present in leukemic bone marrow. In samples with abnormal cytogenetics, the clonal abnormality is present in the SP population [65], although the SP + CD34 + CD38− subpopulation appears to contain cytogenetically normal SCID-repopulating cells in some cases [19]. Efforts to identify cell-surface markers that can differentiate SL-IC from HSC have utilized the interleukin-3 receptor alpha (CD123) [30] and the novel transmembrane glycoprotein CLL-1 [2, 59]. Using a genetic approach, Morrison and colleagues recently demonstrated that loss of *Pten* can functionally separate L-ICs from normal HSCs [68]. This suggests that agents targeting the PI3K pathway downstream of Pten (e.g., mTor inhibitors) might selectively inhibit L-ICs while sparing normal HSCs.

There is substantial variability in NOD/SCID engraftment potential from donor to donor. Some degree of engraftment variability is intrinsic to the NOD/SCID model (even with normal hematopoietic grafts) because of residual immunity in these mice. A body of literature suggests that engraftment variability may also reflect differences in the underlying biology of the leukemia. FAB subtype appears to be a good predictor of engraftability. For unexplained reasons, M3 AML cells typically fail to engraft, while chimerism is highest with M0 and M1 samples and variable with M2, M4, and M5 samples [38, 9, 53]. Samples from patients with adverse prognostic features (e.g., high WBC, poor risk cytogenetics, FLT3-ITD positive, high CD34 expression) tend to engraft better [41, 53], although high WBC and FLT3-ITD were not predictors of engraftment success in another study [47]. Nucleophosmin mutational status also has no apparent impact on the success of engraftment [47]. There does not appear to be a dose–response relationship nor does the source of the graft (bone marrow versus peripheral blood) appear to significantly affect engraftment in NOD/SCID mice [47].

Immunodeficient mice pose additional technical challenges for these experiments. Nearly 70% of NOD/SCID mice develop thymic lymphomas by 40 weeks of age [49]. The high frequency of spontaneous lymphomas and small litter size make it difficult to maintain large colonies. The shortened life span of this strain (mean 8.5 months) also limits the duration of tumor watches, although this obstacle has been largely overcome with NOG mice (life span >90 weeks) [28]. In addition, B- and NK-cell deficiency is not absolute in SCID mice. Although the SCID mutation is less "leaky" in the NOD background, breeders must be screened to verify lack of mature lymphocytes.

Additional technical refinements may lead to higher levels of engraftment, less variability between samples, and a phenotype that more closely resembles human AML. The promise of this model for translational studies might then be more fully realized. Robust xenotransplantation models are an ideal platform to test novel therapies (e.g., small molecule inhibitors, immunotherapy, and other biologics). In one recent example, the sesquiterpene lactone parthenolide was shown to specifically target L-ICs without impairing engraftment of normal HSCs in NOD/SCID mice [23].

Retroviral Transduction/Transplantation

Gene transfer into primary hematopoietic cells using oncoretroviral vectors provides a relatively rapid means of analyzing the role of single genetic changes in the pathogenesis of AML. Genes of interest are cloned into a replication-incompetent retroviral expression vector, most commonly a modified murine stem cell leukemia virus (MSCV). Bone marrow is harvested from donor mice, usually 24–48 h after treatment with 5-fluorouracil (5-FU) to deplete lineage-committed progenitors and stimulate cell cycle entry of more primitive, quiescent stem cells. Transduction is carried out by cocultivation of bone marrow cells with high titer retroviral supernatants with or without polybrene in a cocktail containing recombinant cytokines (e.g., interleukin-3, interleukin-6, stem cell factor) to promote proliferation and survival of primitive cells [32]. Transduced marrow is transplanted into lethally irradiated recipient mice. A reporter gene (e.g., EGFP) is often incorporated to assess transduction efficiency (typically 30–50%) and for lineage-specific tracking of transduced cells in vivo.

Results from these experiments reinforce the paradigm that a single mutation is insufficient to cause AML. For example, retroviral transduction/transplantation of FLT3-ITD alleles into Balb/c hosts caused a myeloproliferative disorder (MPD) characterized by leukocytosis, hepatosplenomegaly, and a median survival of 40 days [32]. Although this syndrome was rapidly fatal, there was no accumulation of immature forms consistent with acute leukemia. Transduction/transplantation of bone marrow cells from PML/RARA transgenic mice with the MSCV FLT3-ITD virus accelerated the onset of acute leukemia with promyelocytic features and increased penetrance compared to transplantation of PML/RARA-expressing cells alone [33].

One important feature of retroviral transduction is that each proviral integration site provides a unique signature that can be used to track the clonality of tumors. For example, retroviral transduction/transplantation with MSCV-NUP98/HOXA9 causes an oligoclonal MPD [36], implying that NUP98/HOXA9 alone is not sufficient for leukemic outgrowth. The disease remained oligoclonal in secondary recipients, demonstrating that self-renewing long-term repopulating cells were targeted by the retrovirus. In contrast, when the MPD evolved to AML, the disease was exclusively mono- or biclonal, suggesting that rare additional cooperating mutations were required for disease progression in this model [36].

The retroviral gene transfer system also lends itself well to asking whether an oncogene can transform hematopoietic cells at all stages of differentiation or whether the targeted cellular compartment restricts transformation potential. For these studies, bone marrow cells are sorted into subpopulations (e.g., HSC, CMP, GMP) prior to retroviral transduction and transplantation. In one example, the BCR/ABL fusion was competent to transform only HSC, whereas MOZ/TIF2, the product of the inv(8)(p11q13), conferred self-renewal potential on committed progenitors [27]. Thus, the leukemic phenotype in this case was dictated by the oncogene (the molecular "seed") rather than the cellular compartment (the "soil") [64].

The relatively high throughput and flexibility of the transduction/transplantation away are well suited to "proof of principle" preclinical therapeutic studies with novel compounds such as small molecule tyrosine kinase inhibitors [56].

Structure/function questions can also be asked by generating and testing panels of mutant constructs on a time scale that would not be feasible with other techniques (e.g., transgenesis). This strategy was used to map the critical tyrosine residues required for induction of MPD by the TEL/PDGFBR fusion [57].

Murine genetic backgrounds can have a confounding influence on the outcome of these experiments. For example, the FLT3-ITD mutations that induce MPD in Balb/c mice cause T-cell lymphomas in a B6C3HF1 background [33]. The hypomorphic *Cdkn2a* allele present in Balb/c mice resulting in reduced p16(INK4a) activity [70] may account for this bias toward myeloid leukemogenesis. Further study is needed to clarify the effects of strain background on the latency, penetrance, and phenotype of AML in all the models discussed in this review (see below). Some investigators have turned to lentivirus-mediated transfer of oncogenes into primary human hematopoietic cells as a strategy to overcome the limitations inherent in using mouse cells to model human AML [3].

Dysregulated transgene expression is a caveat that must be considered when interpreting experiments using retroviral gene transfer. Expression is driven off the strong viral LTR promoter, leading to levels of gene expression that may not be physiologically relevant [50]. This may have important consequences, since careful studies in transgenic models have shown that high-level expression of fusion oncogenes can be toxic and low-level expression may be associated with higher disease penetrance [63].

Finally, the lack of control over proviral integration site makes insertional mutagenesis a potential confounder in these experiments. Retroviral transduction with an MSCV vector expressing only a neomycin registance cassette readily immortalized murine bone marrow cells in association with integration most commonly at the *Evi1* or *Prdm16* loci [16]. Retroviral vectors are also leukemogenic in humans, as illustrated by T-cell leukemias arising in X-SCID patients due to *LMO2* activation by the integrated gene transfer vector [34].

Transgenesis

The entire spectrum of available transgenic tools has been applied to the task of modeling human AML in mice. The relative advantages and disadvantages of each strategy are well illustrated by considering one example: the t(8;21) associated with M2 AML. Despite intensive study, the role of RUNX1/ETO, the fusion protein generated by the t(8;21), is still not well understood. Correlative studies using human samples suggested that RUNX1/ETO is not sufficient to cause AML. Over the past decade, many laboratories have generated transgenic models of RUNX1/ETO to define the role of this protein in the pathogenesis of AML.

The first approach, taken by two laboratories, was to recreate the fusion by targeting human *ETO* into the *Runx1* locus via homologous recombination in embryonic stem cells. In both cases, the heterozygous $Runx1^{+/RUNX1\text{-}ETO}$

mutation proved to be lethal at E13.5 [46, 67]. The animals had pale livers, massive hemorrhage within ventricles of the central nervous system and the pericardial cavity, and a complete absence of definitive fetal liver-derived hematopoiesis [46, 67]. This was a phenocopy of the *Runx1* null mouse [45, 62], providing genetic evidence that RUNX1/ETO acts as a dominant negative inhibitor of RUNX1. The utility of these mutant mice for studies of the mechanism of leukemogenesis is obviously limited because of the embryonic lethal phenotype.

A second-generation conditional model overcame this obstacle. A "floxed" transcriptional stop cassette, placed immediately 5′ of the knocked in RUNX1/ETO cDNA, prevented expression of the fusion protein during development [26]. The heterozygous mice developed normally and were crossed with an Mx1-Cre transgenic line. Treatment of Cre+ $Runx1^{+/Floxed}$ mice with poly I:C activated RUNX1/ETO expression in adult bone marrow. These mice have only subtle hematopoietic defects, reinforcing the paradigm that RUNX1/ETO expression is not sufficient to induce AML. Despite the importance of this model, its complexity (two transgenes, mixed strain backgrounds, interferon induction, Cre-mediated recombination) limit its usefulness as a platform to test potential mutations that can cooperate with RUNX1/ETO to induce AML. In another model, targeted expression of RUNX1/ETO under the control of the *Sca1* promoter avoided embryonic lethality (possibly because both *Runx1* alleles remained intact) and produced MPD at high penetrance [18]. In an elegant refinement of the targeted approach, compound heterozygous mice were generated containing floxed *Runx1* and *Eto* alleles. These mice were crossed to a nestin-Cre transgenic line, resulting in rearrangement between *Runx1* and *Eto* at a frequency of $<10^{-4}$ cells in the brain, kidney, and heart [10]. Although this technical tour de force demonstrated that interchromosomal translocation can be driven in vivo, lack of bone marrow RUNX1/ETO expression limits its utility for the study of leukemogenesis.

Transgenic models of RUNX1/ETO have also been generated using the hMRP8 promoter and a tetracycline inducible system [52, 69]. These animals developed normally with no detectable hematopoietic phenotype, as expected given the necessity of cooperating events for RUNX1/ETO-associated AML. Interpretation of these random integration transgenic models must take into account integration site and copy number effects on transgene expression [21] and the extent to which the heterologous promoter mirrors activity of the endogenous locus.

Modeling "Second Hits"

Genetically engineered mouse models of human AML illustrate that this is not a "single hit" disease. Much of the attention in AML research currently centers on identifying the second and subsequent genetic "hits" that are responsible of disease initiation, progression, relapse, and resistance to treatment. Mouse models provide an extremely useful platform to identify and validate cooperating genetic events that can lead to the development of AML. In general, the approaches involve either inducing second hits or waiting for them to occur spontaneously. In an example of the latter approach, hCG-PML/RARA transgenic mice uniformly develop MPD that evolves to AML with partial penetrance (10–30%) and long latency (>200 days) [22], implying that spontaneous mutations are required for the transition from MPD to AML. Spectral karyotyping of these leukemias revealed an interstitial deletion in chromosome 2 as a common acquired abnormality [71]. The minimally deleted segment contained *Pu.1*, a member of the Ets family of transcription factors, raising the possibility that PU.1 acted as a tumor suppressor to restrict progression from MPD to AML in hCG-PML/RARA transgenic mice. This hypothesis was confirmed by crossing the PML/RARA transgene into a PU.1 haploinsufficient background, resulting in a >5-fold increase in AML incidence [61].

Second hits can be actively induced by combining the approaches discussed above (e.g., retroviral transduction/transplantation of transgenic cells) or by performing genome-wide mutagenesis to screen broadly for potential cooperating mutations. The most commonly employed mutagenesis techniques are the subject of the remainder of this review.

Chemical Mutagenesis

Chemical mutagens are used to create sporadic mutations within the mouse genome. Alkylating agents such as *N*-ethyl-*N*-nitrosurea (ENU) are most commonly employed. ENU is the most potent known mutagen in mice [54]. This agent has been shown to induce mutations that cooperate with the core binding factor rearrangements, resulting in myeloid malignancies [26, 69, 12].

The use of ENU to induce these second hits is problematic because ENU alone causes tumors in most inbred strains of mice. Susceptibility to ENU-induced tumors is highly strain dependent [17]. The most common hematopoietic tumor induced by ENU is thymic lymphoma [17]. Careful histologic and flow cytometric analysis can readily distinguish this "noise" from the AML "signal."

In general, complementation by ENU mutagenesis is a proof of principle experiment designed to demonstrate that a mutation of interest can participate in the development of AML. Identification of the cooperating mutations induced by ENU is difficult because of the non-selectivity of the agent. There are a few known "hot spots" for ENU mutagenesis (e.g., *Kras*) [42]. Until recently, the search for mutated alleles has been restricted to a candidate gene resequencing approach. Unbiased, genome-wide mutational screens may soon be feasible, as sequencing costs on "Next Generation" platforms are falling rapidly.

Insertional Mutagenesis

Endogenous murine retroviruses are oncogenic owing to insertional activation of oncogenes or inactivation of tumor supressors. In the recombinant inbred strains AKXD-23 and BXH-2, somatic mutagenesis by endogenous murine leukemia viruses (MuLV) leads to AML with nearly complete penetrance by 1 year of age [4, 44]. Large-scale integration site cloning and resequencing efforts using these and other strains have resulted in the identification of commonly targeted loci that represent potential leukemia pathogenesis genes [40]. These data have been compiled in the Retroviral Tagged Cancer Gene Database at: http://RTCGD.ncifcrf.gov.

These recombinant inbred strains are powerful tools to employ in the search for cooperating mutations in AML. First, multiple integration sites found within the same AML specimen provide strong a priori evidence that the tagged genes cooperate to cause AML. In a striking example that recapitulated the events leading to the development of leukemia in the X-SCID patients cited above, a mouse leukemia sample was identified that contained integrations at *Lmo2* and *Il2rg*, the therapeutic gene contained in the clinical gene transfer vector [15].

To search for genes that cooperate with a specific mutation, transgenic lines are backcrossed into the BXH-2 background. The tumor latency, penetrance, and common integration sites are then compared to data from control wild-type BXH-2 mice. This strategy has been employed to screen for cooperating genes in several models, including mice transgenic for CBFB/MYH11, NUP98/HOXA9, or haploinsufficient for *Nf1* or *Runx1* [11, 66, 29, 6].

Radiation-Induced mutagenesis

Radiation is leukemogenic in humans and mice [13, 58]. This model system has been used for more than 25 years to search for leukemia genes in mice. After a single 3–5 Gy dose of gamma radiation, susceptible mice develop AML with a latency of >300 days. This phenotype is highly strain dependent, ranging from <10% in most strains to 20–30% in CBA/Ca, C3H/HeJ, and SJL/J mice [51]. Steroid pretreatment increases penetrance to 50–70%, possibly by reducing the frequency of radiation-induced thymic lymphomas [51]. Genetic analysis of these AML mice demonstrated loss of heterozygosity at the band D region on chromosome 2 [24]. In an extraordinary convergence of two independent lines of research, this interval overlapped the region commonly deleted in hCG-PML/RARA leukemias. Indeed, the radiation-induced leukemias with del(2) also lack one copy of *PU.1* and frequently mutate the remaining allele [14].

Despite this track record, it is time consuming and difficult to evaluate genetic changes induced by radiation. Unlike retroviral integration, radiation does not leave a footprint that can be used to identify a mutation. Classically, regions with loss of heterozygosity were mapped using polymorphic microsatellite markers in an F1 intercross. Spectral karyotyping and comparative genomic hybridization are high-throughput, high-resolution techniques that can be performed in inbred strains. The relatively low penetrance and long latency of AML add to the difficulty in working with this model, requiring large cohorts and long follow-up, which is both time consuming and costly.

Conclusions

A large international effort has generated many useful murine models of human AML. Because of space constraints, only representative examples could be cited here. It is likely that mice will remain the dominant model organism for AML research in the foreseeable future.

An important theme that has emerged from these studies is that murine genetic background is a variable that significantly influences leukemia susceptibility and phenotype. The Mouse Phenome Project is an international collaboration that seeks to collect and disseminate data on diverse phenotypes in mice, including cancer susceptibility [8]. This information can be correlated with a growing database of strain-specific polymorphisms to inform our understanding the genetic basis of complex diseases such as AML.

Acknowledgments This study was supported by a grant from the G&P Foundation (T.A.G.). Jan Nolta, Michael Tomasson, Matthew Walter, and Tim Ley provided valuable comments and suggestions.

References

1. Ailles LE, Gerhard B, Kawagoe H, Hogge DE. Growth characteristics of acute myelogenous leukemia progenitors that initiate malignant hematopoiesis in nonobese diabetic/severe combined immunodeficient mice. *Blood*. 1999;94(5): 1761–1772.
2. Bakker AB, van den Oudenrijn S, Bakker AQ, et al. C-type lectin-like molecule-1: a novel myeloid cell surface marker associated with acute myeloid leukemia. *Cancer Res*. 2004;64(22):8443–8450.
3. Barabe F, Kennedy JA, Hope KJ, Dick JE. Modeling the initiation and progression of human acute leukemia in mice. *Science*. 2007;316(5824):600–604.
4. Bedigian HG, Johnson DA, Jenkins NA, Copeland NG, Evans R. Spontaneous and induced leukemias of myeloid origin in recombinant inbred BXH mice. *J Virol*. 1984;51(3):586–594.
5. Berman JN, Kanki JP, Look AT. Zebrafish as a model for myelopoiesis during embryogenesis. *Exp Hematol*. 2005;33(9):997–1006.

6. Blaydes SM, Kogan SC, Truong BT, et al. Retroviral integration at the Epi1 locus cooperates with Nf1 gene loss in the progression to acute myeloid leukemia. *J Virol.* 2001;75(19):9427–9434.
7. Blunt T, Gell D, Fox M, et al. Identification of a nonsense mutation in the carboxyl-terminal region of DNA-dependent protein kinase catalytic subunit in the scid mouse. *Proc Natl Acad Sci USA.* 1996;93(19):10285–10290.
8. Bogue M. Mouse Phenome Project: understanding human biology through mouse genetics and genomics. *J Appl Physiol.* 2003;95(4):1335–1337.
9. Bonnet D, Dick JE. Human acute myeloid leukemia is organized as a hierarchy that originates from a primitive hematopoietic cell. *Nat Med.* 1997. 3(7):730–737.
10. Buchholz F., Refaeli Y, Trumpp A, Bishop JM. Inducible chromosomal translocation of AML1 and ETO genes through Cre/loxP-mediated recombination in the mouse. *EMBO Rep.* 2000;1(2):133–139.
11. Castilla LH, Perrat P, Martinez NJ, et al. Identification of genes that synergize with Cbfb-MYH11 in the pathogenesis of acute myeloid leukemia. *Proc Natl Acad Sci USA.* 2004;101(14):4924–4929.
12. Castilla LH, Garrett L, Adya N, et al. The fusion gene Cbfb-MYH11 blocks myeloid differentiation and predisposes mice to acute myelomonocytic leukaemia. *Nat Genet.* 1999;23(2):144–146.
13. Cleary HJ, Wright E, Plumb M. Specificity of loss of heterozygosity in radiation-induced mouse myeloid and lymphoid leukaemias. *Int J Radiat Biol.* 1999;75(10):1223–1230.
14. Cook WD, McCaw BJ, Herring C, et al. PU.1 is a suppressor of myeloid leukemia, inactivated in mice by gene deletion and mutation of its DNA binding domain. *Blood.* 2004;104(12):3437–3444.
15. Dave UP, Jenkins NA, Copeland NG. Gene therapy insertional mutagenesis insights. *Science.* 2004;303(5656):333.
16. Du Y, Jenkins NA, Copeland NG. Insertional mutagenesis identifies genes that promote the immortalization of primary bone marrow progenitor cells. *Blood.* 2005;106(12):3932–3939.
17. Fenske TS, McMahon C, Edwin D, et al. Identification of candidate alkylator-induced cancer susceptibility genes by whole genome scanning in mice. *Cancer Res.* 2006;66(10). In press.
18. Fenske TS, Pengue G, Mathews V, et al. Stem cell expression of the AML1/ETO fusion protein induces a myeloproliferative disorder in mice. *Proc Natl Acad Sci USA.* 2004;101(42):15184–15189.
19. Feuring-Buske M, Hogge DE. Hoechst 33342 efflux identifies a subpopulation of cytogenetically normal CD34(+)CD38(-) progenitor cells from patients with acute myeloid leukemia. *Blood.* 2001;97(12):3882–3889.
20. Goodell MA, Brose K, Paradis G, Conner AS, Mulligan RC. Isolation and functional properties of murine hematopoietic stem cells that are replicating in vivo. *J Exp Med.* 1996;183(4):1797–1806.
21. Graubert TA, Hug BA, Wesselschmidt R, et al. Stochastic, stage-specific mechanisms account for the variegation of a human globin transgene. *Nucleic Acids Res.* 1998;26(12):2849–2858.
22. Grisolano JL, Wesselschmidt RL, Pelicci PG, Ley TJ. Altered myeloid development and acute leukemia in transgenic mice expressing PML-RAR alpha under control of cathepsin G regulatory sequences. *Blood.* 1997;89(2):376–387.
23. Guzman ML, Rossi RM, Karnischky L, et al. The sesquiterpene lactone parthenolide induces apoptosis of human acute myelogenous leukemia stem and progenitor cells. *Blood.* 2005;105(11):4163–4169.
24. Hayata I, Seki M, Yoshida K, et al. Chromosomal aberrations observed in 52 mouse myeloid leukemias. *Cancer Res.* 1983;43(1):367–373.
25. Hess DA, Meyerose TE, Wirthlin L, et al. Functional characterization of highly purified human hematopoietic repopulating cells isolated according to aldehyde dehydrogenase activity. *Blood.* 2004;104(6):1648–1655.

26. Higuchi M, O'Brien D, Kumaravelu P, Lenny N, Yeoh EJ, Downing JR. Expression of a conditional AML1-ETO oncogene bypasses embryonic lethality and establishes a murine model of human t(8;21) acute myeloid leukemia. *Cancer Cell.* 2002;1(1):63–74.

27. Huntly BJ, Shigematsu H, Deguchi K, et al. MOZ-TIF2, but not BCR-ABL, confers properties of leukemic stem cells to committed murine hematopoietic progenitors. *Cancer Cell.* 2004;6(6):587–596.

28. Ishikawa F, Yoshida S, Saito Y, et al. Chemotherapy-resistant human AML stem cells home to and engraft within the bone-marrow endosteal region. *Nat Biotechnol.* 2007;25(11):1315–1321.

29. Iwasaki M, Kuwata T, Yamazaki Y, et al. Identification of cooperative genes for NUP98-HOXA9 in myeloid leukemogenesis using a mouse model. *Blood.* 2005;105(2):784–793.

30. Jordan CT, Upchurch D, Szilvassy SJ, et al. The interleukin-3 receptor alpha chain is a unique marker for human acute myelogenous leukemia stem cells. *Leukemia.* 2000;14(10):1777–1784.

31. Jung SH, Evans CJ, Uemura C, Banerjee U. The Drosophila lymph gland as a developmental model of hematopoiesis. *Development.* 2005;132(11):2521–2533.

32. Kelly LM, Liu Q, Kutok JL, Williams IR, Boulton CL, Gilliland DG. FLT3 internal tandem duplication mutations associated with human acute myeloid leukemias induce myeloproliferative disease in a murine bone marrow transplant model. *Blood.* 2002;99(1):310–318.

33. Kelly LM, Kutok JL, Williams IR, et al. PML/RARalpha and FLT3-ITD induce an APL-like disease in a mouse model. *Proc Natl Acad Sci USA.* 2002;99(12):8283–8288.

34. Kohn DB, Sadelain M, Glorioso JC. Occurrence of leukaemia following gene therapy of X-linked SCID. *Nat Rev Cancer.* 2003;3(7):477–488.

35. Kollet O, Peled A, Byk T, et al. beta2 microglobulin-deficient (B2m(null)) NOD/SCID mice are excellent recipients for studying human stem cell function. *Blood.* 2000;95(10):3102–3105.

36. Kroon E, Thorsteinsdottir U, Mayotte N, Nakamura T, Sauvageau G. NUP98-HOXA9 expression in hemopoietic stem cells induces chronic and acute myeloid leukemias in mice. *Embo J.* 2001;20(3):350–361.

37. Langenau DM, Traver D, Ferrando AA, et al. Myc-induced T cell leukemia in transgenic zebrafish. *Science.* 2003;299(5608):887–890.

38. Lapidot T, Sirard C, Vormoor J et al. A cell initiating human acute myeloid leukaemia after transplantation into SCID mice. *Nature.* 1994;367(6464):645–648.

39. Larochelle A, Vormoor J, Hanenberg H, et al. Identification of primitive human hematopoietic cells capable of repopulating NOD/SCID mouse bone marrow: implications for gene therapy. *Nat Med.* 1996;2(12):1329–1337.

40. Li J, Shen H, Himmel KL, et al. Leukaemia disease genes: large-scale cloning and pathway predictions. *Nat Genet.* 1999;23(3):348–353.

41. Lumkul R, Gorin NC, Malehorn MT, et al. Human AML cells in NOD/SCID mice: engraftment potential and gene expression. *Leukemia.* 2002;16(9):1818–1826.

42. Malkinson AM. Molecular comparison of human and mouse pulmonary adenocarcinomas. *Exp Lung Res.* 1998;24(4):541–555.

43. Mazurier F, Doedens M, Gan OI, Dick JE. Rapid myeloerythroid repopulation after intrafemoral transplantation of NOD-SCID mice reveals a new class of human stem cells. *Nat Med.* 2003;9(7):959–963.

44. Mucenski ML, Taylor BA, Ihle JN, et al. Identification of a common ecotropic viral integration site, Evi-1, in the DNA of AKXD murine myeloid tumors. *Mol Cell Biol.* 1988;8(1):301–308.

45. Okuda T, van Deursen J, Hiebert SW, Grosveld G, Downing JR. AML1, the target of multiple chromosomal translocations in human leukemia, is essential for normal fetal liver hematopoiesis. *Cell.* 1996;84(2):321–330.

46. Okuda T, Cai Z, Yang S, et al. Expression of a knocked-in AML1-ETO leukemia gene inhibits the establishment of normal definitive hematopoiesis and directly generates dysplastic hematopoietic progenitors. *Blood*. 1998;91(9):3134–3143.

47. Pearce DJ, Taussig D, Zibara K, et al. AML engraftment in the NOD/SCID assay reflects the outcome of AML: implications for our understanding of the heterogeneity of AML. *Blood*. 2006;107(3):1166–1173.

48. Pearce DJ, Taussig D, Simpson C, et al. Characterization of cells with a high aldehyde dehydrogenase activity from cord blood and acute myeloid leukemia samples. *Stem Cells*. 2005;23(6):752–760.

49. Prochazka M, Gaskins HR, Shultz LD, Leiter EH. The nonobese diabetic scid mouse: model for spontaneous thymoma genesis associated with immunodeficiency. *Proc Natl Acad Sci USA*. 1992;89(8):3290–3294.

50. Ren R. Modeling the dosage effect of oncogenes in leukemogenesis. *Curr Opin Hematol*. 2004;11(1):25–34.

51. Resnitzky P, Estrov Z, Haran-Ghera N. High incidence of acute myeloid leukemia in SJL/J mice after X-irradiation and corticosteroids. *Leuk Res*. 1985;9(12):1519–1528.

52. Rhoades KL, Hetherington CJ, Harakawa N, et al. Analysis of the role of AML1-ETO in leukemogenesis, using an inducible transgenic mouse model. *Blood*. 2000;96(6):2108–2115.

53. Rombouts WJ, Martens AC, Ploemacher RE. Identification of variables determining the engraftment potential of human acute myeloid leukemia in the immunodeficient NOD/SCID human chimera model. *Leukemia*. 2000;14(5):889–897.

54. Russell WL, Kelly EM, Hunsicker PR, Bangham JW, Maddux SC, Phipps EL. Specific-locus test shows ethylnitrosourea to be the most potent mutagen in the mouse. *Proc Natl Acad Sci USA*. 1979;76(11):5818–5819.

55. Shultz LD, Banuelos SJ, Leif J, et al. Regulation of human short-term repopulating cell (STRC) engraftment in NOD/SCID mice by host CD122+ cells. *Exp Hematol*. 2003;31(6):551–558.

56. Stover EH, Chen J, Lee BH, et al. The small molecule tyrosine kinase inhibitor AMN107 inhibits TEL-PDGFRbeta and FIP1L1-PDGFRalpha in vitro and in vivo. *Blood*. 2005;106(9):3206–3213.

57. Tomasson MH, Sternberg DW, Williams IR, et al. Fatal myeloproliferation, induced in mice by TEL/PDGFbetaR expression, depends on PDGFbetaR tyrosines 579/581. *J Clin Invest*. 2000;105(4):423–432.

58. Torok S, Borgulya G, Lobmayer P, Jakab Z, Schuler D, Fekete G. Childhood leukaemia incidence in Hungary, 1973-2002. Interpolation model for analysing the possible effects of the Chernobyl accident. *Eur J Epidemiol*. 2005;20(11):899–906.

59. van Rhenen A, van Dongen GA, Kelder A, et al. The novel AML stem cell associated antigen CLL-1 aids in discrimination between normal and leukemic stem cells. *Blood*. 2007;110(7):2659–2666.

60. Vidal M, Cagan RL. Drosophila models for cancer research. *Curr Opin Genet Dev*. 2006;16(1):10–16.

61. Walter MJ, Park JS, Ries RE, et al. Reduced PU.1 expression causes myeloid progenitor expansion and increased leukemia penetrance in mice expressing PML-RARalpha. *Proc Natl Acad Sci USA*. 2005;102(35):12513–12518.

62. Wang Q, Stacy T, Binder M, Marin-Padilla M, Sharpe AH, Speck NA. Disruption of the Cbfa2 gene causes necrosis and hemorrhaging in the central nervous system and blocks definitive hematopoiesis. *Proc Natl Acad Sci USA*. 1996;93(8):3444–3449.

63. Westervelt P, Lane AA, Pollock JL, et al. High-penetrance mouse model of acute promyelocytic leukemia with very low levels of PML-RARalpha expression. *Blood*. 2003;102(5):1857–1865.

64. Westervelt P, Ley TJ. Seed versus soil: the importance of the target cell for transgenic models of human leukemias. *Blood*. 1999;93(7):2143–2148.

65. Wulf GG, Wang RY, Kuehnle I, et al. A leukemic stem cell with intrinsic drug efflux capacity in acute myeloid leukemia. *Blood.* 2001;98(4):1166–1173.
66. Yamashita N, Osato M, Huang L, et al. Haploinsufficiency of Runx1/AML1 promotes myeloid features and leukaemogenesis in BXH2 mice. *Br J Haematol.* 2005;131(4): 495–507.
67. Yergeau DA, Hetherington CJ, Wang Q, et al. Embryonic lethality and impairment of haematopoiesis in mice heterozygous for an AML1-ETO fusion gene. *Nat Genet.* 1997;15(3):303–306.
68. Yilmaz OH, Valdez R, Theisen BK, et al. Pten dependence distinguishes haematopoietic stem cells from leukaemia-initiating cells. *Nature.* 2006;441(7092):475–482.
69. Yuan Y, Zhou L, Miyamoto T, et al. AML1-ETO expression is directly involved in the development of acute myeloid leukemia in the presence of additional mutations. *Proc Natl Acad Sci USA.* 2001;98(18):10398–10403.
70. Zhang S, Ramsay ES, Mock BA. Cdkn2a, the cyclin-dependent kinase inhibitor encoding p16INK4a and p19ARF, is a candidate for the plasmacytoma susceptibility locus, Pctr1. *Proc Natl Acad Sci USA.* 1998;95(5):2429–2434.
71. Zimonjic DB, Pollock JL, Westervelt P, Popescu NC, Ley TJ. Acquired, nonrandom chromosomal abnormalities associated with the development of acute promyelocytic leukemia in transgenic mice. *Proc Natl Acad Sci USA.* 2000;97(24): 13306–13311.

Apoptosis in Leukemias: Regulation and Therapeutic Targeting

Ismael Samudio, Marina Konopleva, Bing Carter, and Michael Andreeff

Abstract Nearly 25 years after the seminal publication of John Foxton Kerr that first described apoptosis, the process of regulated cell death, our understanding of this basic physiological phenomenon is far from complete [39]. From cardiovascular disease to cancer, apoptosis has assumed a central role with broad ranging therapeutic implications that depend on a complete understanding of the molecular events involved in orchestrating cellular demise. More than 145,301 (as of April 2008) published works on this subject have increased our understanding of this process, yet have also identified an incredibly complex regulatory system that is critical for development and is at the core of many diseases, challenging scientists and clinicians to step into its molecular realm and modulate its circuitry for therapeutic purposes. This chapter will review our understanding of the molecular circuitry that controls apoptosis in leukemia and the pharmacological manipulations of this pathway that may yield therapeutic benefit.

Introduction: Apoptosis pathways

Apoptosis can be orchestrated from within the cell (intrinsic), or from the plasma membrane (extrinsic). Phenotypically, the end results of activating the intrinsic or extrinsic pathways of cell death (plasma membrane blebbing, mitochondrial dysfunction, and chromatin lysis) are indistinguishable; however, these end points are achieved via distinct mechanisms. The extrinsic pathway (Fig. 1) is activated by ligation of plasma membrane receptors such as FAS, tumor necrosis factor receptor, and TRAIL receptors [92]. The ligands that activate these receptors comprise a superfamily of secreted peptides that may

I. Samudio (✉)
Section of Molecular Hematology and Therapy, Department of Stem Cell
Transplantation and Cellular Therapy, The University of Texas MD Anderson Cancer
Center, Houston, Texas, USA
e-mail: mandreef@mdanderson.org

L. Nagarajan (ed.), *Acute Myelogenous Leukemia*,
Cancer Treatment and Research 145, DOI 10.1007/978-0-387-69259-3_12,
© Springer Science+Business Media, LLC 2010

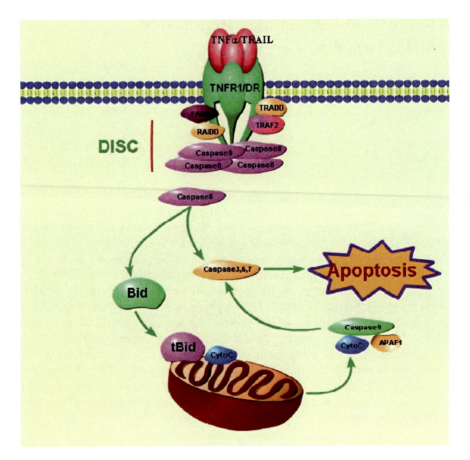

Fig. 1 The extrinsic apoptotic pathway

act in an autocrine, or paracrine fashion, and upon binding their cognate receptor, induce a conformational change that recruits various components of the death inducing signaling complex (DISC) resulting in the activation and processing of the initiator caspases, caspase 8 and/or caspase 10 [21, 64, 79]. Within the DISC, initiator caspases are activated via the "induced proximity" model [70, 85]. In this model, catalytic activation of caspase 8 or caspase 10 occurs by dimerization of the inactive enzymes facilitated by small-scale rearrangements induced by the DISC. The initiator caspases once activated can autocatalytically process themselves, as well as cleave and activate executioner caspases such as caspase 3, caspase 6, or caspase 7. Executioner caspases in turn proteolyze over 100 different proteins involved in various structural and metabolic processes resulting in the characteristic morphological changes associated with apoptosis.

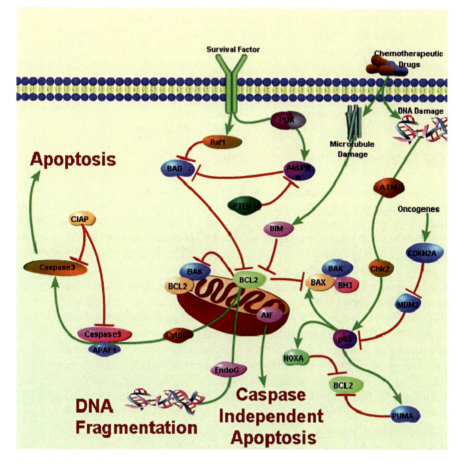

Fig. 2 The intrinsic apoptotic pathway

In contrast to the extrinsic pathway, the intrinsic pathway is activated by mitochondrial dysfunction, independent of alterations in the plasma membrane (Fig. 2). Initiation of the intrinsic pathway involves an increase in the permeability of the outer mitochondrial membrane (MOMP), which facilitates the release of death promoting proteins such as cytochrome C, AIF, EndoG, Omi, and DIABLO that reside in the intermembrane space of this organelle [29]. MOMP appears to be mediated by the formation of a supramolecular pore composed of several mitochondrial proteins like VDAC, Cyclophilin D, and ANT [3, 109]. Of the death-promoting proteins released into the cytosol via MOMP, cytochrome C is essential for the activation of the initiator caspase 9, whereas Omi and DIABLO serve to facilitate the activation of the executioner caspases by caspase 9 [84, 97]. Activation of caspase 9 requires the formation of a multiprotein complex called the apoptosome. The apoptosome consists of cytochrome C, Apaf-1, caspase 9, and ATP, and therefore depends on the intracellular energy stores of the cell [62]. Similar to the DISC-induced activation of caspase 8, caspase 9 is also activated by "induced proximity" within the apoptosome [80]. Interestingly AIF (apoptosis-inducing factor) and EndoG can induce caspase-independent apoptotic cell death by directly mediating chromatinolysis [60, 63], an effect that allows mitochondrial dysfunction to orchestrate cell death even in the absence of caspase activation

The Extrinsic Pathway in Leukemia

The death-inducing ligand TRAIL has received considerable attention as a therapeutic strategy for the treatment of human cancer because of its limited toxicity to normal tissues [116]. Interestingly, albeit the majority of leukemia cells express TRAIL receptors, these samples are notoriously resistant to apoptosis induction by TRAIL [65, 83, 95, 115]. The reasons for this inherent resistance to TRAIL-induced cell death are not completely understood, but may depend on the presence of decoy receptors that bind TRAIL but are not able to induce DISC formation, the expression of FLIP (an inhibitor of caspase 8 activation), or overactivation of NFκB – a near universal consequence of DR ligation [65, 95, 115]. Nevertheless, recent work has demonstrated that a variety of agents can sensitize leukemia cells to apoptosis induction by TRAIL, either by directly affecting the recruitment and assembly of DRs on the cell surface, increasing the expression of DRs, or decreasing the levels of FLIP [91, 99]. One such agent is the novel triterpenoid CDDO. CDDO has been shown to sensitize AML cell lines and primary samples to the cytotoxic effects of TRAIL by decreasing the levels of FLIP [99]. On the other hand, several HDAC inhibitors in clinical use have been shown to sensitize leukemia cells by facilitating the formation of an active DISC and increasing expression of the DRs [91]. Finally, since DR5 is a p53-responsive gene, many forms of genotoxic chemotherapy sensitize leukemia cells with an intact p53 pathway to TRAIL-induced apoptosis via increased expression of DR5 [36, 37, 61]. Currently, two pharmaceutical companies have therapeutic activators of the extrinsic pathway, Genentech with TRAIL itself and Human Genome Sciences with agonistic antibodies to the DRs that mimic the action of TRAIL. First results with agonistic antibodies are encouraging.

The Intrinsic Pathway in Leukemia

The intrinsic apoptotic pathway is critically regulated by pro- and anti-apoptotic members of the Bcl-2 family. Anti-apoptotic members of the Bcl-2 family contain four conserved Bcl-2-homology (BH) domains, designed BH1–BH4, whereas pro-apoptotic members of this family only contain BH1–BH3 [19]. In addition, there is a subset of pro-apoptotic Bcl-2 family members that contain only the BH3 domain and are thus termed "BH3-only" proteins. Of these BH3-only proteins, activators, like Bid and Bim are required to directly activate the pro-apoptotic function of the BH1–BH3 proteins Bax and Bak, whereas sensitizers like Bad, Noxa, or Puma displace Bim or Bid from antiapoptotic Bcl-2 proteins facilitating the subsequent activation of Bax and Bak making BH3 interactions a finely tuned rheostat for apoptotic control [28]. During apoptosis, Bax and Bak oligomerize and form supramolecular pores in the outer mitochondrial membrane leading to the release of pro-apoptotic proteins like cytochrome c and AIF [29]. Interestingly, the expression, intracellular localization, and function of pro- and anti-apoptotic Bcl-2 family members can vary widely depending on cell context and may be temporally regulated throughout development. For instance, the antiapoptotic Bcl-2 protein Mcl-1 has been shown to be a critical regulator of early hematopoietic development [75, 76], and down-regulation of Mcl-1 levels has been shown to be an effective strategy to decrease the viability of multiple myeloma cells [25]. Similarly, the interactions between pro- and anti-apoptotic Bcl-2 proteins display particular selectivity such that the BH3-only proteins Bim, Bmf, and Puma can antagonize Bcl-X_L and Mcl-1, whereas Bid, Bik, and Bad can oppose the anti-apoptotic function of Bcl-X_L, but not of Mcl-1[56]. This is exemplified by the differences in the subcellular location of Bim, which is sequestered to the dynein motor complex and translocated to the mitochondria upon apoptotic stimuli in some experimental systems [81]. However, Bim is already localized at the mitochondria, complexed with anti-apoptotic Bcl-2 family proteins in hematopoietic cells [122], including AML lines [31]. These observations clearly suggest that the mitochondrial levels of expression of various pro- and anti-apoptotic Bcl-2 family members are important determinants of sensitivity of AML cells to apoptotic insults.

Most genotoxic chemotherapeutics activate the intrinsic apoptotic pathway via activation of p53 signaling resulting in the expression of pro-apoptotic target genes such as Noxa, Puma, and Bax, as well as the accumulation of cytoplasmic p53 which functions outside the nucleus as a pro-apoptotic protein by antagonizing the function of Bcl-X_L and directly activating bax [22] or by binding to manganese superoxide dismutase and promoting the formation of reactive oxygen species (ROS) [121]. Additionally, alkylating agents can directly promote MOMP by oxidation of critical cysteine residues of components of the supramolecular pore, in particular ANT [40, 67]. On the other hand, the novel triterpenoid CDDO and its derivatives have been shown to directly induce MOMP, independently of the formation of a supramolecular pore (Fig. 3), possibly by disrupting the impermeability of the inner mitochondrial membrane which leads to matrix swelling with subsequent rupture of the outer membrane [5, 86]. A role of ceramide has also been suggested [49].

These compounds have shown promise in numerous preclinical investigations [54, 102] and in clinical trial in leukemias [53].

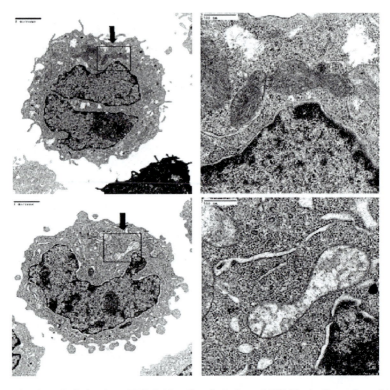

Fig. 3 The CDDO derivative, CDDO-Me, directly induces MOMP via disruption of inner mitochondrial membrane impermeability. *Top* – U937 cells treated with 0.1% DMSO. *Bottom* – U937 cells treated with 300 nM CDDO-Me for 3 h. Images on *left* were captured at 10,000× direct magnification (bar = 2 μm); images on *right* correspond to 50,000× direct magnification of *boxed area* (bar = 500 nm)

Endogenous Inhibitors of Caspase Activity

The *i*nhibitors of *a*poptosis *p*roteins (IAPs) are family of proteins that suppress both mitochondrial and death receptor-mediated apoptosis primarily by inhibiting caspases. Currently, eight IAP members have been identified in humans. Among them, XIAP is the most potent cellular caspase inhibitor. XIAP directly binds to caspase-9 by BIR3 domain and caspases-3 and -7 by BIR1 and BIR2 domains (Deveraux QL, Reed JC. *EMBO J*. 1999;18:5242–5252) and inhibits their activities. The antiapoptotic function of XIAP is supported by multiple in vitro and in vivo studies demonstrating that overexpression of XIAP confers resistance to both mitochondrial and death receptor pathway activation (Wilkinson JC, Duckett CS. *Mol Cell Biol*. 2004;24:7003–7014; Berezovskaya O, Schimmer AD, et al. *Cancer Res*. 2005;65:2378–2386; Tong QS, Zheng LD.

Cancer Gene Ther. 2005;12:509–514). Decreasing XIAP levels with siRNA or antisense oligonucleotide restores chemosensitivity to malignant cells. We have demonstrated in HL-60 cells that knockdown XIAP with antisense oligonucleotide induces caspase activation, apoptosis, and enhances Ara-C-induced cell death [17]. Phenylurea-based small-molecule XIAP inhibitors, which bind to BIR2 domain of XIAP, induce apoptosis in various AML cell lines and patient blasts in a caspase-3-dependent manner [11]. Triptolide, a diterpenoid isolated from a Chinese herb, inhibits XIAP and potently induces AML cell death [13, 15]. Interestingly, XIAP knockout mice looked normal and showed no apparent defect in apoptosis (Harlin H, Reffey SB, et al. *Mol Cell Biol.* 2001;1:3604–3608).

Overexpression of XIAP has been reported by several groups [1, 12, 16, 17, 34, 103, 104]. In addition, we have found that XIAP expression is higher in both AML stem (CD34 + /CD38-) and progenitor (CD34 + /CD38 +) cells than in their normal counterparts by Taq-Man RT-PCR (Andreeff M et al., unpublished results). Clinically, increased XIAP expression is correlated with decreased survival in AML [103, 104]. We have demonstrated that the expression of XIAP is induced by cytokines through the MAPK and the PI3K pathways in AML [12, 16, 17] suggesting that hematopoietic growth factors promote AML cell growth and survival at least in part through upregulating XIAP protein expression. Nuclear factor-κB (NF-κB) also participates in the upregulation of XIAP levels in various malignancies [98]. In AML, NF-κB is constitutively activated [4, 8, 30] which undoubtedly contributes to XIAP overexpression.

Survivin, another member of the IAP family, has an interesting dual function as a caspase inhibitor and as a component of the apoptotic spindle. Inhibition of survivin by survivin AS ODN in leukemia cells results in G_2M block, followed by apoptosis. In CML, survivin levels are entirely dependent on bcr-abl signaling: bcr-abl inhibition with imatinib abrogates survivin expression, while in imatinib-resistant CML there is no effect [13, 15]. In this setting, survivin AS ODN induces pronounced apoptosis [13, 15]. Clinical trials with survivin AS ODN are in progress.

Caspase-Independent Cell Death: The Role of Apoptosis-Inducing Factor (AIF)

Albeit caspases are central players during apoptosis induction, many agents induce apoptosis-like cell death even in the absence of caspase activation or in the presence of pharmacological inhibition of these thiol proteases. In fact, we have reported that in AML cell lines treatment with traditional chemotherapeutic agents like Ara-C, doxorubicin, vincristine, and paclitaxel induce caspase activation and apoptosis, but pharmacological blockade of caspase activity does not prevent cell death induced by any of these agents and did not inhibit, or only partially inhibited, mitochondrial release of apoptosis-inducing factor (AIF) and loss of mitochondrial membrane potential [12, 17]. Caspase inhibition also did not protect AML blasts from chemotherapy-induced cell death in vitro. These results suggest that cytotoxic insults that promote the release of AIF from mitochondria may not require caspase activation to induce cell death. AIF is a bifuctional apoptogenic flavoprotein with NADH oxidase activity resides in the intermembrane space of the mitochondria in healthy cells and has a vital function in oxidative phosphorylation (OXPHOS), possibly due to indirect modulation of OXPHOS complex I [63, 107]. AIF is synthesized in the cytosol as a 67 kDa precursor molecule with a predicted mitochondrial localization sequence (MLS) in its N-terminal prodomain. AIF is matured into a 57 kDa protein that lacks the first 100 amino acids and resides in a soluble form in the mitochondrial intermembrane space. In response to certain pro-apoptotic stimuli that elicit MOMP, AIF is released into the cytoplasm. AIF will then translocate into the nucleus and induce caspase-independent large-scale DNA fragmentation ($\sim$ 50 kbp) and chromatin condensation. AIF contains a DNA-binding domain that is necessary and sufficient to induce apoptosis [117]. In contrast, the NADH oxidase domain is not necessary for apoptotis induction, but is required for the vital role of AIF in modulating OXPHOS [107]. Interestingly, it has been reported that the release of AIF may require caspase activation under some conditions, suggesting that this protein may also amplify caspase-dependent apoptosis [112]. Notably, the synthetic triterpenoid CDDO promotes the release of AIF via perturbations of the permeability of the inner mitochondrial membrane that lead to caspase-independent MOMP [51], suggesting that agents that directly target the mito-chondrial membrane could provide therapeutic efficacy even in the context of caspase dysfunction or inhibition.

Targeting the p53 Pathway to Induce Apoptosis in Leukemia

The transcription factor and tumor suppressor p53 is the most frequently mutated gene in cancer [96]. Activation of the transcriptional activity of p53 by a variety of stress signals leads to the expression of genes, which may orchestrate growth arrest, apoptosis, or both [32]. In addition, p53 has also been shown to function outside the nucleus as a pro-apoptotic protein by antagonizing the function of Bcl-XL and directly activating bax [22] or by binding to manganese superoxide dismutase and promoting the formation of reactive oxygen species (ROS) [121]. The levels of p53 are controlled by the ubiquitin ligase MDM-2 which targets this tumor suppressor protein for pro-teosomal degradation [69]. Interestingly, mutations in p53 have been found in only $\sim$5% of patients with AML [33, 89], but it has been proposed that over-expression of MDM-2 may abrogate the function of p53 in this malignancy

[26, 90]. Indeed, MDM-2 overexpression was observed in up to 53% of leukemias and was associated with unfavorable cytogenetic characteristics and poor prognosis [7, 9, 26]. Recently, potent and selective small-molecule antagonists of MDM-2, Nutlins, have been reported to activate the p53 pathway in cancer cells leading to growth arrest and apoptosis in vivo and in vitro [108]. It has been reported that Nutlins potently induce p53-dependent apoptosis in primary leukemia samples and leukemia cell lines with wild-type p53 [43, 45]. Nutlin activity was positively correlated with baseline MDM2 levels suggesting that the inhibition of MDM-2 may offer considerable clinical benefit for the therapy of leukemia.

MDM-2 inhibition activates transcription-dependent and transcription-independent pro-apoptotic effects of p53 in a cell context-dependent manner [45]. For instance, confocal microscopy analysis revealed that MDM-2 inhibition induced predominantly nuclear accumulation of p53 in the AML cell line MOLM13 (Fig. 2), and in these cells, apoptosis was greatly dependent on de novo protein synthesis. Conversely, MDM-2 inhibition induced mostly cytoplasmic accumulation of p53 in OCI-AML3 cells (Fig. 4), and in these cells inhibition of de novo protein synthesis did not markedly abrogate apoptosis. Nevertheless, in both cell types, induction of the p53 targets Noxa and PUMA, as well as complex formation between p53 and Bcl-xL was observed supporting the notion that p53 modulates the Bcl-2 apoptotic rheostat directly, by behaving as a BH3-only protein, and indirectly by increasing the levels of sensitizer BH3-only proteins. Interestingly, Deng et al. have reported that p53 can directly bind to Bcl-2 and induce a conformational change that exposes the

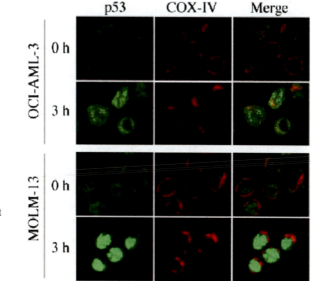

Fig. 4 Subcellular distribution of nutlin 3a-induced p53 is cell context regulated. COX-IV is a mitochondrial OXPHOS protein used to delineate cytoplasmic localization of p53

BH3 domain of this protein, converting Bcl-2 into a pro-apoptotic molecule, adding yet another mechanism by which p53 modulates the Bcl-2 apoptotic rheostat [24]. HDM4 inhibits p53 transcriptionally and can either be inhibited specifically [78, 110] or non-specifically, as it has been shown to be down-regulated by genotoxic stress including chemotherapy [66].

Nutlins potently synergized with doxorubicin, cytarabine, and fludarabine to induce apoptosis in primary leukemia samples [43, 45], suggesting the possi-bility that potentiation of p53-mediated apoptosis may occur through an increase in steady-state p53 protein levels, as effected by nutlin 3a, in combina-tion with further activation of p53 signalling induced by traditional genotoxic chemotherapeutics. Combinations of nutlin with an aurora kinase inhibitor [46], a dual inhibitor of P13 kinase and mTOR ([48], and a CDK1 inhibitor [47] all point to potential clinical combination therapies. Alternatively, it has recently been reported that DNA damage directly activates the BH3-only activator protein Bid via ATM-dependent phosphorylation [38, 123] and decreases the levels of MDM-4 [20, 72], a transcriptional repressor of p53, suggesting that the synergistic nature of nutlin 3a and genotoxic agent combi-nations may be mediated directly at the level of the MOMP, or by enhancement of p53-dependent transcription. Notably, nutlins fail to inhibit the binding of p53 to MDM-4 [77], suggesting that neutralizing the function of this MDM-2 homolog would result in increased transcription-dependent pro-apoptotic activity of the nutlins.

Targeting the Bcl-2 Rheostat to Induce Apoptosis in Leukemia

Bcl-2 overexpression or unchanged levels of Bcl-2 after chemotherapy confers drug resistance to both solid tumors and hematological malignancies [6, 10, 68, 105]. Since the discovery of Bcl-2 in 1984 several approaches have been inves-tigated to decrease bcl-2 levels and increase the efficacy of chemotherapy including antisense oligonucleotides and small molecule BH3 mimetics. Small molecule BH3 mimetics bind to the hydrophobic pocket of antiapoptotic Bcl-2 proteins releasing pro-apoptotic Bcl-2 proteins, which in turn induce apoptosis in leukemia cell lines and primary samples via activation of the mitochondrial pathway of cell death [82]. The scientific literature has reported many naturally occurring as well as synthetic compounds that bind to Bcl-2 and antagonize its prosurvival function. Among these are antimycin A [106], epigallocatechin gallate (EGCG) [58], HA14-1 [111], vitamin E [93], gossypol [42], thiazolidene-diones [94], ABT-737 [50, 75], BH3I-1 and -2 [23], and GX015-070 [52, 82]. Albeit all these various compounds induce apoptosis, bind to Bcl-2, and dis-sociate BH3 peptides from Bcl-2, only ABT-737 demonstrates an absolute requirement for both Bax and Bak to induce apoptosis suggesting this agent induces apoptosis solely through perturbations in the Bcl-2 rheostat. Interest-ingly, ABT-737 was originally postulated to only induce apoptosis in combination with chemotherapy; however, early evidence suggested that this agent could induce apoptosis in a cell context-dependent manner, and this depended on the composition of the Bcl-2 rheostat – particularly the levels of BH3-only activator proteins [18, 73]. Additionally, our observations suggested that the activity of ABT-737 was minimized by the expression of Mcl-1, a finding that at second glance was not surprising since ABT-737 bound Bcl-2, Bcl-xL, and Bcl-w, but failed to interact with Mcl-1 or A1 [50]. This peculiar property resulted from the NMR-guided, structure-based design used to generate ABT-737 where the hydrophobic pocket of Bcl-2 served as the docking target for lead compounds. More interestingly, however, was our observation that ABT-737 also induced apoptosis in primary leukemia samples and cell lines at low nanomolar doses, and this was independent of Bim expression since siRNA ablation of this BH3-only activator failed to rescue leukemia cells from apoptosis induced by this agent [50]. Furthermore, in contrast to the findings in the original report by Oltersdorf et al., we observed that ABT-737 induced cytochrome c release from isolated leukemia cell mitochondria [50]. Taken together, the above observations suggest that leukemia mitochondria, but not mitochondrial from solid tumors, are poised to activate free Bax and Bak, even in the absence of BH3-only activator proteins, to induce apoptosis via MOMP.

Another peculiar property of ABT-737 was observed in the context of Bcl-2 phosphorylation. Studies in our laboratory have identified Bcl-2 phosphorylation in most primary AML samples, and this was associated with increased antiapoptotic potential and resistance to ABT-737-induced apoptosis. Since ERK is a Bcl-2 kinase, we hypothesized that the ERK and MEK1 inhibitor PD98059 would sensitize cells expressing phosphorylated Bcl-2 to the apoptotic effects of ABT-737. As predicted, PD98059 potently synergized with ABT-737 to induce apoptosis in leukemia cells. Unexpectedly, however, it was observed that PD98059 in addition to decreasing Bcl-2 phosphorylation also decreased the levels of Mcl-1, suggesting that multiple effects were responsible to the powerful synergy in between ABT-737 and ERK/MEK1 inhibition [50]. In support for the criticality of Mcl-1 in preventing ABT-737-induced apoptosis, it was observed that ABT-737 inhibited clonogenic growth in primary AML samples, and their relative sensitivity to this agent was highly correlated to Mcl-1, but not to Bcl-2 protein expression. Moreover, multicolor flow cytometric analysis of primary AML samples treated with ABT-737 or cytarabine revealed that ABT-737, but not cytarabine, preferentially induced apoptosis in the leukemia stem cell compartment – a critical requirement for maximal therapeutic efficacy of any drug (Fig. 5). In contrast, no inhibition was observed of colony formation by normal bone marrow cells treated with ABT-737, and this was associated with high expression of Mcl-1 in these cells. Clinically, this identifies leukemia as a bona fide tumor target for ABT-737, as long as Mcl-1 is not the dominant antiapoptotic Bcl-2 protein.

We have also demonstrated that the combination of nutlin 3a and ABT-737 synergistically induces mitochondrial apoptosis in leukemia cells, via in part an interplay of cell cycle modulation with the bcl-2 apoptotic rheostat [44]. While

Fig. 5 ABT-737, but not Ara-C, induces apoptosis in the CD34+/38-/123+ leukemia stem cell compartment

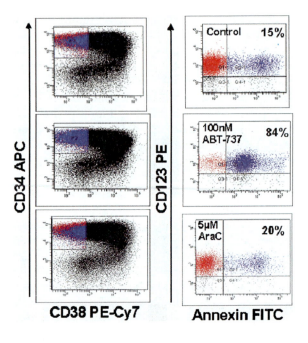

nutlin-3a induced p53-mediated apoptosis predominantly in S and G2/M cells, ABT-737 induced apoptosis predominantly in G1, the cell cycle phase with the lowest Bcl-2 protein levels and Bcl-2/Bax ratios. The complementary effects of nutlin-3a and ABT-737 in different cell cycle phases, as well as the induction of pro-apoptotic Bcl-2 family proteins in response to nutlin 3a, account in large part for their synergistic activity. Our data suggest that combined targeting of Mdm2 and Bcl-2 proteins could offer considerable therapeutic promise in AML [44]. ABT-737 lowers the apoptotic threshold and should therefore be synergistic or at least additive with other agents. One potential choice is the combination with Gleevec (imatinib) in CML [55] and with the FLT3-ITD inhibitor Sorafenib [119, 120].

Another Bcl-2 "inhibitor" with a detailed preclinical and also recent clinical history [87] is Obatoclax Mesylate (GX-015-070). This agent was shown to target the BH3 region, although probably at a lower affinity than ABT-737. A clinical trial has been reported in AML with inconclusive results [87]. No on-target effects on platelets (which contain large amounts of Bcl-X_L and are highly sensitive to ABT-767 and ABT-263) have been observed and no dose-dependent anti-leukemia effects. As Obatoclax targets Mcl-1, unlike ABT-737, potential synergistic effects with ABT-737 could be exploited and Mcl-1-dependent malignancies could be targeted. Another group of compounds, Gossypol and Apo-Gossypol, target the BH3 hydrophobic grove, but other established mechanisms make these compounds less specific, which is also borne out in the reported toxicities of Gossypol [35, 41, 113].

Interestingly, recent observations suggest that the FLT3 inhibitor sorafenib can activate the expression of multiple pro-apoptotic Bcl-2 proteins such as Bim, Bad, Bax and Bak [119], while simultaneously decreasing the levels of Mcl-1 and XIAP, changes that render AML cells exquisitively sensitive to ABT-737. These observations support the combinatorial use of sorafenib with BH3 mimetics as a novel therapeutic strategy in AML. Promising results of a phase I study in hematological malignancies utilizing ABT-263, the orally available clinical analog of ABT-737, are evolving and Sorafenib showed high activity as single agent in FLT3-ITD AML [120]. Optimized, mechanism-based combinations of targeted agents should result in "synthetic lethality" overcoming mechanisms of resistance of their combinatorial partner agents.

Targeting XIAP to Potentiate Apoptosis in Leukemia

Giving its role in opposing apoptosis and promoting chemoresistance, as well as its overexpression in malignant cells over normal cells, XIAP has attracted great attention as a potential therapeutic target for cancer. A phase I/II clinic trial with a XIAP antisense oligonucleotide [57] alone or in combination with chemotherapies is ongoing in various solid tumors and AML. Various small molecules that inhibit BIR3 or BIR2 domains of XIAP are under preclinical development [59, 74, 88, 100, 101, 114]. Among them, polyphenylurea-based small molecular BIR2 inhibitors of XIAP induce apoptosis in various AML cell lines and patient blasts in a caspase-3-dependent manner [12]. By screening a library of molecules isolated from Chinese herbs, Nikolovska-Coleska et al. identified embelin, a natural product from the Japanese Ardisia plant that binds the surface groove in the BIR3 domain of XIAP. Its binding affinity is similar to the SMAC peptide and induces apoptosis of malignant prostate cancer cells at low micromolar concentrations [71]. Our group has found that triptolide inhibits XIAP mRNA and protein and potently induces AML cell death at low nanomolar concentrations [13, 15]. Curiously, triptolide can also activate p53 resulting in increased expression of DR5 and potent synergism with TRAIL to induce apoptosis of AML cell lines and primary samples [14]. A clinical trial with a water-soluble derivative of triptolide in solid tumors is being conducted in France. The potent effect of triptolide in leukemia cells in culture warrants its clinical evaluation in AML.

Conclusions

The findings presented here support hypothesis-driven therapeutic concepts to maximize the induction of apoptosis in leukemia cells. For example, the inactivation of p53 by HDM2 is frequent in leukemias and can be overcome by agents that disrupt the HDM2-p53 interaction. HDM4 which inhibits p53 through a different mechanism will also need consideration. Second, the inactivation of certain pro-apoptotic by anti-apoptotic Bcl-2 family members is very frequent in leukemias and both protein–protein interactions are presently being investigated in targeted clinical trials. The inhibitor-of-apoptosis (IAP) family proteins are likely to contribute to apoptosis resistance, and clinical trials targeting XIAP and survivin are ongoing (Schimmer, Andreeff 2009, In Press). The notion that p53 modulates the Bcl-2 apoptotic rheostat directly, by behaving as a BH3-only protein, and indirectly by increasing the levels of sensitizer BH3-only proteins, provides a rationale to combine HDM2 inhibitors with BH3 mimetics. Ongoing work has indeed revealed that nutlins potently synergize with ABT-737 to induce apoptosis in AML cell lines and primary samples. Second, the finding that inhibition of MAPK signaling decreased the levels of Mcl-1 and abrogated Bcl-2 phosphorylation resulting in increased sensitivity to ABT-737 demonstrates that the true "Achilles heel" of leukemia may be exploited by targeting both signal transduction and apoptotic programs simultaneously ("synthetic lethality"). Third, XIAP and survivin inhibitors have progressed to clinical trials to maximize the contribution of caspase activation to cell death. While still incomplete, our present understanding of the apoptotic circuitry of leukemia cells is illustrated in Fig. 6. Activation of the intrinsic apoptosis pathway, particularly the induction of MOMP, is a central event in promoting cell death of leukemia cells, whether by activated p53 signaling, BH3

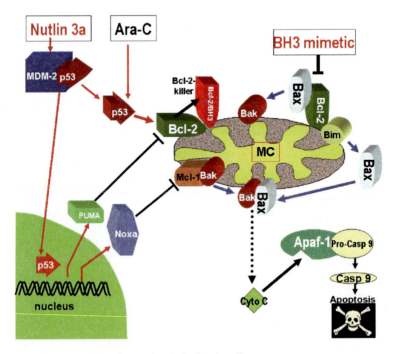

Fig. 6 Targeting the apoptotic circuitry in leukemia cells

mimetics, or IAP inhibition alone or in combination strategies with kinase inhibitors and classical apoptosis inducers, such as chemotherapeutic agents. Critical knowledge will be gained in carefully analyzed clinical trials. The expected surprises will extend our knowledge of the apoptotic machinery of leukemia cells, lead to refinements of the basic concepts discussed here, and the development of new agents that will need to exert pro-apoptotic activity against leukemic stem cells in hypoxic microenvironments that takes into account biochemical and molecular realities not yet fully appreciated in our current concepts [2, 27, 118].

References

1. Adida C, Recher, C, Raffoux E, et al. Expression and prognostic significance of survivin in de novo acute myeloid leukaemia. *Br J Haematol.* 2000;111:196–203.
2. Andreeff M, Zeng Z, Tabe Y, et al. Microenvironment and leukemia: Ying and Yang. *Ann Hematol.* 2008;87:S94–S98.
3. Asakura T, Ohkawa K. Chemotherapeutic agents that induce mitochondrial apoptosis. *Curr Cancer Drug Targets.* 2004;4:577–590.
4. Baumgartner B, Weber M, Quirling M, et al. Increased IkappaB kinase activity is associated with activated NF-kappaB in acute myeloid blasts. *Leukemia.* 2002;16;2062–2071.
5. Brookes PS, Morse K, Ray D, et al. The triterpenoid 2-cyano-3,12-dioxooleana-1,9-dien-28-oic acid and its derivatives elicit human lymphoid cell apoptosis through a novel pathway involving the unregulated mitochondrial permeability transition pore. *Cancer Res.* 2007;67;1793–1802.
6. Buchholz TA, Davis DW, McConkey DJ, et al. Chemotherapy-induced apoptosis and Bcl-2 levels correlate with breast cancer response to chemotherapy. *Cancer J.* 2003;9;33–41.
7. Bueso-Ramos CE, Manshouri T, Haidar MA, Huh YO, Keating MJ, Albitar M. Multiple patterns of MDM-2 deregulation in human leukemias: implications in leukemogenesis and prognosis. *Leuk Lymphoma.* 1995;17;13–18.
8. Bueso-Ramos CE, Rocha FC, Shishodia S, et al. Expression of constitutively active nuclear-kappa B RelA transcription factor in blasts of acute myeloid leukemia. *Hum Pathol.* 2004;35:246–253.
9. Bueso-Ramos CE, Yang Y, deLeon E, McCown P, Stass SA, Albitar M. The human MDM-2 oncogene is overexpressed in leukemias. *Blood.* 1993;82:2617–2623.
10. Campos L, Rouault JP, Sabido O, et al. High expression of bcl-2 protein in acute myeloid leukemia cells is associated with poor response to chemotherapy. *Blood.* 1993;81:3091–3096.
11. Carter BZ, Gronda M, Wang Z, et al. Small-molecule XIAP inhibitors derepress downstream effector caspases and induce apoptosis of acute myeloid leukemia cells. *Blood.* 2005;105:4043–4050.
12. Carter BZ, Kornblau SM, Tsao T, et al. Caspase-independent cell death in AML: caspase inhibition in vitro with pan-caspase inhibitors or in vivo by XIAP or survivin does not affect cell survival or prognosis. *Blood.* 2003;102:4179–4186.
13. Carter BZ, Mak D, Schober WD, et al. Regulation of survivin expression through bcr-abl/MAPK cascade: targeting survivin overcomes Imatinib resistance and increases Imatinib sensitivity in Imatinib responsive CML cells. *Blood.* 2006:107:1555–1563.

14. Carter BZ, Mak DH, Schober WD, et al. Triptolide induces caspase-dependent cell death mediated via the mitochondrial pathway in leukemic cells. *Blood*. 2006;108:630–637.

15. Carter BZ, Mak DH, Schober WD, et al. Triptolide sensitizes AML cells to TRAIL-induced apoptosis via decrease of XIAP and p53-mediated increase of DR5. *Blood*. 2008;111:3742–3750.

16. Carter BZ, Milella M, Altieri DC, Andreeff M. Cytokine-regulated expression of survivin in myeloid leukemia. *Blood*. 2001;97:2784–2790.

17. Carter BZ, Milella M, Tsao T, et al. Regulation and targeting of antiapoptotic XIAP in acute myeloid leukemia. *Leukemia*. 2003;17:2081–2089.

18. Certo M, Moore VG, Nishino M, et al. Mitochondria primed by death signals determine cellular addiction to antiapoptotic BCL-2 family members. *Cancer Cell*. 2006;9:351–365.

19. Chan SL, Yu VC. Proteins of the bcl-2 family in apoptosis signalling: from mechanistic insights to therapeutic opportunities. *Clin Exp Pharmacol Physiol*. 2004;31:119–128.

20. Chen L, Gilkes DM, Pan Y, Lane WS, Chen J. ATM and Chk2-dependent phosphorylation of MDMX contribute to p53 activation after DNA damage. *EMBO J*. 2005;24:3411–3422.

21. Chen M, Wang J. Initiator caspases in apoptosis signaling pathways. *Apoptosis*. 2002;7:313–319.

22. Chipuk JE, Kuwana T, Bouchier-Hayes L, Droin NM, Newmeyer DD, Schuler M, Green DR. Direct activation of Bax by p53 mediates mitochondrial membrane permeabilization and apoptosis. *Science*. 2004;303:1010–1014.

23. Degterev A, Lugovskoy A, Cardone M, et al. Identification of small-molecule inhibitors of interaction between the BH3 domain and Bcl-xL. *Nat Cell Biol*. 2001;3:173–182.

24. Deng X, Gao F, Flagg T, Anderson J, May WS. Bcl2's Flexible loop domain regulates p53 binding and survival. *Mol Cell Biol*. 2006;26:4421–4434.

25. Derenne S, Monia B, Dean NM, et al. Antisense strategy shows that Mcl-1 rather than Bcl-2 or Bcl-x(L) is an essential survival protein of human myeloma cells. *Blood*. 2002;100:194–199.

26. Faderl S, Kantarjian HM, Estey E, et al. The prognostic significance of p16(INK4a)/p14(ARF) locus deletion and MDM-2 protein expression in adult acute myelogenous leukemia. *Cancer*. 2000;89:1976–1982.

27. Fiegl M, Samudio I, Clise-Dwyer K, Burks JK, Mnjoyan Z, Andreeff M. CXCR4 expression and biological activity in acute myeloid leukemia are dependent on oxygen partial pressure. *Blood*. 2009;113:1504–1512.

28. Fleischer A, Rebollo A, Ayllon V. BH3-only proteins: the lords of death. *Arch Immunol Ther Exp (Warsz.)*. 2003;51:9–17.

29. Green DR, Reed JC. Mitochondria and apoptosis. *Science*. 1998;281:1309–1312.

30. Guzman ML, Neering SJ, Upchurch D, et al. Nuclear factor-kappaB is constitutively activated in primitive human acute myelogenous leukemia cells. *Blood*. 2001;98:2301–2307.

31. Harada H, Quearry B, Ruiz-Vela A, Korsmeyer SJ. Survival factor-induced extracellular signal-regulated kinase phosphorylates BIM, inhibiting its association with BAX and proapoptotic activity. *Proc Natl Acad Sci USA*. 2004;101:15313–15317.

32. Harms K, Nozell S, Chen X. The common and distinct target genes of the p53 family transcription factors. *Cell Mol Life Sci*. 2004;61:822–842.

33. Hu G, Zhang W, Deisseroth AB. P53 gene mutations in acute myelogenous leukaemia. *Br J Haematol*. 1992;81:489–494.

34. Invernizzi R, Travaglino E, Lunghi M, et al. Survivin expression in acute leukemias and myelodysplastic syndromes. *Leuk Lymphoma*. 2004;45:2229–2237.

35. Jia L, Coward LC, Kerstner-Wood CD, et al. Comparison of pharmacokinetic and metabolic profiling among gossypol, apogossypol and apogossypol hexaacetate. *Cancer Chemother Pharmacol*. 2008;61:63–73.

36. Johnston JB, Kabore AF, Strutinsky J, et al. Role of the TRAIL/APO2-L death receptors in chlorambucil- and fludarabine-induced apoptosis in chronic lymphocytic leukemia. *Oncogene*. 2003;22:8356–8369.

37. Jones DT, Ganeshaguru K, Mitchell WA, et al. Cytotoxic drugs enhance the ex vivo sensitivity of malignant cells from a subset of acute myeloid leukaemia patients to apoptosis induction by tumour necrosis factor receptor-related apoptosis-inducing ligand. *Br J Haematol.* 2003;121:713–720.
38. Kamer I, Sarig R, Zaltsman Y, et al. Proapoptotic BID is an ATM effector in the DNA-damage response. *Cell.* 2005;122:593–603.
39. Kerr JF, Wyllie AH, Currie AR. Apoptosis: a basic biological phenomenon with wide-ranging implications in tissue kinetics. *Br J Cancer.* 1972;26:239–257.
40. Kim TS, Yun BY, Kim IY. Induction of the mitochondrial permeability transition by selenium compounds mediated by oxidation of the protein thiol groups and generation of the superoxide. *Biochem Pharmacol.* 2003;66:2301–2311.
41. Kitada S, Kress CL, Krajewska M, Jia L, Pellecchia M, Reed JC. Bcl-2 antagonist apogossypol (NSC736630) displays single-agent activity in Bcl-2-transgenic mice and has superior efficacy with less toxicity compared with gossypol (NSC19048). *Blood.* 2008;111:3211–3219.
42. Kitada S, Leone M, Sareth S, Zhai D, Reed JC, Pellecchia M. Discovery, characterization, and structure-activity relationships studies of proapoptotic polyphenols targeting B-cell lymphocyte/leukemia-2 proteins. *J Med Chem.* 2003;46:4259–4264.
43. Kojima K, Konopleva M, McQueen T, O'Brien S, Plunkett W, Andreeff M. Mdm2 inhibitor Nutlin-3a induces p53-mediated apoptosis by transcription-dependent and transcription-independent mechanisms and may overcome Atm-mediated resistance to fludarabine in chronic lymphocytic leukemia. *Blood.* 2006a;108:993–1000.
44. Kojima K, Konopleva M, Samudio IJ, Schober WD, Bornmann WG, Andreeff M. Concomitant inhibition of MDM2 and Bcl-2 protein function synergistically induce mitochondrial apoptosis in AML. *Cell Cycle.* 2006b;5:2778–2786.
45. Kojima K, Konopleva M, Samudio IJ, et al. MDM2 antagonists induce p53-dependent apoptosis in AML: implications for leukemia therapy. *Blood.* 2005;106:3150–3159.
46. Kojima K, Konopleva M, Tsao T, Nakakuma H, Andreeff M. Concomitant inhibition of Mdm2-p53 interaction and Aurora kinases activates the p53-dependent postmitotic checkpoints and synergistically induces p53-mediated mitochondrial apoptosis along with reduced endoreduplication in acute myelogenous leukemia. *Blood.* 2008a;112: 2886–2895.
47. Kojima K, Shimanuki M, Shikami M, Andreeff M, Nakakuma H. The CDK1 inhibitor RO-3306 enhances p53-mediated Bax activation and mitochondrial apoptosis in AML. *Cancer Sci.* 2009;100:1128–1136.
48. Kojima K, Shimanuki M, Shikami M, et al. The dual PI3 kinase/mTOR inhibitor PI-103 prevents p53 induction by Mdm2 inhibition but enhances p53-mediated mitochondrial apoptosis in p53 wild-type AML. *Leukemia.* 2008b;22:1728–1736.
49. Kong G, Wang D, Wang H, et al. Synthetic triterpenoids have cytotoxicity in pediatric acute lymphoblastic leukemia cell lines but cytotoxicity is independent of induced ceramide increase in MOLT-4 cells. *Leukemia.* 2008;22:1258–1262.
50. Konopleva M, Contractor R, Tsao T, et al. Mechanisms of apoptosis sensitivity and resistance to the BH3 mimetic ABT-737 in acute myeloid leukemia. *Cancer Cell.* 2006;10:375–388.
51. Konopleva M, Tsao T, Estrov Z, et al. The synthetic triterpenoid 2-cyano-3,12-dioxoo-leana-1,9-dien-28-oic acid induces caspase-dependent and -independent apoptosis in acute myelogenous leukemia. *Cancer Res.* 2004;64:7927–7935.
52. Konopleva M, Watt J, Contractor R, et al. Mechanisms of antileukemic activity of the novel Bcl-2 homology domain-3 mimetic GX15-070 (obatoclax). *Cancer Res.* 2008;68:3413–3420.
53. Koschmieder S, D'Alo F, Radomska H, et al. CDDO induces granulocytic differentiation of myeloid leukemic blasts through translational up-regulation of p42 CCAAT enhancer binding protein alpha. *Blood.* 2007;110:3695–3705.

54. Kress CL, Konopleva M, Martinez-Garcia V, et al. Triterpenoids display single agent anti-tumor activity in a transgenic mouse model of chronic lymphocytic leukemia and small B cell lymphoma. *PLoS ONE.* 2007;2:e559.
55. Kuroda J, Kimura S, Andreeff M, et al. ABT-737 is a useful component of combinatory chemotherapies for chronic myeloid leukaemias with diverse drug-resistance mechanisms. *Br J Haematol.* 2007;140:181–190.
56. Kuwana T, Bouchier-Hayes L, Chipuk JE, et al. BH3 domains of BH3-only proteins differentially regulate Bax-mediated mitochondrial membrane permeabilization both directly and indirectly. *Mol Cell.* 2005;17:525–535.
57. Lacasse EC, Kandimalla ER, Winocour P, et al. Application of XIAP antisense to cancer and other proliferative disorders: development of AEG35156/GEM640. *Ann NY Acad Sci.* 2005;1058:215–234.
58. Leone M, Zhai D, Sareth S, Kitada S, Reed JC, Pellecchia M. Cancer prevention by tea polyphenols is linked to their direct inhibition of antiapoptotic Bcl-2-family proteins. *Cancer Res.* 2003;63:8118–8121.
59. Li L, Thomas RM, Suzuki H, De Brabander JK, Wang X, Harran PG. A small molecule Smac mimic potentiates T. *Science.* 2004;305:1471–1474.
60. Li LY, Luo X, Wang X. Endonuclease G is an apoptotic DNase when released from mitochondria. *Nature.* 2001;412:95–99.
61. Liu Q, Hilsenbeck S, Gazitt Y. Arsenic trioxide-induced apoptosis in myeloma cells: p53-dependent G1 or G2/M cell cycle arrest, activation of caspase-8 or caspase-9, and synergy with APO2/TRAIL. *Blood.* 2003;101:4078–4087.
62. Liu X, Kim CN, Yang J, Jemmerson R, Wang X. Induction of apoptotic program in cell-free extracts: requirement for dATP and cytochrome c. *Cell.* 1996;86:147–157.
63. Lorenzo HK, Susin SA, Penninger J, Kroemer G. Apoptosis inducing factor (AIF): a phylogenetically old, caspase-independent effector of cell death. *Cell Death Differ.* 1999;6:516–524.
64. MacFarlane M. TRAIL-induced signalling and apoptosis. *Toxicol Lett.* 2003;139:89–97.
65. MacFarlane M, Harper N, Snowden RT, et al. Mechanisms of resistance to TRAIL-induced apoptosis in primary B cell chronic lymphocytic leukaemia. *Oncogene.* 2002;21:6809–6818.
66. Markey M, Berberich SJ. Full-length hdmX transcripts decrease following genotoxic stress. *Oncogene.* 2008;27:6657–6666.
67. McStay GP, Clarke SJ, Halestrap AP. Role of critical thiol groups on the matrix surface of the adenine nucleotide translocase in the mechanism of the mitochondrial permeability transition pore. *Biochem J.* 2002;367:541–548.
68. Miyashita T, Reed JC. Bcl-2 oncoprotein blocks chemotherapy-induced apoptosis in a human leukemia cell line. *Blood.* 1993;81:151–157.
69. Moll UM, Petrenko O. The MDM2-p53 interaction. *Mol Cancer Res.* 2003;1:1001–1008.
70. Muzio M, Stockwell BR, Stennicke HR, Salvesen GS, Dixit VM. An induced proximity model for caspase-8 activation. *J Biol Chem.* 1998;273:2926–2930.
71. Nikolovska-Coleska Z, Xu L, Hu Z, et al. Discovery of embelin as a cell-permeable, small-molecular weight inhibitor of XIAP through structure-based computational screening of a traditional herbal medicine three-dimensional structure database. *J Med Chem.* 2004;47:2430–2440.
72. Okamoto K, Kashima K, Pereg Y, et al. DNA damage-induced phosphorylation of MdmX at serine 367 activates p53 by targeting MdmX for Mdm2-dependent degradation. *Mol Cell Biol.* 2005;25:9608–9620.
73. Oltersdorf T, Elmore SW, Shoemaker AR, et al. An inhibitor of Bcl-2 family proteins induces regression of solid tumours. *Nature.* 2005;435:677–681.
74. Oost TK, Sun C, Armstrong RC, et al. Discovery of potent antagonists of the antiapoptotic protein XIAP for the treatment of cancer. *J Med Chem.* 2004;47:4417–4426.
75. Opferman JT, Iwasaki H, Ong CC, Suh H, Mizuno S, Akashi K, Korsmeyer SJ. Obligate role of anti-apoptotic MCL-1 in the survival of hematopoietic stem cells. *Science.* 2005;307:1101–1104.

76. Opferman JT, Letai A, Beard C, Sorcinelli MD, Ong CC, Korsmeyer SJ. Development and maintenance of B and T lymphocytes requires antiapoptotic MCL-1. *Nature.* 2003;426:671–676.

77. Patton JT, Mayo LD, Singhi AD, Gudkov AV, Stark GR, Jackson MW. Levels of HdmX expression dictate the sensitivity of normal and transformed cells to Nutlin-3. *Cancer Res.* 2006;66:3169–3176.

78. Pazgier M, Liu M, Zou G, et al. Structural basis for high-affinity peptide inhibition of p53 interactions with MDM2 and MDMX. *Proc Natl Acad Sci USA.* 2009;106:4665–4670.

79. Peter ME, Krammer PH. The CD95(APO-1/Fas) DISC and beyond. *Cell Death Differ.* 2003;10:26–35.

80. Pop C, Timmer J, Sperandio S, Salvesen GS. The apoptosome activates caspase-9 by dimerization. *Mol Cell.* 2006;22:269–275.

81. Puthalakath H, Huang DC, O'Reilly LA, King SM, Strasser A. The proapoptotic activity of the Bcl-2 family member Bim is regulated by interaction with the dynein motor complex. *Mol Cell.* 1999;3:287–296.

82. Reed JC, Pellecchia M. Apoptosis-based therapies for hematologic malignancies. *Blood.* 2005;106:408–418.

83. Riccioni R, Pasquini L, Mariani G, et al. TRAIL decoy receptors mediate resistance of acute myeloid leukemia cells to TRAIL. *Haematologica.* 2005;90:612–624.

84. Saelens X, Festjens N, Vande WL, van Gurp M, van Loo G, Vandenabeele P. Toxic proteins released from mitochondria in cell death. *Oncogene.* 2004;23:2861–2874.

85. Salvesen GS, Dixit VM. Caspase activation: the induced-proximity model. *Proc Natl Acad Sci USA.* 1999;96:10964–10967.

86. Samudio I, Konopleva M, Pelicano H, et al. A novel mechanism of action of methyl-2-cyano-3,12 dioxoolean-1,9 diene-28-oate (CDDO-Me): direct permeabilization of the inner mitochondrial membrane to inhibit electron transport and induce apoptosis. *Mol Pharmacol.* 2006;5, 1182–1193.

87. Schimmer AD, Kantarjian H, O'Brien S, et al. A phase I study of the pan Bcl-2 family inhibitor obatoclax mesylate in patients with refractory hematologic malignancies. *Clin Cancer Res.* 2008. In press.

88. Schimmer AD, Welsh K, Pinilla C, et al. Small-molecule antagonists of apoptosis suppressor XIAP exhibit broad antitumor activity. *Cancer Cell.* 2004;5:25–35.

89. Schottelius A, Brennscheidt U, Ludwig WD, Mertelsmann RH, Herrmann F, Lubbert M. Mechanisms of p53 alteration in acute leukemias. *Leukemia.* 1994;8:1673–1681.

90. Seliger B, Papadileris S, Vogel D, et al. Analysis of the p53 and MDM-2 gene in acute myeloid leukemia. *Eur J Haematol.* 1996;57:230–240.

91. Shankar S, Singh TR, Fandy TE, Luetrakul T, Ross DD, Srivastava RK. Interactive effects of histone deacetylase inhibitors and TRAIL on apoptosis in human leukemia cells: involvement of both death receptor and mitochondrial pathways. *Int J Mol Med.* 2005;16:1125–1138.

92. Sheikh MS, Huang Y. Death receptors as targets of cancer therapeutics. *Curr Cancer Drug Targets.* 2004;4:97–104.

93. Shiau CW, Huang JW, Wang DS, et al. alpha-Tocopheryl succinate induces apoptosis in prostate cancer cells in part through inhibition of Bcl-xL/Bcl-2 function. *J Biol Chem.* 2006;281:11819–11825.

94. Shiau CW, Yang CC, Kulp SK, et al. Thiazolidinediones mediate apoptosis in prostate cancer cells in part through inhibition of Bcl-xL/Bcl-2 functions independently of PPAR-gamma. *Cancer Res.* 2005;65:1561–1569.

95. Shiiki K, Yoshikawa H, Kinoshita H, et al. Potential mechanisms of resistance to TRAIL/Apo2L-induced apoptosis in human promyelocytic leukemia HL-60 cells during granulocytic differentiation. *Cell Death Differ.* 2000;7:939–946.

96. Soussi T, Lozano G. p53 mutation heterogeneity in cancer. *Biochem Biophys Res Commun.* 2005;331:834–842.

97. Srinivasula SM, Datta P, Fan XJ, Fernandes-Alnemri T, Huang Z, Alnemri ES. Molecular determinants of the caspase-promoting activity of Smac/DIABLO and its role in the death receptor pathway. *J Biol Chem.* 2000;275:36152–36157.

98. Stehlik C, de Martin R, Kumabashiri I, Schmid JA, Binder BR, Lipp J. Nuclear factor (NF)-kappaB-regulated X-chromosome-linked iap gene expression protects endothelial cells from tumor necrosis factor alpha-induced apoptosis. *J Exp Med.* 1998;188:211–216.

99. Suh WS, Kim YS, Schimmer AD, et al. Synthetic triterpenoids activate a pathway for apoptosis in AML cells involving downregulation of FLIP and sensitization to TRAIL. *Leukemia.* 2003;17:2122–2129.

100. Sun H, Nikolovska-Coleska Z, Chen J, et al. Structure-based design, synthesis and biochemical testing of novel and potent Smac peptido-mimetics. *Bioorg Med Chem Lett.* 2005;15:793–797.

101. Sun H, Nikolovska-Coleska Z, Yang CY, et al. Structure-based design of potent, conformationally constrained Smac mimetics. *J Am Chem Soc.* 2004;126:16686–16687.

102. Tabe Y, Konopleva M, Kondo Y, et al. PPARgamma-active triterpenoid CDDO enhances ATRA-induced differentiation in APL. *Cancer Biol Ther.* 2007;6:1967–1977.

103. Tamm I, Kornblau SM, Segall H, et al. Expression and prognostic significance of IAP-family genes in human cancers and myeloid leukemias. *Clin Cancer Res.* 2000;6:1796–1803.

104. Tamm I, Richter S, Oltersdorf D, et al. High expression levels of x-linked inhibitor of apoptosis protein and survivin correlate with poor overall survival in childhood de novo acute myeloid leukemia. *Clin Cancer Res.* 2004;10:3737–3744.

105. Tothova E, Fricova M, Stecova N, Kafkova A, Elbertova A. High expression of Bcl-2 protein in acute myeloid leukemia cells is associated with poor response to chemotherapy. *Neoplasma.* 2002;49:141–144.

106. Tzung SP, Kim KM, Basanez G, et al. Antimycin A mimics a cell-death-inducing Bcl-2 homology domain 3. *Nat Cell Biol.* 2001;3:183–191.

107. Vahsen N, Cande C, Briere JJ, et al. AIF deficiency compromises oxidative phosphorylation. *EMBO J.* 2004;23:4679–4689.

108. Vassilev LT, Vu BT, Graves B, et al. In vivo activation of the p53 pathway by small-molecule antagonists of MDM2. *Science.* 2004;303:844–848.

109. Verrier F, Deniaud A, Lebras M, et al. Dynamic evolution of the adenine nucleotide translocase interactome during chemotherapy-induced apoptosis. *Oncogene.* 2004;23:8049–8064.

110. Wade M, Wahl GM. Targeting Mdm2 and Mdmx in cancer therapy: better living through medicinal chemistry? *Mol Cancer Res.* 2009;7:1–11.

111. Wang JL, Liu D, Zhang ZJ, et al. Structure-based discovery of an organic compound that binds Bcl-2 protein and induces apoptosis of tumor cells. *Proc Natl Acad Sci USA.* 2000;97:7124–7129.

112. Waterhouse NJ, Sedelies KA, Sutton VR, et al. Functional dissociation of DeltaPsim and cytochrome c release defines the contribution of mitochondria upstream of caspase activation during granzyme B-induced apoptosis. *Cell Death Differ.* 2006;13:607–618.

113. Wei J, Rega MF, Kitada S, et al. Synthesis and evaluation of Apogossypol atropisomers as potential Bcl-xL antagonists. *Cancer Lett.* 2009;273:107–113.

114. Wu TY, Wagner KW, Bursulaya B, Schultz PG, Deveraux QL. Development and characterization of nonpeptidic small molecule inhibitors of the XIAP/caspase-3 interaction. *Chem Biol.* 2003;10:759–767.

115. Wuchter C, Krappmann D, Cai Z, et al. In vitro susceptibility to TRAIL-induced apoptosis of acute leukemia cells in the context of TRAIL receptor gene expression and constitutive NF-kappa B activity. *Leukemia.* 2001;15:921–928.

116. Yagita H, Takeda K, Hayakawa Y, Smyth MJ, Okumura K. TRAIL and its receptors as targets for cancer therapy. *Cancer Sci.* 2004;95:777–783.

117. Ye H, Cande C, Stephanou NC, et al. DNA binding is required for the apoptogenic action of apoptosis inducing factor. *Nat Struct Biol.* 2002;9:680–684.
118. Zeng Z, Shi Y-X, Samudio IJ, et al. Targeting the leukemia microenvironment by CXCR4 inhibition overcomes resistance to kinase inhibitors and chemotherapy in AML. *Blood.* 2009;113(24):6215–6224.
119. Zhang W, Konopleva M, Ruvolo VR, et al. Sorafenib induces apoptosis of AML cells via Bim-mediated activation of the intrinsic apoptotic pathway. *Leukemia.* 2008a;22:808–818.
120. Zhang WG, Konopleva M, Shi YX, et al. Mutant FLT3: a direct target of sorafenib in acute myelogenous leukemia. *J Natl Cancer Inst.* 2008b;100:184–198.
121. Zhao Y, Chaiswing L, Velez JM, et al. p53 translocation to mitochondria precedes its nuclear translocation and targets mitochondrial oxidative defense protein-manganese superoxide dismutase. *Cancer Res.* 2005;65:3745–3750.
122. Zhu Y, Swanson BJ, Wang M, et al. Constitutive association of the proapoptotic protein Bim with Bcl-2-related proteins on mitochondria in T cells. *Proc Natl Acad Sci USA.* 2004;101:7681–7686.
123. Zinkel SS, Hurov KE, Ong C, Abtahi FM, Gross A, Korsmeyer SJ. A role for proa-poptotic BID in the DNA-damage response. *Cell.* 2005;122:579–591.

Acute Promyelocytic Leukemia: A Paradigm for Differentiation Therapy

David Grimwade, Anita R. Mistry, Ellen Solomon, and Fabien Guidez

Abstract Acute promyelocytic leukemia(APL) is characterized by the t(15;17) chromosomal translocation leading to the formation of the PML-RARα oncoprotein. This leukemia has attracted considerable interest in recent years, being the first in which therapies that specifically target the underlying molecular lesion, i.e., all-*trans* retinoic acid (ATRA) and arsenic trioxide (ATO), leading to induction of differentiation and apoptosis have been successfully used in clinical practice. The advent of ATRA therapy has transformed APL from being a disease with a poor outlook to one of the most prognostically favorable subsets of acute myeloid leukemia. Further improvements in outcome may be achieved with the use of ATO, which achieves high rates of remission in the relatively small proportion of patients now relapsing following standard first-line therapy with ATRA and anthracycline-based chemotherapy. Moreover, recent studies have suggested that ATO and ATRA, or even ATO alone, used as front-line treatment of *PML-RARA-* associated APL can induce long-term remissions. This raises the possibility that some patients can be cured using differentiation therapies alone, without the need for chemotherapy, thereby potentially reducing treatment-related toxicity. It is clear that the success of such an approach is critically dependent upon molecular diagnostics and monitoring for minimal residual disease (MRD) to distinguish those patients who can potentially be cured with differentiation therapy from those requiring additional myelosuppressive agents. This represents an exciting new phase in the treatment of acute leukemia, highlighting the potential of molecularly targeted and MRD-directed therapies to achieve an individualized approach to patient management.

D. Grimwade (✉)
Department of Medical and Molecular Genetics, King's College London School of Medicine, London, UK
e-mail: david.grimwade@genetics.kcl.ac.uk

L. Nagarajan (ed.), *Acute Myelogenous Leukemia*,
Cancer Treatment and Research 145, DOI 10.1007/978-0-387-69259-3_13,
© Springer Science+Business Media, LLC 2010

Clinical Features of Acute Promyelocytic Leukemia

Acute promyelocytic leukemia (APL) is one of the commonest forms of acute myeloid leukemia (AML) accounting for around 10% of cases arising in children and younger adults [30]. APL can be readily identified, based upon the distinct morphological appearance of the leukemic cells, with two major subtypes being recognized in the French-American-British (FAB) classification of AML [4]. The majority of cases (~75%) present with the classical form of the disease (FAB M3) with a marrow replaced by heavily granulated promyelocytic blasts, including characteristic cells crammed with bundles of Auer rods ("faggot cells"). This hypergranular form of APL is typically accompanied by cytopenias with few or no abnormal cells in the peripheral blood. Approximately a quarter of cases present with the hypogranular (microgranular) variant form of the disease (FAB M3v), which is associated with leucocytosis due to circulating blasts characterized by bilobed nuclei, which appear less heavily granulated than those of classical APL by light microscopy. Despite the morphological differences, the majority of cases harbor a common molecular lesion, i.e., the *PML-RARA* fusion gene, whose presence predicts a favorable response to molecularly targeted therapies in the form of all-*trans* retinoic acid (ATRA) and arsenic trioxide (ATO) [29].

A key clinical feature that needs to be taken into account in the initial management of APL is the severe bleeding tendency, reflecting to varying degrees triggering of the coagulation cascade with disseminated intravascular coagulation (DIC), increased fibrinolysis and proteolysis, compounded by thrombocytopenia [19]. The coagulation defect is exacerbated by chemotherapy through disruption of APL blasts and their granular contents; indeed cerebral hemorrhage remains one of the commonest causes of death, leading to the demise of up to 10% of patients. In contrast to the effects of chemotherapy, ATRA has been found to have an ameliorating effect on the coagulopathy; hence prompt initiation of ATRA therapy as soon as the diagnosis is suspected is of paramount importance [57]. Moreover, randomized trials have demonstrated that the addition of ATRA to conventional chemotherapy leads to a substantial reduction in rates of relapse associated with improved overall survival [21, 70]. Over the last 15 years it has also been recognized that APL is exquisitely sensitive to anthracyclines, which correlates with the low levels of P-glycoprotein expression observed in this subset of AML [60, 35]. These data have led to the adoption of extended courses of ATRA combined with anthracycline-based chemotherapy as the standard treatment approach to *PML-RARA*-associated APL, leading to the cure of approximately 70% of patients [65].

Molecular Pathogenesis of APL

While ATRA and ATO have ultimately proved to be highly effective molecularly targeted therapies, ironically both agents were discovered to be particularly efficacious in APL by physicians working in China long before the molecular basis of the disease was determined [14]. Indeed, it was the particular sensitivity to retinoic acid (RA) that prompted some investigators to focus upon and ultimately identify the gene encoding Retinoic Acid Receptor Alpha (RARα) located at 17q21 as the translocation target in APL [13] (reviewed in [59]). Seven fusion partners of *RARA* have been identified to date (Fig. 1). The vast majority of cases involve the *PML* (*P* ro*M* yelocytic *L* eukemia) gene as a result of the t(15;17)(q22;q21) chromosomal translocation. Other less common fusion partners include the PLZF (*P* romyelocytic *L* eukemia *Z* inc *F* inger) (~1% APL) and nucleophosmin (*NPM1*) genes (~0.5% APL) due to t(11;17)(q23;q21) and t(5;17)(q35;q12-21), respectively. *Nu* clear *M* itotic *A* pparatus protein (NuMA) and *S* ignal *T* ransducer/*A* ctivator of *T* ranscription 5b (STAT5b) have been identified as the fusion partner in cases with t(11;17)(q13;q21) and interstitial 17q deletion, respectively, but have only been reported in single patients to date. Recently, the number of APL fusion partners has been further extended with the discovery of cases involving the *PRKAR1A* (encoding the regulatory subunit type 1-alpha of protein kinase A) and *FIP1L1* (Fip1-like1) genes (Fig. 1). Precise molecular diagnosis of APL is essential, since the nature of the fusion partner has an important bearing upon disease biology, particularly the response to molecularly targeted therapies. APL involving PML, NPM1, FIP1L1 and

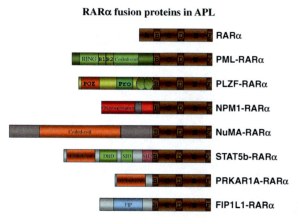

Fig. 1 Fusion proteins underlying the pathogenesis of acute promyelocytic leukemia. Schematic representation of RARα and the APL-associated RARα fusion proteins. RARα functional domains (A to F) are as indicated. Different colors are used to represent functional regions of the PML, PLZF, NPM1, NuMA, STAT5b, PRKAR1A and FIP1L1 proteins. PML regions labeled as RING, B1, and B2 represent cysteine–histidine rich domains. Circled Zn^{++} and PRO symbols represent Krüppel-like zinc finger motifs and proline-rich regions in the PLZF moiety, respectively. Labeled coiled-coil regions in PML, NuMA, and STAT5b and oligomerization and POZ domains in NPM1 and PLZF, respectively, represent protein–protein interaction motifs which are present in the N-termini of all the RARα chimeras. Relative positions of STAT5b DNA-binding (DBD), SH3, and SH2 domains are also indicated. The fusions involving PRKAR1A and FIP1L1 generated by a 17q rearrangement [9] and t(4;17)(q12;q21) [41] respectively, are the latest to have been described to date

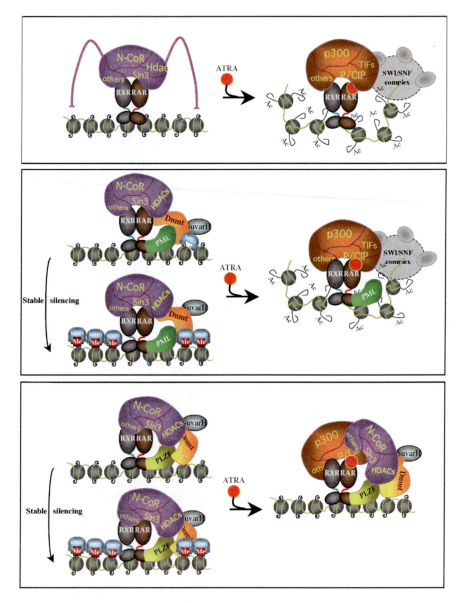

Fig. 2 RARα/RXR heterodimers and APL fusion proteins function as retinoic acid concentration-dependent transcription factors. In the absence of ligand (retinoic acid, RA), RARα heterodimerizes with RXRα and acts as a transcriptional repressor by recruiting nuclear receptor corepressors, including N-coR, Sin3, and in turn histone deacetylases (HDACs) (upper panel, left). The presence of RA, at physiological concentrations, induces an allosteric change in the receptor leading to the release of the corepressor complex and the recruitment of the co-activator complex thus leading to the activation of transcription of genes required for cellular differentiation and growth inhibition (*upper panel, right*). At this concentration of the ligand, both PML-RARα and PLZF-RARα are potent transcriptional repressors in view of

NuMA is retinoid responsive, while that involving PLZF or STAT5b responds poorly to ATRA. Sensitivity to ATO has only been demonstrated in cases with an underlying PML-RARα fusion, indeed those with PLZF-RARα have been shown to be resistant to this agent (reviewed in [59]).

The consistent association between APL and rearrangement of the *RARA* locus highlights the importance of deregulation of RARα which functions as a ligand- (i.e., retinoic acid, RA) dependent transcription factor involved in myeloid differentiation, in determining the disease phenotype. Indeed, genomic break points consistently occur within intron 2, such that the same domains of RARα (B through F) are retained within the APL fusion proteins (see Fig. 1); these include the DNA-binding domain (C) and the E-region which includes the ligand-binding domain (LBD) and regions involved in binding to coactivator or corepressor complexes in a ligand-dependent fashion (as discussed in more detail below). The E-region also provides an interface for heterodimerization with a second family of nuclear hormone receptors, retinoid-X-receptors (RXRs), required for high-affinity binding to the specific response elements (retinoic acid response elements, RAREs) within regulatory regions of retinoid target genes [10, 61]. Both RARs and RXRs act to transduce the retinoid signal and are activated by 9-*cis* retinoic acid (9-*cis* RA); but in addition, RARs are specifically activated by ATRA [43].

Unliganded RARα is capable of binding DNA response elements with high affinity leading to transcriptional repression due to interaction with corepressor molecules, i.e., SMRT (silencing mediator for RAR and TR) and N-CoR (nuclear receptor corepressor) [42, 62]. The corepressors in turn recruit the histone deacetylase (HDAC)-containing Sin3a complex, leading to deacetylation of core histones and chromatin condensation [62]. The chromatin in this state is inaccessible to transcriptional activators and the basal transcription machinery, effectively silencing retinoid target genes (Fig. 2). Conversely, binding of ligand to the receptor induces a conformational change, favoring recruitment of coactivator complexes to previously inaccessible residues accompanied by dissociation of corepressors, leading to transcriptional activation at retinoid target genes. The coactivator complexes act in concert to initiate transcription in a number of ways, including acetylation of core histones (CBP, ACTR),

◄───

Fig. 2 (continued) an increased and aberrant affinity for nuclear corepressors and HDACs (*middle and bottom panels, left*). At pharmacological doses of RA, while PML-RARα can be freed from corepressor interactions thus directly mediating trans-activation of RARα target genes (middle panel, right), corepressor binding to the N-terminal POZ domain PLZF-RARα persists correlating with the retinoid insensitivity of this subset of APL (*bottom panel, right*). Recent studies have shown that the PML-RARα protein induces promoter DNA hypermethylation at CpG dinucleotides by direct recruitment of DNA methyltransferase (DNMT) enzymes and by forming stable complexes with the methyl binding protein MBD1, which docks to methylated CpG sites to establish a silenced chromatin state. The presence of SUV39H1, which induces trimethylation of histone H3 (H3K9), further reinforces the epigenetic silencing mediated by APL fusion proteins

unwinding of DNA mediated by helicases (Tripl/Sugl), and interaction with components of the basal transcription machinery (TIF-1) [62]. Hence RARα has the potential to exert a dual function; with unliganded receptor acting as a negative regulator of granulocytic differentiation, while RA-bound RARα stimulates this process. The APL fusion proteins such as PML-RARα effectively act in a similar fashion, but operating at a higher threshold concentration of RA. While at physiological levels of RA the wild-type RARα receptor functions as a transcriptional activator to induce myeloid differentiation, this concentration of ligand (10^{-9}–10^{-8}M) is insufficient to displace the corepressor complexes including SMRT/N-CoR, Sin3a, and HDACs from APL fusion proteins, leading to chromatin condensation at regulatory elements of target genes [34, 31, 49, 28]. Epigenetic silencing is further compounded by recruitment of DNA methyltransferases (DNMTs) [15], MBD1 [72], and the histone methyltransferase SUV39H1, responsible for trimethylation of lysine 9 of histone H3 [7] (Fig. 2). A number of potential RA-RARα target genes which could be aberrantly regulated by APL fusion proteins have been identified. These include the *HOX* gene cluster and genes encoding CCAAT enhancer binding proteins β and ε (C/EBPβ, C/EBPε), interferon regulatory factor 1 (IRF1), STAT1α, c-MYC, and $p21^{WAF/CIP}$ [17] (reviewed in [59]). The PML-RARα oncoprotein not only retains binding activity for classical RAREs but can also bind to novel DNA response elements, thereby providing a potential gain of function for the fusion protein leading to the repression of an extended range of target genes that could potentially contribute to the block in differentiation and the mechanisms underlying leukemic transformation [39, 77]. Recruitment of repressor complexes by the APL fusion proteins may be further enhanced by the oligomerization capacity conferred by dimerization domains within each of the respective fusion partners (see Fig. 1), which appears to play an important role in leukemogenesis [77, 69, 44, 50, 58].

Mechanisms Underlying ATRA Response in APL

While APL fusion proteins block myeloid differentiation at physiological levels of RA (see above); paradoxically they also have the capacity to mediate myeloid differentiation at pharmacological concentrations of ATRA (10^{-7}–10^{-6}M). This effect is critically dependent upon displacement of corepressor complexes from the APL fusion protein in favor of coactivators (Fig. 2). While binding of ligand leads to dissociation of corepressor complexes from the C-terminal RARα moiety of the APL fusion proteins; a number of studies have shown that the resistance to ATRA therapy that characterizes APL with the PLZF-RARα fusion is correlated with persistent retinoid-insensitive binding of corepressor complexes to the POZ repressor domain of the PLZF moiety of the fusion protein (Fig. 2). Indeed, subsequent *in vitro* studies have indicated that primary leukemic blasts with the PLZF-RARα fusion can be induced to differentiate if ATRA is combined with the HDAC inhibitor trichostatin A [31, 40].

In addition to mediating release of corepressor molecules, ATRA has been shown to induce degradation of the PML-RARα oncoprotein through caspase-dependent cleavage within the PML moiety and through the ubiquitin/proteosome system (reviewed in [59]). This has a number of effects which could contribute to differentiation. Cleaved PML-RARα can no longer interact with wild-type PML, thereby allowing the multiprotein nuclear body structures (PML nuclear bodies), which have been implicated in cellular growth control and whose disruption is a hallmark of PML-RARα associated APL, to re-form. Additionally, it can be envisaged that the cleaved PML-RARα protein could activate RA-RARα target genes. Furthermore, degradation of PML-RARα leads to release of sequestered RXR, which could interact with wild-type RARα and other nuclear hormone receptors.

Differentiation of APL blasts is followed by upregulation of the death ligand TRAIL (tumor necrosis factor-related apoptosis-inducing ligand) leading to induction of apoptosis [1]. TRAIL is believed to act through the death-signaling receptors DR4 and DR5. Interestingly chemotherapeutic agents can induce DR5. Hence in patients treated with ATRA and chemotherapy the TRAIL pathway could be activated at multiple points, potentially contributing to the success of current combination treatment approaches.

Mechanism of ATO Activity in APL

ATO has wide-ranging effects on a variety of biological pathways and enzyme systems through cross-linking of proteins containing thiol (-SH) groups (reviewed in [59]). Clinical efficacy of this agent in PML-RARα-associated APL seems to reflect a combination of concentration-dependent responses involving partial differentiation and induction of apoptosis (Fig. 3).

At lower concentrations (0.1–0.5 μM), differentiation is favored, which can result in clinical complications relating to induction of hyperleucocytosis and differentiation syndrome (see below). ATO reverses a number of processes implicated in the pathogenesis of PML-RARα-associated APL, which could contribute to release of the block in differentiation (Fig. 3). In particular, it triggers MAP kinase-dependent phosphorylation of SMRT, leading to dissociation of corepressor/HDAC complexes from the PML-RARα fusion protein thereby abrogating its transcriptional repressor function [36]. Moreover, ATO induces degradation of the fusion protein, a process that takes place in the reconstituted PML nuclear bodies (see below).

At higher concentrations (0.5–2 μM), ATO can induce apoptosis of APL blasts. The precise mechanisms have not been elucidated, but a number of pathways have been implicated (reviewed in [59]). These include a direct effect on the transition pore of the mitochondrion, altering transmembrane potentials, leading to release of cytochrome C and culminating in triggering of the caspase cascade. In addition, arsenic may induce apoptosis through more indirect effects on the mitochondrion involving an increase in reactive oxygen species,

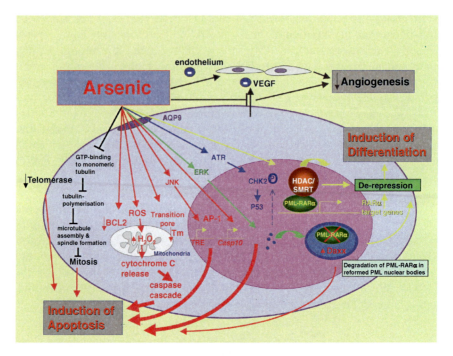

Fig. 3 Potential mechanisms implicated in apoptotic and differentiation responses to arsenic trioxide in acute promyelocytic leukemia

through a number of mechanisms including inhibition of glutathione peroxidase which converts hydrogen peroxide to water (reviewed in [59]) and upregulation of NADPH oxidase [11]. Further studies have focused upon the modulatory effects of ATO on components of signal transduction pathways, including MAP kinases p38, JNK, and ERK [12, 33, 71, 26]. It has been appreciated for several years that treatment of APL blasts with ATO restores nuclear architecture, leading to relocalization of wild-type PML protein to multi- protein nuclear body structures (PML nuclear bodies) that are disrupted in the presence of the PML-RARα oncoprotein. The pathways through which ATO mediates such effects are now being elucidated, with a recent study providing evidence that ATO induces phosphorylation of ERK, inducing downstream phosphorylation of PML, promoting conjugation with SUMO1 (for *S* mall *U* biquitin-related *MO* difier) which in turn leads to relocalization of PML to form nuclear bodies. This process is accompanied by recruitment of other nuclear body constituents such as Daxx, thereby promoting apoptosis [33]. ATO has also been shown to activate ATR (ataxia telangiectasia mutated and Rad3-related kinase) leading to activation of Chk2 inducing P53-dependent apoptosis. Activation of Chk2 is further enhanced by restoration of PML function following the degradation of PML-RARα mediated by ATO [38].

Interestingly, a more recent study has suggested that the particular sensitivity of APL to ATO correlates with expression level of the transmembrane protein aquaglyceroporin 9 (AQP9) which may be involved in uptake of the drug [48]. AQP9 expression is upregulated by ATRA leading to increased intracellular levels of ATO, which may contribute to the apparent clinical benefit of combination therapy with these agents [48].

It has become clear over the last few years that ATO not only has very far reaching effects on a wide range of cellular pathways but also influences the marrow micro-environment (see Fig. 3). As such, the extent to which any given mechanism contributes to mediating the differentiation and pro-apoptotic responses observed in APL or indeed apoptotic responses in other forms of hematological malignancy remains to be established.

Establishment of ATRA in the Treatment of APL

Early Experience with ATRA as Single-Agent Therapy

Use of ATRA in APL was the first successful clinical application of differentiation therapy, leading to induction of morphological remission in virtually all patients with newly diagnosed PML-RARα-associated disease [8, 37]. However, unfortunately it soon became apparent that use of ATRA as single-agent therapy is insufficient to maintain remission, with disease relapse typically occurring within a few months. A number of mechanisms have been implicated in the development of secondary resistance to retinoid therapy [59, 74, 23]. Of key importance is the decline in plasma levels that occurs over the first few weeks of therapy due to increased drug metabolism through the induction of cytochrome P450, interaction with lipid hydroperoxides, and possibly upregulation of *MDR1*, but also due to induction of cellular retinoic acid binding proteins (e.g., CRABPII), such that intracellular RA concentrations fall below the threshold required to target the PML-RARα oncoprotein. A further mechanism of resistance that has been observed in patients subject to prolonged ATRA therapy involves the emergence of leukemic clones harboring mutations in the LBD of the PML-RARα fusion protein which impair RA binding (reviewed in [23]). The invariable occurrence of disease relapse in patients receiving continuous daily ATRA therapy also implies that this treatment approach does not effectively target the leukemic stem cell population.

"Differentiation Syndrome" – The "Downside" of Differentiation Therapy

In approximately a third of newly diagnosed APL patients receiving ATRA as a single agent, treatment is complicated by a potentially fatal "differentiation syndrome" [22]; this typically presents with unexplained fever and dyspnea, with evidence of fluid retention with pulmonary infiltrates. The syndrome usually develops within the first 2 weeks of therapy and is commonly, but not necessarily, accompanied by a rising leukocyte count. In most cases a fatal outcome is averted by prompt initiation of steroid therapy. The syndrome is believed to be due to cytokine release and modulation of adhesion molecules on the surface of APL blasts and vascular endothelial cells [59]. The syndrome may in part relate to induction of APL blast aggregation. This phenomenon may be observed *in vitro* following exposure of the APL cell line NB4 to ATRA and is correlated with upregulation of cellular adhesion molecules including LFA1 and ICAM2 [54, 45]. Interestingly, these aggregates rapidly disperse following exposure to steroids, which may contribute to the clinical efficacy of these agents in this situation. Early introduction of chemotherapy has significantly reduced the incidence of differentiation syndrome, which now complicates remission induction in less than 10% of cases. Patients presenting with elevated leukocyte counts tend to be at highest risk; this group commonly treated with prophylactic steroids, although there is no clear evidence that this is beneficial (reviewed in [57]).

Optimizing ATRA Therapy to Improve Outcome in APL

Since ATRA therapy proved to be insufficient to maintain long-term remission in APL, subsequent studies investigated the impact of combining ATRA with chemotherapy. Randomized studies showed that addition of ATRA significantly reduces risk of relapse and improves overall survival compared to use of chemotherapy alone [21, 70]. Subsequent randomized trials investigated the best way to schedule ATRA in relation to chemotherapy. Giving ATRA as a short 5-day course (with the aim of reducing the coagulopathy) [60] and as a more extended course (to induce morphological remission) [20] prior to commencing chemotherapy were both associated with a poorer outcome with higher risk of relapse, as compared to patients receiving ATRA as an extended course commenced simultaneously with induction chemotherapy. The latter approach has been adopted in the ATRA and anthracycline-dominated protocols developed by the Italian GIMEMA and Spanish PETHEMA groups that are specific to *PML-RARA+* APL and lead to cure of approximately 70% of patients [3, 63].

Experience with Arsenic Compounds in the Management of APL

Approximately 10–15% of APL patients relapse following first-line therapy with ATRA and anthracycline-based chemotherapy. For such patients, ATRA cannot be relied upon to induce a further remission due to acquired resistance, which can in some instances be due to relapse with subclones harboring mutations within the LBD of the PML-RARα oncoprotein (reviewed in [23]).

However, ATO has been shown to be an extremely active agent in relapsed APL, achieving high rates of remission (approximately 80%) [67, 68] and has been confirmed to be effective in cases with LBD mutations [53]. ATO also carries the distinct advantage that it does not induce myelosuppression and can therefore be given largely on an outpatient basis. Moreover, in the majority of cases, clinical response to ATO is accompanied by molecular remission [67, 68]; this is in stark contrast to ATRA, with which achievement of molecular remission is exceptional at any stage of the disease. As such, ATO is now widely considered as the first-line treatment approach in relapsed disease [64]. However, the drug is associated with a number of recognized adverse effects, the most common being induction of hyperleucocytosis and differentiation syndrome [16]. The latter, which is clinically identical to the syndrome induced by ATRA, can occur in up to a third of patients, is managed in a similar fashion, and typically responds to steroids. Some concerns have been raised with respect to the use of ATO due a high incidence of severe hepatotoxicity observed in an early Chinese study and reports of cardiac toxicity with QT prolongation and fatal cardiac arrhythmia. However, there has been limited hepatotoxicity and no fatal cardiac arrhythmias with the use of proprietary drug and it is possible that the reported adverse effects reflect the nature of the drug preparation used or the patient population subject to study. The drug can, however, cause neurological toxicity, particularly peripheral neuropathy which can be dose limiting. Although most patients with relapsed disease achieve a further remission with ATO, the rate of subsequent relapse is relatively high [16] and hence this agent is typically used as "a bridge to transplantation" [47]. In patients achieving molecular remission in whom it is possible to harvest PCR negative hematopoietic stem cells, an autologous transplant may be the preferred option to consolidate remission. Eligible patients with evidence of persistent disease as determined by molecular monitoring and/or those in whom PCR negative stem cells could not be harvested may benefit from an allogeneic transplant (reviewed in [64]).

Following the impressive results obtained with ATO in relapsed disease, a number of studies have investigated arsenic compounds as a component of consolidation, but also as first-line therapy in newly diagnosed APL. These include tetra-arsenic tetra-sulfide (As_4S_4) [52] and arsenic trioxide (ATO) [25, 55]. Shen and colleagues reported use of ATRA and ATO for remission induction of APL, suggesting a benefit for the combination over either alone [66]. Studies performed in India and Iran have reported exciting results with the use of ATO as single-agent therapy, with high CR rates, associated with durable molecular remissions. These findings have been extended by Estey and colleagues at the MD Anderson Hospital, Houston, using proprietary ATO in combination with ATRA to treat APL patients with low presenting leukocyte count ($< 10 \times 10^9$/l). A similar approach has been adopted for patients presenting with higher WBC, who in addition received anti-CD33-targeted therapy with gemtuzumab ozogamicin on induction [18]. Patients were monitored regularly by polymerase chain reaction to identify those with evidence of minimal residual disease requiring additional therapy, for example, with gemtuzumab ozogamicin, anthracycline chemotherapy, and possibly transplantation to maintain molecular remission. Results were extremely encouraging, including those obtained in older patients, who normally have a high risk of induction death with conventional treatment protocols involving chemotherapy. Taken together, these studies present an exciting prospect that a substantial proportion of APL patients may be curable with differentiation therapy alone, allowing chemotherapy to be dispensed with.

The Future of Differentiation Therapy in APL and Beyond

Over the past decade it has become established that APL fusion proteins function as transcriptional repressors recruiting HDACs and DNMTs leading to silencing of genes implicated in myeloid differentiation. This suggests that demethylating agents and histone deacetylase inhibitors (HDACIs) could be of value in the management of APL patients. There is some clinical evidence for efficacy of HDACIs in this disease, with phenylbutyrate treatment leading to remission in a patient in clinical relapse [73]; while sodium valproate may have contributed to prolonged remission in a patient who received minimal therapy [56]. However, the response to HDACIs in patients with relapsed disease has generally been disappointing [76]. Nevertheless, targeting epigenetic changes provides a promising strategy for extending differentiation therapies to other subsets of leukemia, with early studies showing encouraging results for combination therapy with an HDACI and demethylating agent in relapsed and newly diagnosed high-risk AML [24, 27]. Interestingly, the efficacy of this approach extends beyond the molecularly-defined subgroups of AML characterized by chimeric fusion proteins that are recognized to recruit corepressor complexes, such as AML1-ETO.

Investigation of pathways which could potentially be exploited to overcome retinoid resistance in APL has also provided a further exciting avenue to explore the potential of a differentiation approach to the treatment of other forms of AML. Agents that specifically target RXR (the so-called "rexinoids") have been shown to synergize with cyclic AMP (cAMP) to induce differentiation of APL blasts expressing PML-RARα with a RA-resistance conferring LBD mutation [5]. Similarly, the combination of rexinoid and cAMP was found to effectively bypass the RA resistance conferred by the PLZF-RARα oncoprotein to induce differentiation of blasts from a patient with t(11;17)-associated APL [2]. This pathway may have been implicated in the clinical response observed in a multiply relapsed APL patient treated with theophylline, which serves to increase intracellular cAMP levels [32]. This approach may be applicable to other subsets of leukemia, with combinations of rexinoid and cAMP agonists inducing apoptosis in a wide range of AMLs, with efficacy correlated with upregulation of TRAIL and its cognate receptors [2]. Development of additional strategies for the induction of differentiation may arise from greater understanding of the regulatory networks at the transcriptional and proteomic level that underlie the response to ATRA and ATO in APL [75].

While ATRA and ATO have highlighted the dramatic potential of differentiation therapies, it is sobering to think that these highly successful molecularly targeted therapies were identified by chance rather than by design. Nevertheless, results of studies conducted in China [66], India [55], Iran [25], and the United States [18] using arsenic-based compounds as first-line therapy in newly diagnosed APL represent an exciting advance in the management of acute leukemia indicating some patients may be curable with differentiation therapies alone, without a requirement for chemotherapy. This novel approach could provide a number of benefits, potentially reducing risk of induction death in APL due to hemorrhage and enabling treatment to be largely administered on an outpatient basis. Moreover, elimination of chemotherapy from the treatment regimen could reduce treatment-related toxicity; indeed it has become apparent that some APL patients develop secondary myelodysplasia/ AML as a complication of their initial chemotherapy [46, 51]. Use of this differentiation therapy "chemotherapy-free" approach is a particularly attractive proposition for the management of older patients with APL. Based on these encouraging preliminary data, "differentiation therapy" is now being evaluated in randomized controlled trials against conventional ATRA and anthracycline-based protocols. However, it is clear that the success of this novel treatment strategy will be critically dependent upon reliable molecular monitoring to identify the subgroup of patients who require more conventional therapy [18].

While our understanding of the molecular pathogenesis of APL and its response to targeted therapies has advanced dramatically over the past decade, it remains uncertain as to whether sensitivity to differentiation therapies is dictated solely by the underlying molecular lesion and the mechanisms by which myeloid development is blocked, or whether the nature of the target progenitor subject to leukemic transformation has a critical bearing upon the feasibility of this approach. Certainly use of ATRA and ATO in APL represents a paradigm for differentiation therapy, transforming APL from a disease with poor prognosis to one in which most patients can expect to be cured. This has provided considerable impetus to take on the significant challenge of widening the scope of this novel treatment strategy to improve outcomes of other subsets of leukemia in the future.

Acknowledgments DG and FG gratefully acknowledge grant support from the Leukaemia Research Fund of Great Britain. DG is also supported by the European LeukemiaNet.

References

1. Altucci L, Rossin A, Raffelsberger W, Reitmair A, Chomienne C, Gronemeyer H. Retinoic acid-induced apoptosis in leukemia cells is mediated by paracrine action of tumor-selective death ligand TRAIL. *Nat Med*. 2001;7:680–686.

2. Altucci L, et al. Rexinoid-triggered differentiation and tumor-selective apoptosis of acute myeloid leukemia by protein kinase A-mediated desubordination of retinoid X receptor. *Cancer Res.* 2005;65:8754–8765.

3. Avvisati G, et al. AIDA (all-trans retinoic acid + idarubicin) in newly diagnosed acute promyelocytic leukemia: a Gruppo Italiano Malattie Ematologiche Maligne dell'Adulto (GIMEMA) pilot study. *Blood.* 1996;88:1390–1398.

4. Bennett JM, et al. Proposed revised criteria for the classification of acute myeloid leukemia. A report of the French-American-British Cooperative Group. *Ann Intern Med.* 1985;103:620–625.

5. Benoit G, et al. RAR-independent RXR signaling induces t(15;17) leukemia cell maturation. *Embo J.* 1999;18:7011–7018.

6. Burnett AK, Grimwade D, Solomon E, Wheatley K, Goldstone AH. Presenting white blood cell count and kinetics of molecular remission predict prognosis in acute promyelocytic leukemia treated with all-trans retinoic acid: result of the Randomized MRC Trial. *Blood.* 1999;93:4131–4143.

7. Carbone R, et al. Recruitment of the histone methyltransferase SUV39H1 and its role in the oncogenic properties of the leukemia-associated PML-retinoic acid receptor fusion protein. *Mol Cell Biol.* 2006;26:1288–1296.

8. Castaigne S, et al. All-trans retinoic acid as a differentiation therapy for acute promyelocytic leukemia. I. Clinical results. *Blood.* 1990;76:1704–1709.

9. Catalano A, et al. The PRKAR1A gene is fused to RARA in a new variant acute promyelocytic leukemia. *Blood.* 2007;110:4073–4076.

10. Chambon P. A decade of molecular biology of retinoic acid receptors. *Faseb J.* 1996;10:940–954.

11. Chou WC, Jie C, Kenedy AA, Jones RJ, Trush MA, Dang CV. Role of NADPH oxidase in arsenic-induced reactive oxygen species formation and cytotoxicity in myeloid leukemia cells. *Proc Natl Acad Sci USA.* 2004;101:4578–4583.

12. Davison K, Mann KK, Waxman S, Miller WH, Jr. JNK activation is a mediator of arsenic trioxide-induced apoptosis in acute promyelocytic leukemia cells. *Blood.* 2004;103:3496–3502.

13. de Thé H, Chomienne C, Lanotte M, Degos L, Dejean A. The t(15;17) translocation of acute promyelocytic leukaemia fuses the retinoic acid receptor alpha gene to a novel transcribed locus. *Nature.* 1990;347:558–561.

14. Degos L. The history of acute promyelocytic leukaemia. *Br J Haematol.* 2003;122:539–553.

15. Di Croce L, et al. Methyltransferase recruitment and DNA hypermethylation of target promoters by an oncogenic transcription factor. *Science.* 2002;295:1079–1082.

16. Douer D, Tallman MS. Arsenic trioxide: new clinical experience with an old medication in hematologic malignancies. *J Clin Oncol.* 2005;23:2396–2410.

17. Duprez E, Wagner K, Koch H, Tenen DG. C/EBPbeta: a major PML-RARA-responsive gene in retinoic acid-induced differentiation of APL cells. *Embo J.* 2003;22:5806–5816.

18. Estey E, et al. Use of all-trans retinoic acid plus arsenic trioxide as an alternative to chemotherapy in untreated acute promyelocytic leukemia. *Blood.* 2006;107:3469–3473.

19. Falanga A, Rickles FR. Pathogenesis and management of the bleeding diathesis in acute promyelocytic leukaemia. *Best Pract Res Clin Haematol.* 2003;16:463–482.

20. Fenaux P, et al. A randomized comparison of all transretinoic acid (ATRA) followed by chemotherapy and ATRA plus chemotherapy and the role of maintenance therapy in newly diagnosed acute promyelocytic leukemia. The European APL Group. *Blood.* 1999;94:1192–1200.

21. Fenaux P, et al. Effect of all transretinoic acid in newly diagnosed acute promyelocytic leukemia. Results of a multicenter randomized trial. European APL 91 Group. *Blood.* 1993;82:3241–3249.

22. Frankel SR, Eardley A, Lauwers G, Weiss M, Warrell RP, Jr. The "retinoic acid syndrome" in acute promyelocytic leukemia. *Ann Intern Med.* 1992;117:292–296.

23. Gallagher RE. Retinoic acid resistance in acute promyelocytic leukemia. *Leukemia*. 2002;16:1940–1958.

24. Garcia-Manero G, et al. Phase I/II study of the combination of 5-aza-2′ -deoxycytidine with valproic acid in patients with leukemia. *Blood*. 2006;108:3271–3279.

25. Ghavamzadeh A, et al. Treatment of acute promyelocytic leukemia with arsenic trioxide without ATRA and/or chemotherapy. *Ann Oncol*. 2006;17:131–134.

26. Giafis N, et al. Role of the p38 mitogen-activated protein kinase pathway in the generation of arsenic trioxide-dependent cellular responses. *Cancer Res*. 2006;66:6763–6771.

27. Gore SD, et al. Combined DNA methyltransferase and histone deacetylase inhibition in the treatment of myeloid neoplasms. *Cancer Res*. 2006;66:6361–6369.

28. Grignani F, et al. Fusion proteins of the retinoic acid receptor-alpha recruit histone deacetylase in promyelocytic leukaemia. *Nature*. 1998;391:815–818.

29. Grimwade D, et al. Characterization of acute promyelocytic leukemia cases lacking the classic t(15;17): results of the European Working Party. Groupe Francais de Cytogenetique Hematologique, Groupe de Francais d'Hematologie Cellulaire, UK Cancer Cytogenetics Group and BIOMED 1 European Community-Concerted Action "Molecular Cytogenetic Diagnosis in Haematological Malignancies". *Blood*. 2000;96: 1297–1308.

30. Grimwade D, et al. The importance of diagnostic cytogenetics on outcome in AML: analysis of 1,612 patients entered into the MRC AML 10 trial. The Medical Research Council Adult and Children's Leukaemia Working Parties. *Blood*. 1998;92:2322–2333.

31. Guidez F, Ivins S, Zhu J, Söderström M, Waxman S, Zelent A. Reduced retinoic acid-sensitivities of nuclear receptor corepressor binding to PML- and PLZF-RARalpha underlie molecular pathogenesis and treatment of acute promyelocytic leukemia. *Blood*. 1998;91:2634–2642.

32. Guillemin MC, et al. In vivo activation of cAMP signaling induces growth arrest and differentiation in acute promyelocytic leukemia. *J Exp Med*. 2002;196:1373–1380.

33. Hayakawa F, Privalsky ML. Phosphorylation of PML by mitogen-activated protein kinases plays a key role in arsenic trioxide-mediated apoptosis. *Cancer Cell*. 2004;5:389–401.

34. He LZ, et al. Distinct interactions of PML-RARalpha and PLZF-RARalpha with co-repressors determine differential responses to RA in APL. *Nat Genet*. 1998;18:126–135.

35. Head D, et al. Effect of aggressive daunomycin therapy on survival in acute promyelocytic leukemia. *Blood*. 1995;86:1717–1728.

36. Hong SH, Yang Z, Privalsky ML. Arsenic trioxide is a potent inhibitor of the interaction of SMRT corepressor with its transcription factor partners, including the PML-retinoic acid receptor alpha oncoprotein found in human acute promyelocytic leukemia. *Mol Cell Biol*. 2001;21:7172–7182.

37. Huang ME, et al. Use of all-trans retinoic acid in the treatment of acute promyelocytic leukemia. *Blood*. 1988;72:567–572.

38. Joe Y, et al. ATR, PML, and CHK2 Play a Role in Arsenic Trioxide-induced Apoptosis. *J Biol Chem*. 2006;281:28764–28771.

39. Kamashev D, Vitoux D, de Thé H. PML-RARA-RXR oligomers mediate retinoid and rexinoid/cAMP cross-talk in acute promyelocytic leukemia cell differentiation. *J Exp Med*. 2004;199:1163–1174.

40. Kitamura K, Hoshi S, Koike M, Kiyoi H, Saito H, Naoe T. Histone deacetylase inhibitor but not arsenic trioxide differentiates acute promyelocytic leukaemia cells with t(11;17) in combination with all-trans retinoic acid. *Br J Haematol*. 2000;108:696–702.

41. Kondo T, Mori A, Darmanin S, Hashino S, Tanaka J, Asaka M. The seventh pathogenic fusion gene FIP1L1-RARA was isolated from a t(4;17)-positive acute promyelocytic leukemia. *Haematologica*. 2008;93:1414–1416.

42. Kurokawa R, et al. Polarity-specific activities of retinoic acid receptors determined by a co-repressor. *Nature*. 1995;377:451–454.

43. Kurokawa R, et al. Regulation of retinoid signalling by receptor polarity and allosteric control of ligand binding. *Nature*. 1994;371:528–531.

44. Kwok C, Zeisig BB, Dong S, So CW. Forced homo-oligomerization of RARalpha leads to transformation of primary hematopoietic cells. *Cancer Cell.* 2006;9:95–108.
45. Larson RS, Brown DC, Sklar LA. Retinoic acid induces aggregation of the acute promyelocytic leukemia cell line NB-4 by utilization of LFA-1 and ICAM-2. *Blood.* 1997;90:2747–2756.
46. Latagliata R, et al. Therapy-related myelodysplastic syndrome-acute myelogenous leukemia in patients treated for acute promyelocytic leukemia: an emerging problem. *Blood.* 2002;99:822–824.
47. Leoni F, et al. Arsenic trioxide therapy for relapsed acute promyelocytic leukemia: a bridge to transplantation. *Haematologica.* 2002;87:485–489.
48. Leung J, Pang A, Yuen WH, Kwong YL, Tse EW. Relationship of expression of aquaglyceroporin 9 with arsenic uptake and sensitivity in leukemia cells. *Blood.* 2007;109:740–746.
49. Lin RJ, Nagy L, Inoue S, Shao W, Miller WH, Evans RM. Role of the histone deacetylase complex in acute promyelocytic leukaemia. *Nature.* 1998;391:811–814.
50. Lin RJ, Evans RM. Acquisition of oncogenic potential by RAR chimeras in acute promyelocytic leukemia through formation of homodimers. *Mol Cell.* 2000;5:821–830.
51. Lobe I, et al. Myelodysplastic syndrome after acute promyelocytic leukemia: the European APL group experience. *Leukemia.* 2003;17:1600–1604.
52. Lu DP, et al. Tetra-arsenic tetra-sulfide for the treatment of acute promyelocytic leukemia: a pilot report. *Blood.* 2002;99:3136–3143.
53. Marasca R, et al. Missense mutations in the PML/RARalpha ligand binding domain in ATRA-resistant As(2)O(3) sensitive relapsed acute promyelocytic leukemia. *Haematologica.* 1999;84:963–968.
54. Marchetti M, Falanga A, Giovanelli S, Oldani E, Barbui T. All-trans-retinoic acid increases adhesion to endothelium of the human promyelocytic leukaemia cell line NB4. *Br J Haematol.* 1996;93:360–366.
55. Mathews V, et al. Single-agent arsenic trioxide in the treatment of newly diagnosed acute promyelocytic leukemia: durable remissions with minimal toxicity. *Blood.* 2006;107:2627–2632.
56. McMullin MF, Nugent E, Thompson A, Hull D, Jones FG, Grimwade D. Prolonged molecular remission in PML-RARalpha-positive acute promyelocytic leukemia treated with minimal chemotherapy followed by maintenance including the histone deacetylase inhibitor sodium valproate. *Leukemia.* 2005;19: 1676–1677.
57. Milligan DW, et al. Guidelines on the management of acute myeloid leukaemia in adults. *Br J Haematol.* 2006;135:450–474.
58. Minucci S, et al. Oligomerization of RAR and AML1 transcription factors as a novel mechanism of oncogenic activation. *Mol Cell.* 2000;5:811–820.
59. Mistry AR, Pedersen EW, Solomon E, Grimwade D. The molecular pathogenesis of acute promyelocytic leukaemia: implications for the clinical management of the disease. *Blood Rev.* 2003;17:71–97.
60. Paietta E, et al. Significantly lower P-glycoprotein expression in acute promyelocytic leukemia than in other types of acute myeloid leukemia: immunological, molecular and functional analyses. *Leukemia.* 1994;8:968–973.
61. Rosenfeld MG, Glass CK. Coregulator codes of transcriptional regulation by nuclear receptors. *J Biol Chem.* 2001;276:36865–36868.
62. Rosenfeld MG, Lunyak VV, Glass CK. Sensors and signals: a coactivator/corepressor/epigenetic code for integrating signal-dependent programs of transcriptional response. *Genes Dev.* 2006;20:1405–1428.
63. Sanz MA, et al. A modified AIDA protocol with anthracycline-based consolidation results in high antileukemic efficacy and reduced toxicity in newly diagnosed PML/RARalpha-positive acute promyelocytic leukemia. PETHEMA group. *Blood.* 1999;94:3015–3021.
64. Sanz MA, Fenaux P, Lo Coco F. Arsenic trioxide in the treatment of acute promyelocytic leukemia. A review of current evidence. *Haematologica.* 2005;90:1231–1235.

65. Sanz MA, Tallman MS, Lo-Coco F. Tricks of the trade for the appropriate management of newly diagnosed acute promyelocytic leukemia. *Blood.* 2005;105:3019–3025.
66. Shen ZX, et al. All-trans retinoic acid/As2O3 combination yields a high quality remission and survival in newly diagnosed acute promyelocytic leukemia. *Proc Natl Acad Sci U S A.* 2004;101:5328–5335.
67. Soignet SL, et al. Complete remission after treatment of acute promyelocytic leukemia with arsenic trioxide. *N Engl J Med.* 1998;339:1341–1348.
68. Soignet SL, et al. United States multicenter study of arsenic trioxide in relapsed acute promyelocytic leukemia. *J Clin Oncol.* 2001;19:3852–3860.
69. Sternsdorf T, et al. Forced retinoic acid receptor alpha homodimers prime mice for APL-like leukemia. *Cancer Cell.* 2006;9:81–94.
70. Tallman MS, et al. All-trans-retinoic acid in acute promyelocytic leukemia. *N Engl J Med.* 1997;337:1021–1028.
71. Verma A, et al. Activation of Rac1 and the p38 mitogen-activated protein kinase pathway in response to arsenic trioxide. *J Biol Chem.* 2002;277:44988–44995.
72. Villa R, et al. The methyl-CpG binding protein MBD1 is required for PML-RARalpha function. *Proc Natl Acad Sci USA.* 2006;103:1400–1405.
73. Warrell RP, Jr, He LZ, Richon V, Calleja E, Pandolfi PP. Therapeutic targeting of transcription in acute promyelocytic leukemia by use of an inhibitor of histone deacetylase. *J Natl Cancer Inst.* 1998;90:1621–1625.
74. Warrell RP. Retinoid resistance in acute promyelocytic leukemia:new mechanisms, strategies and implications. *Blood.* 1993;82:2175–2181.
75. Zheng PZ, et al. Systems analysis of transcriptome and proteome in retinoic acid/arsenic trioxide-induced cell differentiation/apoptosis of promyelocytic leukemia. *Proc Natl Acad Sci U S A.* 2005;102:7653–7658.
76. Zhou DC, Kim SH, Ding W, Schultz C, Warrell RP, Gallagher RE. Frequent mutations in the ligand-binding domain of PML-RARalpha after multiple relapses of acute promyelocytic leukemia: analysis for functional relationship to response to all-trans retinoic acid and histone deacetylase inhibitors in vitro and in vivo. *Blood.* 2002;99:1356–1363.
77. Zhou J, Pérès L, Honoré N, Nasr R, Zhu J, de Thé H. Dimerization-induced corepressor binding and relaxed DNA-binding specificity are critical for PML/RARA induced immortalization. *Proc Natl Acad Sci USA.* 2006;103:9238–9243.

Immunotherapy of AML

Gheath Alatrash and Jeffrey J. Molldrem

Abstract The applications of chemotherapy for the treatment of AML have been unchanged over the past three decades, with only 30% of patients demonstrating disease-free survival (DFS) [118]. Despite achieving CR following induction chemotherapy, the majority of patients relapse and succumb to their disease [6]. In view of the limitations encountered by cytarabine/anthracycline based regimes, attention has shifted to immunotherapy as a means to treat AML and provide significant long-term DFS. This chapter will discuss the role of the immune system and recent advances in immunotherapy for the treatment of AML, focusing on cellular and non-cellular approaches.

Introduction: Biology of Anti-leukemia Immunity

The immune system plays a major role in the prevention of cancer by recognizing and eliminating malignantly transformed cells, a process known as immune-surveillance [28, 110], and following tumor establishment by inhibiting and controlling metastases. Immunodeficiency states, whether genetic [92], virally acquired [108], or iatrogenic (e.g., post-organ transplantation) [58], have clearly been correlated with a higher incidence of both solid and hematologic malignancies. The emergence of a clinically significant malignant disorder thus can be attributed to a failure of the immune system to mount an appropriate anti-tumor response and eradicate the malignant clone. Defects in the host immune system, which may be directly caused by tumor-secreted molecule [17], as well as a myriad of evasion mechanisms employed by tumor cells, can lead to further disease progression and ultimately to the death of the tumor-bearing host. Understanding the interactions between tumor and immune cells is critical to the success of immunotherapy in the treatment of cancer.

J.J. Molldrem (✉)
Division of Cancer Medicine, University of Texas MD, Anderson Cancer Center,
Houston, TX 77030, USA

L. Nagarajan (ed.), *Acute Myelogenous Leukemia,*
Cancer Treatment and Research 145, DOI 10.1007/978-0-387-69259-3_14,
© Springer Science+Business Media, LLC 2010

In addition to being morphologically different from normal bone marrow blasts, leukemia blasts are further distinguished by their distinct patterns of antigen expression, thereby providing a rationale for applying immunotherapy to patients with AML. Anti-leukemia effects of HSCT were first reported in murine models in the late 1950s [4] and subsequently in humans in the 1960s [81]. Other approaches were attempted in the mid-late 1970s by Gutterman et al. [49] and Mathe et al. [82], who used BCG and killed AML blasts, respectively, to enhance anti-leukemia immunity. Although responses were noted using these therapies, failure to reproduce results in subsequent studies and the morbidity associated with HSCT's graft-versus-host disease (GVHD), initially called "secondary syndrome" [81] posed major limitations. Advances in our understanding of anti-leukemia immunity, identification of the specific cell types and mechanisms that lead to graft-versus-leukemia (GVL) effects of HSCT, and the discovery of leukemia-associated antigens (LAA) have enabled us to employ novel immunotherapeutic approaches to the therapy of AML.

The success of allogeneic HSCT and DLI in the treatment of leukemia underscores the ability of the immune system to control and eradicate hematological malignancies. Immunotherapeutic approaches to cancer therapy can be broadly divided into two categories: cellular or non-cellular. The major cell types which have been used in cellular immunotherapy are T cells, natural killer (NK) cells, and dendritic cells (DC). These cells can be stimulated in vitro or engineered to express tumor-specific receptors prior to administration to patients. Examples of cellular immunotherapy that has shown efficacy in AML include HSCT, DLI, and adoptive cellular immunotherapy. On the other hand, vaccines, tumor antibodies, and immune modulators (i.e., cytokines) constitute the non-cellular approaches to leukemia immunotherapy. One of the major limitations to cellular and non-cellular immunotherapy has been the anti-self immune responses which are generated following their administration. An example is GVHD which causes significant morbidity and mortality, despite adequate prophylaxis. Nevertheless, observations that distinct immune mechanisms lead to GVHD and GVL emphasize the prospect to develop targeted therapy that would exploit the full potential of GVL while minimizing GVHD. The interactions between effector cells, helper cells, co-stimulatory molecules, and leukemia targets highlight the complexity of anti-leukemia immunotherapy; more importantly, they point to the multiple pathways of the immune system that can potentially be intervened on, in order to generate potent, yet highly specific, anti-leukemia immune responses.

Cellular Therapy

Hematopoietic Stem Cell Transplant (HSCT)

Although the efficacy of allogeneic HSCT in treatment of AML has been clearly demonstrated, much controversy currently exists regarding the timing of transplantation and the patient population that should receive this toxic regimen. Multiple randomized control trials comparing allogeneic HSCT, autologous HSCT, and chemotherapy have been conducted over the past decade [18, 20, 63, 103, 116]. Although the results vary slightly, in general, studies have consistently demonstrated improved DFS with allogeneic HSCT, compared to autologous HSCT and chemotherapy. However, there does not appear to be significant differences in overall survival (OS), and patients receiving allogeneic HSCT have significantly higher treatment-related mortality rates. Transplant-related morbidity and mortality associated with GVHD, which can occur in up to 50% of patients receiving allogeneic grafts despite adequate prophylaxis, are major limitations to HSCT. GVL and GVHD are mediated by donor T-lymphocytes which target both tumor cells and normal tissues [33], although NK cells are also thought to play a role in these phenomena [94]. Attempts at reducing GVHD by using T-cell-depleted grafts led to a successful reduction in GVHD incidence, at the cost of increased disease relapse [76]. Although distinguishing GVL from GVHD remains a difficult task, the two processes may be separated as they are mediated by specific T-cell populations [23, 25, 34, 42, 59]. Significant GVHD occurs in 50% of patients with DLI, 90% of whom achieve significant disease response; however, 55% of patients without GVHD also achieve disease response [43, 66].

In order to further identify the subgroups of AML patients that would benefit most from allogeneic HSCT, numerous investigators have probed the role of cytogenetic abnormalities and have demonstrated a correlation with outcomes following transplant. AML has been divided into three risk categories based on cytogenetic mutations: favorable (e.g., inv(16)/t(16;16), t(8;21), and t(15;17)); adverse risk (e.g., -5/del(5q),-7del(7q), t(6;9), t(9;22), 3q26 abnormality, or complex karyotype [more than three abnormalities]); and intermediate risk (e.g., normal karyotype, $-Y$, $+8$, $+21$, or others) [19, 48]. In view of the benefits observed with chemotherapy in the favorable and intermediate risk groups [20, 107, 116], allogeneic HSCT is often recommended for patients in the adverse risk group following first CR, although some patients in the intermediate-risk group are routinely offered allogeneic HSCT [113, 116, 129].

Donor Lymphocyte Infusion (DLI)

Since its initial use in the 1980s, DLI has shown great promise in the therapy of hematologic malignancies, especially in patients who relapse following HSCT. The exact mechanisms through which DLI exerts its clinical effects are not yet fully delineated; however, current data point to T cells as the primary mediators of DLI's anti-leukemia activities. One proposed mechanism involves normalization of T-cell clonality following DLI. To examine the influence of DLI on the T-cell compartment of recipients, Claret et al. [23] analyzed the T-cell receptor (TCR) repertoire in four patients with relapsed chronic myeloid leukemia (CML) who achieved a CR after infusion of $CD4^+$ lymphocytes from HLA-identical sibling donors. TCR repertoire in patients was examined over a 1-yr period before and after DLI and results demonstrated abnormal TCR patterns persisting during the initial period following DLI but thereafter normalizing; 1 yr after DLI, all patients demonstrated almost complete normalization of V-beta repertoire. Wu et al. [127] analyzed sera from three patients with relapsed CML who achieved a complete molecular remission after infusion of donor T cells and noted that sera from these patients recognized 13 distinct gene products represented in a CML-derived cDNA library, none of which were recognized by sera from healthy donors or patients with chronic GVHD. These conclusions imply that the responses to DLI may be mediated by the development of coordinated T- and B-cell immunity.

The greatest success with DLI has been encountered in CML, although some responses were reported in patients with AML. In a retrospective analysis of 84 CML, 23 AML and 22 ALL patients who received DLI, the European Group for Blood and Marrow Transplantation reported CR in 73% of patients with relapsed CML, 29% of patients with AML, and zero percent of patients with ALL [67]. These results were confirmed by another retrospective analysis of 140 allogeneic transplant patients from 25 North American programs. A response rate of 60% was observed for patients with CML, in contrast to 15 and 18% CR rates in patients with AML or ALL, respectively.

Similar to HSCT, one of the major limitations encountered is GVHD, which has been reported in more than 50% of patients following DLI [24, 73]. Multiple approaches have been attempted to reduce GVHD while harnessing the benefits of GVL following DLI. Mackinnon et al. [74] administered escalating doses of donor leukocytes to 22 patients with relapsed CML, ranging between 1×10^5 and 5×10^8. Nineteen of the 22 patients achieved remission at doses between 1×10^7 and 5×10^8. A direct correlation was demonstrated between the incidence of GVHD and the dose of T cells administered. To determine the effects of $CD8^+$ cells on GVL and GVHD, Giralt et al. [41] administered $CD8^+$-depleted DLI to 10 patients with CML who relapsed following HSCT. A mean of $0.9 \pm 0.3 \times 10^8$ mononuclear cells/kg were infused, containing $0.6 \pm 0.4 \times 10^6$ $CD3^+CD8^+$ cells/kg were administered. Hematologic and cytogenetic remissions were noted in six patients, while only three patients had evidence of GVHD (two patients developed $\geq$ grade II acute GVHD and one patient developed mild chronic GVHD). Alyea et al. [2] administered escalating doses (0.3, 1.0, or 1.5×10^8 cells/kg) of $CD4^+$ donor T cells after ex vivo depletion of $CD8^+$ lymphocytes to 40 patients with relapsed hematologic malignancies (CML and MM) after allogeneic HSCT. GVHD increased in parallel to the increasing doses of $CD4^+$ cells. Although all patients in this trial who developed GVHD demonstrated tumor regression, evidence of GVHD was not a prerequisite for achieving a response. Together, these studies demonstrate evidence that relatively low numbers of CD8-depleted donor lymphocytes can induce CR and that $CD4^+$ donor cells may play a role in GVHD. Although these data were primarily generated from patients with CML and MM, the effects of T-cell depletion of DLI on GVHD and GVL is likely independent of the underlying disease, and therefore may be reasonably extrapolated to patients with AML.

Selective T-Cell Therapy

T cells appear to be major mediators of HSCT's GVL therapeutic effects as demonstrated by (1) higher disease relapse rates in patients receiving autologous (vs. allogeneic) [38] or T-cell depleted grafts [76, 80] and (2) efficacy of DLI in disease control in patients with CML and AML who have relapsed following HSCT [24, 67] (discussed previously). In order for T cells to mount appropriate anti-leukemia responses, they must first recognize antigens expressed by malignantly transformed clones and subsequently receive co-stimulatory signaling. The delivery of a strong TCR signal in the absence of co-stimulation or stimulation with a low-affinity ligand in the presence of co-stimulation leads to a state of tolerance and anergy [109]. Thus, generation and expansion of cytotoxic T-lymphocytes (CTL) with specific anti-leukemia activities in vitro remain technically difficult, especially due to the poor immunogenicity of the leukemic cell. Nevertheless, numerous approaches have been undertaken to utilize leukemia-specific CTL in AML, some of which will be discussed in this section.

Studies using in vitro expanded LAA-specific CTL have shown promising pre-clinical results for AML therapy. The minor histocompatibility antigens (mHAg) HA-1 and HA-2 are polymorphic antigens selectively expressed by hematopoietic cells, which in the context of HLA-matched HSCT, were shown to be involved in both GvL and GvHD [32]. Moreover, HA-1 and HA-2 were targeted by donor CTL clones that showed restricted specificity to recipient hematopoietic tissue [27]. Additionally, the emergence of HA-1 and HA-2-specific CTL has been correlated with clinical remission in CML and MM [79]. These findings predict that adoptive immunotherapy with mHAg-specific CTL may lead to clinical efficacy. Many studies have shown the successful in vitro generation of mHAg-specific CTL by co-culture of donor lymphocytes with donor-derived DC pulsed with mHAg peptide or retrovirally transduced with the mHAg [95, 96]. Subsequent adoptive therapy with human CTL specific for HA-1 demonstrated anti-leukemia efficacy against human lymphoblastic leukemia in nonobese diabetic severe combined immunodeficiency (NOD/SCID) mouse model [51]. In a phase I/II study using in vitro generated CTLs for treatment of patients with relapsed leukemia after allogeneic stem cell transplantation, Falkenburg et al. demonstrated efficacy of these CTLs against recipient blasts and reported clinical responses following CTL administration [78]. Since recipient blasts express mHAgs, CTL recognizing blast mHAg may have been the major contributor to the observed clinical efficacy. Further studies are underway to determine the efficacy mHAg-specific CTL in treatment of leukemia.

Given that AML cells secrete immunosuppressive factors in order to evade immunity [16, 17] and considering the critical role played by co-stimulatory signaling of T cells in generating antigen-specific CTL responses [109], various methods to enhance T-cell co-stimulation have been undertaken. Using a poorly immunogenic murine myeloid leukemia cell line lacking B7, the ligand for the T-cell co-stimulatory receptor CD28, Boyer et al. [12] genetically engineered blasts to express murine B7, and subsequently used them as stimulators for $CD8^+$ T cells. A large number of CTL was obtained and administered to leukemia-bearing mice following HSCT. A survival advantage was noted in the mice which received the CTL, and immune responses were detected up to 3 months post-infusion. Similar studies have been conducted using human cells, whereby primary AML blasts transduced simultaneously with B7.1 and interleukin (IL)-2 led to the expansion of activated $CD8^+$ cells and increased secretion of Th1 cytokines (interferon [IFN]-γ, tumor necrosis factor (TNF)-α and IL-2) [21]. Presently, clinical trials are underway to examine the efficacy of the administration of irradiated autologous tumor cells modified to express B7-1 and IL-2 in AML patients who achieved first remission following HSCT [21].

TCR modification has provided yet another approach to generate antigen-specific CTL. CTL expressing TCR specific for Wilms' tumor antigen (WT1) have been used to target AML cells. The Wilms' tumor gene encodes a zinc finger transcription factor involved in apoptosis, cell proliferation, and organ development [57]. Traditionally, the Wilms' gene is associated with the pediatric renal cancer which bears its name (i.e., Wilms' tumor). However, it is also overexpressed in many tumors including lymphoid and myeloid leukemias and has been strongly linked to leukemogenesis [123]. A number of putative HLA-binding motifs have been documented within the WT1 protein, some of which bind relevant HLA allele and elicit a peptide-specific CTL response [5]. Xue et al. [128] modified the TCR in a leukemia mouse model to target WT1 and demonstrated the ability of these cells to kill leukemia cells and produce cytokines in vitro. Subsequent administration of TCR-transduced T cells to NOD/SCID mice harboring human leukemia cells showed elimination of leukemia in animals that received the WT1-specific CTL, compared with controls. Tsuji et al. [124] transduced TCR genes obtained from WT1 peptide-specific CTL into polyclonally activated Th1 and CTL. TCR gene-modified CTL and Th1 cells showed cytotoxicity, as well as IFN-γ and IL-2 production, in response to WT1-expressing leukemia targets. Thus, TCR gene-modification may provide the means to generate leukemia-specific CTL for use in AML immunotherapy.

Despite these promising results, there are major obstacles to T-cell-based therapy that must be overcome. Clinical results with CTL were demonstrated primarily following the infusion of a large number of CTL to patients. However, the generation of antigen-specific CTL is a complex and time-consuming process, thereby limiting the application of CTL therapy to rapidly progressive malignancies such as AML. Furthermore, the long-time interval required for the generation of antigen-specific CTL may lead to alterations in CTL function and proliferation, which can in turn interfere with the in vivo activities of CTL following infusion. The development of more efficient techniques to generate antigen-specific CTL may improve the clinical outcomes of CTL therapy in AML.

Natural Killer Cells

NK cells are large granular lymphocytes which make up 3–15% of peripheral blood lymphocytes, 5% of spleen cells, and approximately 25% of liver lymphocytes [119, 120]. NK cells express IgG Fc-receptor IIIA (CD16) and the neural cell adhesion molecule (NCAM, CD56), but lack CD3 which is found on T cells in association with the T-cell receptor (TCR) [87]. A distinguishing feature of NK cells is the ability to lyse target cells without prior sensitization, and when target and effector are HLA-mismatched [104]. This spontaneous cytotoxicity of NK cells makes them ideal immune cells, functioning in the first-line defense against pathogens as part of the innate immune system and in immunesurveillance [60]. In contrast, T cells must first be exposed to specific antigens for their activation, rendering a latent period of 7–10 days before T cells develop their initial or primary reactivity [55]. Even though NK cell killing is antigen unrestricted, numerous activating and inhibitory receptors have been identified on NK cells, the net of which determines whether or not an immune response is generated. Among the many receptors identified on NK cells which regulate their activities, killer immunoglobulin receptors (KIRs) [125] are members of the immunoglobulin superfamily which recognize antigens at the HLA-A, -B, or -C loci, while natural killer group 2 NKG2/CD94 receptors recognize antigens of the nonclassical HLA-E, -F, or -G loci [11, 13, 69]. The identification of these receptors, along with reports demonstrating normal receptor expression, cytokine production, and potent anti-leukemia activities of NK cells from AML patients [112], has led to the exploitation of the potential for NK cells in AML immunotherapy.

The role of NK cells in anti-leukemia immunotherapy is highlighted by haploidentical HSCT. In a multivariate analysis by Ruggeri et al. [105] that examined the role of NK cell alloreactivity on patient outcomes, KIR ligand incompatibility in the GVH direction was the sole independent predictor of survival in AML. This was subsequently confirmed by a mouse xenograft model for AML, in which infusion of alloreactive NK cells demonstrated reduced GVHD, while maintaining sufficient GVL activity. After failing to demonstrate anti-tumor activity of autologous NK cell therapy [89], Miller et al. [88] administered subcutaneous IL-2 following haploidentical HSCT to patients with solid and hematologic malignancies including 19 cases of poor-prognosis AML and demonstrated that adoptively transferred human NK cells were able to expand in vivo. In addition, induction of complete hematologic remission was shown in 5 of 19 poor-prognosis patients with AML. Similar to T cells, NK cell receptors modulation, with the goal of increasing target specificity, has been attempted with promising results. Schirrmann et al. [106] transfected the humanized chimeric immunoglobulin TCR targeting CD33, and AML marker, into a human NK cell line, YT. These cells subsequently demonstrated the ability to lyse the human AML cell line KG1 more effectively in contrast to untransfected cells. Most of the NK cell immunotherapy studies for AML has been performed in vitro and in animal models, and thus remains investigational at this time.

NKT Cells

CD1d is a non-classical MHC molecule composed of a monomorphic glyco-protein heavy chain non-covalently associated with β2-microglobulin. It is expressed on cells of hematopoietic origin, including circulating T and B cells, as well as APCs (i.e., monocytes, macrophages, and some subsets of DC) [30, 101, 114]. CD1d was found to be highly expressed in cells from patients with myeloid and lymphoid leukemia [31, 86]. The immunotherapeutic potential for CD1d relies on its ability to bind lipid and glycolipid antigens and present them to NKT cells [36, 44, 100]. The glycolipid antigen α-galactosylceramide (α-GalCer), which binds CD1d, has demonstrated effective activation of NKT cells [53] and anti-tumor activity against a broad spectrum of malignancies in murine models and humans [40, 54, 61, 97, 121]. Although to date most trials using α-GalCer have been conducted in patients with solid tumors [40, 61], imatinib-treated CML-chronic phase patients in complete cytogenetic response were shown to have NKT cells capable of producing IFN-γ, in contrast to patients in partial remission [111]. These data, in addition to prior reports demonstrating CD1d expression by AML cells [86], provide promising evidence for use of α-GalCer/α-GalCer-pulsed DC for therapy of AML.

Antibodies

Antibody-based immunotherapy is primarily effective against malignancies that express lineage-specific surface antigens. CD33 is a cell-surface glycopro-tein which is mostly expressed on cells from myeloid/monocytic lineage. It is found in 80% of AML blasts but is absent on pluripotent stem cells [99]. Gemtuzumab ozogamicin (Mylotarg^{TM}), the humanized IgG4 anti-CD33 anti-body conjugated to a derivative of the anti-tumor antibiotic calicheamicin, has shown promising results in the treatment of relapsed AML. Phase II studies conducted in 277 adult AML patients in first relapse who received Gemtuzu-mab at 9 mg/m^2 demonstrated an OR rate of 26%. However, these promising results were tarnished by long duration of pancytopenia, grade III and IV liver function test abnormalities, and a significant risk of developing veno-occlusive disease (VOD) [68]. In view of these toxicities and in vitro studies that demon-strated a re-expression of CD33 antigenic sites on the cell surface of blasts cells after exposure to Gemtuzumab [126], Taksin et al. conducted a safety/efficacy trial of fractionated doses of Gemtuzumab (3 mg/m^2 on days 1, 4, and 7) administered to 57 patients with AML in first relapse [117]. CR and CRp (complete remission with incomplete platelet recovery) were reported in 26 and 7% of patients, respectively, with a much lower toxicity profile compared with the original dose of 9 mg/m^2. Gemtuzumab is FDA approved for use in patients aged 60 or older with CD33+ AML in first relapse who are not considered candidates for cytotoxic chemotherapy. Ongoing trials are currently evaluating the role of Gemtuzumab as part of induction regimens and in combination with cytotoxic chemotherapy [65, 122]. Other Abs currently under investigation for use in AML include anti-FMS-like tyrosine kinase 3 (FLT3) Ab, anti-PML Ab, anti-CD44 Ab, antibodies conjugated to radioactive isotopes, and anti-TRAIL receptor therapies [62, 71, 93].

Cytokines

The application of type-I IFNs (IFNs-α/β) and IL-2 to AML therapy has yielded disappointing results. Although in vitro activity of type-I IFNs against AML have been reported [9], there are few clinical studies of IFN therapy in AML. In one of the largest trials, the United Kingdom Medical Research Council AML-11 trial [45], 362 patients who achieved remission following induction with three different chemotherapy regimens were randomized to receive a 12-month maintenance treatment with low-dose IFN-α. No benefit was seen with respect to relapse risk, DFS, or OS. Similarly, numerous clinical trials have been conducted to determine the efficacy of IL-2 in AML patients in remission; none of which have demonstrated improved disease control following IL-2 therapy [7, 8, 39, 85]. In contrast, conflicting data have been reported with use of IL-2 in the setting of relapsed or refractory AML. Two studies by Meloni et al. [83, 84] demonstrated efficacy for IL-2 in the therapy of patients with a low disease burden in the bone marrow. Maraninchi et al. [77] reported significant anti-leukemia biological effects in a similar patient population with use of high-dose IL-2. Nevertheless, the severe toxicity and limited remission rates make IL-2 a less valuable therapeutic option for AML patients with advanced disease.

Vaccine Therapy

The discovery of LAAs and specific T-cell populations which effectively target and kill LAA-expressing malignant cells has led to major advances in the development of targeted immunotherapy (Fig 1.). T-lymphocytes which preferentially target LAA-expressing cells can more efficiently eliminate tumor cells, and thereby decrease non-specific interactions with normal tissue that can lead to GVHD. Unlike CML, where 95% of patients bear the Bcr-Abl gene mutation which provides a therapeutic target, there is no known high-frequency disease-defining mutation in AML, and this poses some obstacles to developing targeted therapy for AML. In addition, since the bone marrow origin of AML is known to demolish the host's immune system, immunotherapy aimed at boosting the host immunity would be appropriate as an adjunct to chemotherapy, as part of a post-remission treatment modality. This is a reasonable approach considering that most patients with AML relapse despite achieving CR following induction. To date, the LAA which have been identified in AML include the following: myeloperoxidase, proteinase 3 (P3) [26], WT1 [37], MUC1 epithelial tumor antigen [14], the anti-apoptotic protein Survivin [1], and the PML-RARa

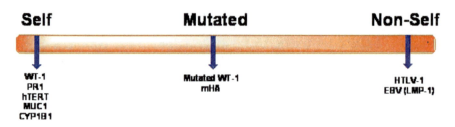

Fig. 1 Cartoon depicting Leukemia-Associated Antigens recognized by specific T cell subsets

fusion peptide which is detected in acute promyelocytic leukemia [22]. In addition, the mHAg HA1 and HA2 (discussed previously) have been shown to induce an HLAA*0201-restricted allogeneic cytotoxic response against leukemia cells and are therefore potential targets for immunotherapy [79]. These LAA are variably expressed during myelopoiesis and are distributed in different compartments within the cell (Fig. 2). The development of immunotherapeutic modalities targeting some of these antigens has demonstrated encouraging results. The following section will highlight some of the leukemia vaccines which have been applied to AML.

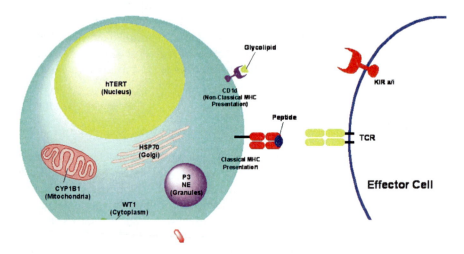

Fig. 2 LAAs are expressed at varying levels, localize to different subcellular compartments

WT1

WT1 has shown promise in AML when used as a part of a vaccine regimen. Early clinical studies have shown increased frequency of WT1-specific T cells in AML and myelodysplastic syndrome (MDS) patients following vaccination with WT1 peptide and direct correlation with clinical response [98]. A phase I/II trial reported CR in an HLA-A2 patient with relapsed AML who received four biweekly and then monthly vaccinations with WT1 peptide plus the T-helper protein keyhole limpet hemocyanin (KLH) and GM-CSF for a total of 15 vaccinations [75]. More recently, Scheibenbogen et al. reported the results of a phase II trial of 16 HLA-A2-positive patients with AML and 1 patient with MDS who received up to 18 vaccinations (median 8) of WT1 peptide with KLH and GM-CSF. Twelve patients had elevated blast counts at study entry and five patients were in CR with a high relapse risk. In patients with elevated blast counts, six demonstrated evidence of anti-leukemia activity; one patient achieved CR for 12 months. Furthermore, tetramer and intracellular cytokine staining demonstrated WT1-specific T-cell responses in peripheral blood and bone marrow [64].

PR1

PR1 is an HLA-A201 restricted nonomer peptide (VLQELNVTV) that is derived from the differentiation stage-specific neutral serine proteases P3 and neutrophil elastase (NE), which share 54% amino acid (aa) sequence homology and are normally stored in primary azurophil granules of myeloid progenitor cells [91]. The pre–pro forms of both proteins contain a 25 aa leader sequence which traffics them to the endoplasmic reticulum (ER) for processing and enzyme activation [72, 102]. P3 and NE are aberrantly expressed in myeloid leukemia (2–5-fold higher versus normal cells) and rheumatologic disorders such as Wegener's granulomatosis and small vessels vasculitis [10, 15, 26, 35]. The leukemogenic and immunogenic properties of these proteins, the latter demonstrated by their essential role in generating autoimmunity characteristic of the aforementioned rheumatologic diseases, make them ideal targets for the development of anti-leukemia immunotherapy. PR1-specific CTL that recognize and kill PR1 expressing HLA-A2 CML cells have been correlated with clinical responses to IFN-α2b therapy in 11 of 12 patients with CML; in contrast, PR1-specific CTLs were absent in all non-responders ($n = 7$) [90, 91]. Similarly, PR1-specific CTLs were detected in six of eight patients with CML receiving allogeneic HSCT, while they were absent in patients who failed to respond to allogeneic HSCT and in those who received cytotoxic chemotherapy. A phase I/II vaccine study in patients with refractory or relapsed myeloid leukemia has been conducted which combined PR1 peptide and GM-CSF in 15 CML and AML patients with progressive disease. PR1-specific CTL, measured using PR1/HLA-A2 tetramers, were detected in eight patients, five of whom obtained a clinical response. In addition, durable remission was reported in two patients with refractory AML [56].

RHAMM/CD168

The receptor for hyaluronic acid (HA)-mediated motility (RHAMM/CD168) has also been used as a target for vaccine therapy for AML. RHAMM is a glycophosphatidylinositol (GPI)-anchored receptor that is involved in cell motility [29]. In addition, it is oncogenic when overexpressed, is critical for *ras*-mediated transformation [50], and has been reported in blasts of more than 80% of patients with AML, MDS, and multiple myeloma (MM) [47]. In a phase I/II vaccine study, clinical and immunological responses were noted following administration of RHAMM R3 peptide emulsified with incomplete Freund's adjuvant and GM-CSF to patients with AML, MDS, and MM overexpressing RHAMM/CD168 [46]. Furthermore, a reduction of the RHAMM/CD168 antigen by real-time RT-PCR and a decreased number of CD33+ cells were also noted following vaccine administration.

Dendritic Cells (DCs)-Based Vaccines

DCs are the most potent antigen presenting cell (APC) and are unique in their capacity to sensitize naïve, unprimed T cells [3, 52], as they express high levels of major histocompatibility complex (MHC) class I and class II and can provide secondary signals essential to T-cell activation. Studies using DCs which have been incubated with leukemia lysates [70, 115] or transfected with genes encoding over-expressed antigen (e.g., survivin) [130] have yielded promising results for AML therapy, although no clinical trials have yet been conducted using these techniques. The well-established anti-cancer properties of DCs justify the rationale for conducting additional investigations for their use in AML immunotherapy.

Conclusion

Although the use of immunotherapy in AML is yet to be the standard of care in frontline settings, the promising results seen with DLI, HSCT, vaccines, and various other immune-modulating therapies continue to provide the impetus to further apply immunotherapy in AML. The identification of LAAs and tumor-specific immune targets, as well as our improved understanding of the biology of the immune system and its interactions with tumor cells, will enable us to enhance upon the current applications of HSCT, vaccines, and cytokine therapies in AML. Ongoing studies which aim to increase anti-leukemia specificity while reducing treatment-associated toxicity may provide the platform for future applications of immunotherapy to AML, as well as for the treatment of other malignancies.

References

1. Adida C, Haioun C, Gaulard P, et al. Prognostic significance of survivin expression in diffuse large B-cell lymphomas. *Blood.* 2000;96:1921–1925.
2. Alyea EP, Soiffer RJ, Canning C, et al. Toxicity and efficacy of defined doses of CD4(+) donor lymphocytes for treatment of relapse after allogeneic bone marrow transplant. *Blood.* 1998;91:3671–3680.

3. Banchereau J, Steinman RM. Dendritic cells and the control of immunity. *Nature.* 1998;392:245–252.

4. Barnes DW, Loutit JF. Treatment of murine leukaemia with x-rays and homologous bone marrow. II. *Br J Haematol.* 1957;3:241–252.

5. Bellantuono I, Gao L, Parry S, et al. Two distinct HLA-A0201-presented epitopes of the Wilms tumor antigen 1 can function as targets for leukemia-reactive CTL. *Blood.* 2002;100:3835–3837.

6. Bennett JM, Young ML, Andersen JW, et al. Long-term survival in acute myeloid leukemia: the Eastern Cooperative Oncology Group experience. *Cancer.* 1997;80:2205–2209.

7. Blaise D, Attal M, Pico JL, et al. The use of a sequential high dose recombinant interleukin 2 regimen after autologous bone marrow transplantation does not improve the disease free survival of patients with acute leukemia transplanted in first complete remission. *Leuk Lymphoma.* 1997;25:469–478.

8. Blaise D, Attal M, Reiffers J, et al. Randomized study of recombinant interleukin-2 after autologous bone marrow transplantation for acute leukemia in first complete remission. *Eur Cytokine Netw.* 2000;11:91–98.

9. Borden EC, Hogan TF, Voelkel JG. Comparative antiproliferative activity in vitro of natural interferons alpha and beta for diploid and transformed human cells. *Cancer Res.* 1982;42:4948–4953.

10. Borregaard N, Cowland JB. Granules of the human neutrophilic polymorphonuclear leukocyte. *Blood.* 1997;89:3503–3521.

11. Borrego F, Masilamani M, Marusina AI, Tang X, Coligan JE. The CD94/NKG2 family of receptors: from molecules and cells to clinical relevance. *Immunol Res.* 2006;35:263–278.

12. Boyer MW, Vallera DA, Taylor PA, et al. The role of B7 costimulation by murine acute myeloid leukemia in the generation and function of a CD8+ T-cell line with potent in vivo graft-versus-leukemia properties. *Blood.* 1997;89:3477–3485.

13. Braud VM, Allan DS, O'Callaghan CA, et al. HLA-E binds to natural killer cell receptors CD94/NKG2A, B and C. *Nature.* 1998;391:795–799.

14. Brossart P, Schneider A, Dill P, et al. The epithelial tumor antigen MUC1 is expressed in hematological malignancies and is recognized by MUC1-specific cytotoxic T-lymphocytes. *Cancer Res.* 2001;61:6846–6850.

15. Brouwer E, Stegeman CA, Huitema MG, Limburg PC, Kallenberg CG. T cell reactivity to proteinase 3 and myeloperoxidase in patients with Wegener's granulomatosis (WG). *Clin Exp Immunol.* 1994;98:448–453.

16. Buggins AG, Lea N, Gaken J, et al. Effect of costimulation and the microenvironment on antigen presentation by leukemic cells. *Blood.* 1999;94:3479–3490.

17. Buggins AG, Milojkovic D, Arno MJ, Lea NC, Mufti GJ, Thomas NS, Hirst WJ. Microenvironment produced by acute myeloid leukemia cells prevents T cell activation and proliferation by inhibition of NF-kappaB, c-Myc, and pRb pathways. *J Immunol.* 2001;167:6021–6030.

18. Burnett AK, Wheatley K, Goldstone AH, et al. The value of allogeneic bone marrow transplant in patients with acute myeloid leukaemia at differing risk of relapse: results of the UK MRC AML 10 trial. *Br J Haematol.* 2002;118:385–400.

19. Byrd JC, Mrozek K, Dodge RK, et al. Pretreatment cytogenetic abnormalities are predictive of induction success, cumulative incidence of relapse, and overall survival in adult patients with de novo acute myeloid leukemia: results from Cancer and Leukemia Group B (CALGB 8461). *Blood.* 2002;100:4325–4336.

20. Cassileth PA, Harrington DP, Appelbaum FR, et al. Chemotherapy compared with autologous or allogeneic bone marrow transplantation in the management of acute myeloid leukemia in first remission [see comment]. *N Engl J Med* 1649;339:1649–1656.

21. Chan L, Hardwick NR, Guinn BA, et al. An immune edited tumour versus a tumour edited immune system: Prospects for immune therapy of acute myeloid leukaemia. *Cancer Immunol Immunother.* 2006;55:1017–1024.

22. Chomienne C, Ballerini P, Balitrand N, et al. The retinoic acid receptor alpha gene is rearranged in retinoic acid-sensitive promyelocytic leukemias. *Leukemia.* 1990;4:802–807.
23. Claret EJ, Alyea EP, Orsini E, et al. Characterization of T cell repertoire in patients with graft-versus-leukemia after donor lymphocyte infusion. *J Clin Invest.* 1997;100:855–866.
24. Collins RH, Jr, Shpilberg O, Drobyski WR, et al. Donor leukocyte infusions in 140 patients with relapsed malignancy after allogeneic bone marrow transplantation. *J Clin Oncol.* 1997;15:433–444.
25. Datta AR, Barrett AJ, Jiang YZ, et al. Distinct T cell populations distinguish chronic myeloid leukaemia cells from lymphocytes in the same individual: a model for separating GVHD from GVL reactions. *Bone Marrow Transplant.* 1994;14:517–524.
26. Dengler R, Munstermann U, al-Batran S, et al. Immunocytochemical and flow cytometric detection of proteinase 3 (myeloblastin) in normal and leukaemic myeloid cells. *Br J Haematol.* 1995;89:250–257.
27. Dickinson AM, Wang XN, Sviland L, et al. In situ dissection of the graft-versus-host activities of cytotoxic T cells specific for minor histocompatibility antigens. *Nat Med.* 2002;8:410–414.
28. Dunn GP, Bruce AT, Ikeda H, Old LJ, Schreiber RD. Cancer immunoediting: from immunosurveillance to tumor escape. *Nat Immunol.* 2002;3:991–998.
29. Entwistle J, Zhang S, Yang B, et al. Characterization of the murine gene encoding the hyaluronan receptor RHAMM. *Gene* 1995;163:233–238.
30. Exley M, Garcia J, Wilson SB, et al. CD1d structure and regulation on human thymocytes, peripheral blood T cells, B cells and monocytes. *Immunology.* 2000;100:37–47.
31. Fais F, Morabito F, Stelitano C, et al. CD1d is expressed on B-chronic lymphocytic leukemia cells and mediates alpha-galactosylceramide presentation to natural killer T lymphocytes. *Int J Cancer.* 2004;109:402–411.
32. Falkenburg JH, van de Corput L, Marijt EW, Willemze R. Minor histocompatibility antigens in human stem cell transplantation. *Exp Hematol* 2003;31:743–751.
33. Ferrara JL, Deeg HJ. Graft-versus-host disease. *N Engl J Med.* 1991;324:667–674.
34. Fowler DH, Breglio J, Nagel G, Eckhaus MA, Gress RE. Allospecific CD8+ Tc1 and Tc2 populations in graft-versus-leukemia effect and graft-versus-host disease. *J Immunol.* 1996;157:4811–4821.
35. Franssen CF, Stegeman CA, Kallenberg CG, et al. Antiproteinase 3- and antimyeloperoxidase-associated vasculitis. *Kidney Int.* 2000;57:2195–2206.
36. Fujii S, Shimizu K, Kronenberg M, Steinman RM. Prolonged IFN-gamma-producing NKT response induced with alpha-galactosylceramide-loaded DCs. *Nat Immunol.* 2002;3:867–874.
37. Gaiger A, Reese V, Disis ML, Cheever MA. Immunity to WT1 in the animal model and in patients with acute myeloid leukemia. *Blood.* 2000;96:1480–1489.
38. Gale RP, Horowitz MM, Ash RC, et al. Identical-twin bone marrow transplants for leukemia. *Ann Intern Med.* 1994;120:646–652.
39. Ganser A, Heil G, Seipelt G, et al. Intensive chemotherapy with idarubicin, ara-C, etoposide, and m-AMSA followed by immunotherapy with interleukin-2 for myelodysplastic syndromes and high-risk Acute Myeloid Leukemia (AML). *Ann Hematol.* 2000;79:30–35.
40. Giaccone G, Punt CJ, Ando Y, et al. A phase I study of the natural killer T-cell ligand alpha-galactosylceramide (KRN7000) in patients with solid tumors. *Clin Cancer Res.* 2002;8:3702–3709.
41. Giralt S, Hester J, Huh Y, et al. CD8-depleted donor lymphocyte infusion as treatment for relapsed chronic myelogenous leukemia after allogeneic bone marrow transplantation. *Blood.* 1995;86:4337–4343.
42. Giralt SA, Champlin RE. Leukemia relapse after allogeneic bone marrow transplantation: a review. *Blood.* 1994;84:3603–3612.
43. Giralt SA, Kolb HJ. Donor lymphocyte infusions. *Curr Opin Oncol.* 1996;8:96–102.

44. Godfrey DI, Hammond KJ, Poulton LD, Smyth MJ, Baxter AG. NKT cells: facts, functions and fallacies. *Immunol Today*. 2000;21:573–583.
45. Goldstone AH, Burnett AK, Wheatley K, Smith AG, Hutchinson RM, Clark RE. Attempts to improve treatment outcomes in acute myeloid leukemia (AML) in older patients: the results of the United Kingdom Medical Research Council AML11 trial. *Blood*. 2001;98:1302–1311.
46. Greiner J, GiannopoulosK, Li L, et al. RHAMM/CD168-R3 Peptide vaccination of HLA-A2+ patients with acute myeloid leukemia (AML), myelodysplastic syndrome (MDS) and multiple myeloma (MM). American Society of Hematology Annual Meeting, 2005;106. Abstract 2781.
47. Greiner J, Li L, Ringhoffer M, Barth TF, et al. Identification and characterization of epitopes of the receptor for hyaluronic acid-mediated motility (RHAMM/CD168) recognized by CD8+ T cells of HLA-A2-positive patients with acute myeloid leukemia. *Blood*. 2005;106:938–945.
48. Grimwade D, Walker H, Oliver F, et al. The importance of diagnostic cytogenetics on outcome in AML: analysis of 1,612 patients entered into the MRC AML 10 trial. The Medical Research Council Adult and Children's Leukaemia Working Parties. *Blood*. 1998;92:2322–2333.
49. Gutterman JU, Hersh EM, Rodriguez V, et al. Chemoimmunotherapy of adult acute leukaemia. Prolongation of remission in myeloblastic leukaemia with B.C.G. *Lancet*. 1405;2:1405–1409.
50. Hall CL, Yang B, Yang X, et al. Overexpression of the hyaluronan receptor RHAMM is transforming and is also required for H-ras transformation. *Cell*. 1995;82:19–26.
51. Hambach L, Nijmeijer BA, Aghai Z, et al. Human cytotoxic T lymphocytes specific for a single minor histocompatibility antigen HA-1 are effective against human lymphoblastic leukaemia in NOD/scid mice. *Leukemia* 2006;20:371–374.
52. Hart DN. Dendritic cells: unique leukocyte populations which control the primary immune response. *Blood*. 1997;90:3245–3287.
53. Hayakawa Y, Godfrey DI, Smyth MJ. Alpha-galactosylceramide: potential immunomodulatory activity and future application. *Curr Med Chem*. 2004;11:241–252.
54. Hayakawa Y, Rovero S, Forni G, Smyth MJ. Alpha-galactosylceramide (KRN7000) suppression of chemical- and oncogene-dependent carcinogenesis. *Proc Natl Acad Sci USA*. 2003;100:9464–9469.
55. Herberman RB, Ortaldo JR. Natural killer cells: their roles in defenses against disease. *Science*. 1981;214:24–30.
56. Heslop HE, Stevenson FK, Molldrem JJ. Immunotherapy of hematologic malignancy. *Hematol Am Soc Hematol Educ Program*. 2003;331–349.
57. Hewitt SM, Hamada S, McDonnell TJ, Rauscher FJ, Saunders GF. Regulation of the proto-oncogenes bcl-2 and c-myc by the Wilms' tumor suppressor gene WT1. *Cancer Res*. 1995;55:5386–5389.
58. Hiesse C, Larue JR, Kriaa F, et al. Incidence and type of malignancies occurring after renal transplantation in conventionally and in cyclosporine-treated recipients: single-center analysis of a 20-year period in 1600 patients. *Transplant Proc*. 1995;27:2450–2451.
59. Horowitz MM, Gale RP, Sondel PM, et al. Graft-versus-leukemia reactions after bone marrow transplantation. *Blood*. 1990;75:555–562.
60. Imai K, Matsuyama S, Miyake S, Suga K, Nakachi K. Natural cytotoxic activity of peripheral-blood lymphocytes and cancer incidence: an 11-year follow-up study of a general population. *Lancet*. 1795;356:1795–1799.
61. Ishikawa A, Motohashi S, Ishikawa E, et al. A phase I study of alpha-galactosylceramide (KRN7000)-pulsed dendritic cells in patients with advanced and recurrent non-small cell lung cancer. *Clin Cancer Res*. 1910;11:1910–1917.
62. Kaufmann SH, Steensma DP. On the TRAIL of a new therapy for leukemia. *Leukemia*. 2005;19:2195–2202.

63. Keating S, de Witte T, Suciu S, et al. The influence of HLA-matched sibling donor availability on treatment outcome for patients with AML: an analysis of the AML 8A study of the EORTC Leukaemia Cooperative Group and GIMEMA. European Organization for Research and Treatment of Cancer. Gruppo Italiano Malattie Ematologiche Maligne dell'Adulto. *Br J Haematol.* 1998;102:1344–1353.

64. Keilholz U, Letsch A, Asemissen A, et al. Clinical and immune responses of WT1-peptide vaccination in patients with acute myeloid leukemia. *ASCO Ann Meet Proc.* 2006;24:2511.

65. Kell WJ, Burnett AK, Chopra R, et al. A feasibility study of simultaneous administration of gemtuzumab ozogamicin with intensive chemotherapy in induction and consolidation in younger patients with acute myeloid leukemia. *Blood.* 2003;102:4277–4283.

66. Kolb HJ, Holler E. Adoptive immunotherapy with donor lymphocyte transfusions. *Curr Opin Oncol.* 1997;9:139–145.

67. Kolb HJ, Schattenberg A, Goldman JM, et al. Graft-versus-leukemia effect of donor lymphocyte transfusions in marrow grafted patients. *Blood.* 1995;86:2041–2050.

68. Larson RA, Sievers EL, Stadtmauer EA, et al. Final report of the efficacy and safety of gemtuzumab ozogamicin (Mylotarg) in patients with CD33-positive acute myeloid leukemia in first recurrence. *Cancer.* 2005;104:1442–1452.

69. Lazetic S, Chang C, Houchins JP, Lanier LL, Phillips JH. Human natural killer cell receptors involved in MHC class I recognition are disulfide-linked heterodimers of CD94 and NKG2 subunits. *J Immunol.* 1996;157:4741–4745.

70. Lee JJ, Kook H, Park MS, et al. Immunotherapy using autologous monocyte-derived dendritic cells pulsed with leukemic cell lysates for acute myeloid leukemia relapse after autologous peripheral blood stem cell transplantation. *J Clin Apher.* 2004;19:66–70.

71. Li Y, Li H, Wang MN, et al. Suppression of leukemia expressing wild-type or ITD-mutant FLT3 receptor by a fully human anti-FLT3 neutralizing antibody. *Blood.* 2004;104:1137–1144.

72. Lindmark A, Gullberg U, Osson I. Processing and intracellular transport of cathepsin G and neutrophil elastase in the leukemic myeloid cell line U-937-modulation by brefeldin A, ammonium chloride, and monensin. *J Leukoc Biol.* 1994;55:50–57.

73. Lokhorst HM, Schattenberg A, Cornelissen JJ, Thomas LL, Verdonck LF. Donor leukocyte infusions are effective in relapsed multiple myeloma after allogeneic bone marrow transplantation. *Blood.* 1997;90:4206–4211.

74. Mackinnon S, Papadopoulos EB, Carabasi MH, et al. Adoptive immunotherapy evaluating escalating doses of donor leukocytes for relapse of chronic myeloid leukemia after bone marrow transplantation: separation of graft-versus-leukemia responses from graft-versus-host disease. *Blood.* 1995;86:1261–1268.

75. Mailander V, Scheibenbogen C, Thiel E, Letsch A, Blau IW, Keilholz U. Complete remission in a patient with recurrent acute myeloid leukemia induced by vaccination with WT1 peptide in the absence of hematological or renal toxicity. *Leukemia.* 2004;18:165–166.

76. Maraninchi D, Gluckman E, Blaise D, et al. Impact of T-cell depletion on outcome of allogeneic bone-marrow transplantation for standard-risk leukaemias. *Lancet.* 1987;2:175–178.

77. Maraninchi D, Vey N, Viens P, et al. A phase II study of interleukin-2 in 49 patients with relapsed or refractory acute leukemia. *Leuk Lymphoma.* 1998;31:343–349.

78. Marijt E, Wafelman A, van der Hoorn M, et al. Phase I/II feasibility study evaluating the generation of leukemia-reactive cytotoxic T lymphocyte lines for treatment of patients with relapsed leukemia after allogeneic stem cell transplantation. *Haematologica.* 2007;92:72–80.

79. Marijt WA, Heemskerk MH, Kloosterboer FM, et al. Hematopoiesis-restricted minor histocompatibility antigens HA-1- or HA-2-specific T cells can induce complete remissions of relapsed leukemia. *Proc Natl Acad Sci USA.* 2003;100:2742–2747.

80. Marmont AM, Horowitz MM, Gale RP, et al. T-cell depletion of HLA-identical transplants in leukemia. *Blood.* 1991;78:2120–2130.

81. Mathe G, Amiel JL, Schwarzenberg L, et al. Successful allogenic bone marrow transplantation in man: chimerism, induced specific tolerance and possible anti-leukemic effects. *Blood.* 1965;25:179–196.

82. Mathe G, Schwarzenberg L, Delgado M, De Vassal F. Active immunotherapy trials on acute lymphoid leukemia lymphosarcoma and acute myeloid leukemia. *Eur J Cancer.* 1977;13:445–455.

83. Meloni G, Foa R, Vignetti M, et al. Interleukin-2 may induce prolonged remissions in advanced acute myelogenous leukemia. *Blood.* 1994;84:2158–2163.

84. Meloni G, Vignetti M, Andrizzi C, Capria S, Foa R, Mandelli F. Interleukin-2 for the treatment of advanced acute myelogenous leukemia patients with limited disease: updated experience with 20 cases. *Leuk Lymphoma.* 1996;21:429–435.

85. Meloni G, Vignetti M, Pogliani E, et al. Interleukin-2 therapy in relapsed acute myelogenous leukemia. *Cancer J Sci Am.* 1997;3 (suppl 1):S43–47.

86. Metelitsa LS, Weinberg KI, Emanuel PD, Seeger RC. Expression of CD1d by myelomonocytic leukemias provides a target for cytotoxic NKT cells. *Leukemia.* 2003;17:1068–1077.

87. Miller JS. The biology of natural killer cells in cancer, infection, and pregnancy. *Exp Hematol.* 2001;29:1157–1168.

88. Miller JS, Soignier Y, Panoskaltsis-Mortari A, et al. Successful adoptive transfer and in vivo expansion of human haploidentical NK cells in patients with cancer. *Blood.* 2005;105:3051–3057.

89. Miller JS, Tessmer-Tuck J, Pierson BA, et al. Low dose subcutaneous interleukin-2 after autologous transplantation generates sustained in vivo natural killer cell activity. *Biol Blood Marrow Transplant.* 1997;3:34–44.

90. Molldrem JJ, Lee PP, Wang C, Champlin RE, Davis MM. A PR1-human leukocyte antigen-A2 tetramer can be used to isolate low-frequency cytotoxic T lymphocytes from healthy donors that selectively lyse chronic myelogenous leukemia. *Cancer Res.* 1999;59:2675–2681.

91. Molldrem JJ, Lee PP, Wang C, et al. Evidence that specific T lymphocytes may participate in the elimination of chronic myelogenous leukemia. *Nat Med.* 2000;6:1018–1023.

92. Mueller BU, Pizzo PA. Cancer in children with primary or secondary immunodeficiencies [see comment]. *J Pediatr.* 1995;126:1–10.

93. Mulford DA, Jurcic JG. Antibody-based treatment of acute myeloid leukaemia. *Expert Opin Biol Ther.* 2004;4:95–105.

94. Murphy WJ, Longo DL. The potential role of NK cells in the separation of graft-versus-tumor effects from graft-versus-host disease after allogeneic bone marrow transplantation. *Immunol Rev.* 1997;157:167–176.

95. Mutis T, Ghoreschi K, Schrama E, et al. Efficient induction of minor histocompatibility antigen HA-1-specific cytotoxic T-cells using dendritic cells retrovirally transduced with HA-1-coding cDNA. *Biol Blood Marrow Transplant.* 2002;8:412–419.

96. Mutis T, Verdijk R, Schrama E, Esendam B, Brand A, Goulmy E. Feasibility of immunotherapy of relapsed leukemia with ex vivo-generated cytotoxic T lymphocytes specific for hematopoietic system-restricted minor histocompatibility antigens. *Blood.* 1999;93:2336–2341.

97. Nakagawa R, Motoki K, Ueno H, et al. Treatment of hepatic metastasis of the colon26 adenocarcinoma with an alpha-galactosylceramide, KRN7000. *Cancer Res.* 1998;58:1202–1207.

98. Oka Y, Tsuboi A, Taguchi T, et al. Induction of WT1 (Wilms' tumor gene)-specific cytotoxic T lymphocytes by WT1 peptide vaccine and the resultant cancer regression. *Proc Natl Acad Sci USA.* 2004;101:13885–13890.

99. Peiper SC, Ashmun RA, Look AT. Molecular cloning, expression, and chromosomal localization of a human gene encoding the CD33 myeloid differentiation antigen. *Blood.* 1988;72:314–321.
100. Porcelli SA, Modlin RL. The CD1 system: antigen-presenting molecules for T cell recognition of lipids and glycolipids. *Annu Rev Immunol.* 17:297–329.
101. Pulendran B, Lingappa J, Kennedy MK, et al. Developmental pathways of dendritic cells in vivo: distinct function, phenotype, and localization of dendritic cell subsets in FLT3 ligand-treated mice. *J Immunol.* 1997;159:2222–2231.
102. Rao NV, Rao GV, Marshall BC, Hoidal JR. Biosynthesis and processing of proteinase 3 in U937 cells. Processing pathways are distinct from those of cathepsin G. *J Biol Chem.* 1996;271:2972–2978.
103. Reiffers J, Stoppa AM, Attal M, et al. Allogeneic vs autologous stem cell transplantation vs chemotherapy in patients with acute myeloid leukemia in first remission: the BGMT 87 study. *Leukemia.* 1996;10:1874–1882.
104. Robertson MJ, Ritz J. Biology and clinical relevance of human natural killer cells. *Blood.* 1990;76:2421–2438.
105. Ruggeri L, Capanni M, Urbani E, et al. Effectiveness of donor natural killer cell alloreactivity in mismatched hematopoietic transplants. *Science.* 2002;295:2097–2100.
106. Schirrmann T, Pecher G. Specific targeting of CD33(+) leukemia cells by a natural killer cell line modified with a chimeric receptor. *Leuk Res.* 2005;29:301–306.
107. Schlenk RF, Benner A, Krauter J, et al. Individual patient data-based meta-analysis of patients aged 16 to 60 years with core binding factor acute myeloid leukemia: a survey of the German Acute Myeloid Leukemia Intergroup. *J Clin Oncol.* 2004;22:3741–3750.
108. Schulz TF, Boshoff CH, Weiss RA. HIV infection and neoplasia [see comment]. *Lancet.* 1996;348:587–591.
109. Schwartz RH. T cell anergy. *Annu Rev Immunol.* 2003;21:305–334.
110. Shankaran V, Ikeda H, Bruce AT, et al. IFNgamma and lymphocytes prevent primary tumour development and shape tumour immunogenicity. *Nature.* 2001;410:1107–1111.
111. Shimizu K, Hidaka M, Kadowaki N, et al. Evaluation of the function of human invariant NKT cells from cancer patients using alpha-galactosylceramide-loaded murine dendritic cells. *J Immunol.* 2006;177:3484–3492.
112. Siegler U, Kalberer CP, Nowbakht P, Sendelov S, Meyer-Monard S, Wodnar-Filipowicz A. Activated natural killer cells from patients with acute myeloid leukemia are cytotoxic against autologous leukemic blasts in NOD/SCID mice. *Leukemia.* 2005;19:2215–2222.
113. Slovak ML, Kopecky KJ, Cassileth PA, et al. Karyotypic analysis predicts outcome of preremission and postremission therapy in adult acute myeloid leukemia: a Southwest Oncology Group/Eastern Cooperative Oncology Group Study. *Blood.* 2000;96:4075–4083.
114. Spada FM, Borriello F, Sugita M, Watts GF, Koezuka Y, Porcelli SA. Low expression level but potent antigen presenting function of CD1d on monocyte lineage cells. *Eur J Immunol.* 2000;30:3468–3477.
115. Spisek R, Chevallier P, Morineau N, et al. Induction of leukemia-specific cytotoxic response by cross-presentation of late-apoptotic leukemic blasts by autologous dendritic cells of nonleukemic origin. *Cancer Res.* 2002;62:2861–2868.
116. Suciu S, Mandelli F, de Witte T, et al. Allogeneic compared with autologous stem cell transplantation in the treatment of patients younger than 46 years with acute myeloid leukemia (AML) in first complete remission (CR1): an intention-to-treat analysis of the EORTC/GIMEMAAML-10 trial. *Blood.* 2003;102:1232–1240.
117. Taksin AL, Legrand O, Raffoux E, et al. High efficacy and safety profile of fractionated doses of Mylotarg as induction therapy in patients with relapsed acute myeloblastic leukemia: a prospective study of the alfa group. *Leukemia.* 2007;21:66–71.

118. Tallman MS, Gilliland DG, Rowe JM. Drug therapy for acute myeloid leukemia. *Blood.* 2005;106:1154–1163.

119. Timonen T, Ortaldo JR, Herberman RB. Characteristics of human large granular lymphocytes and relationship to natural killer and K cells. *J Exp Med.* 1981;153:569–582.

120. Timonen T, Saksela E. Isolation of human NK cells by density gradient centrifugation. *J Immunol Methods.* 1980;36:285–291.

121. Toura I, Kawano T, Akutsu Y, Nakayama T, Ochiai T, Taniguchi M. Cutting edge: inhibition of experimental tumor metastasis by dendritic cells pulsed with alpha-galactosylceramide. *J Immunol.* 1999;163:2387–2391.

122. Tsimberidou AM, Giles FJ, Estey E, O'Brien S, Keating MJ, Kantarjian HM. The role of gemtuzumab ozogamicin in acute leukaemia therapy. *Br J Haematol.* 2006;132:398–409.

123. Tsuboi A, Oka Y, Ogawa H, et al. Constitutive expression of the Wilms' tumor gene WT1 inhibits the differentiation of myeloid progenitor cells but promotes their proliferation in response to granulocyte-colony stimulating factor (G-CSF). *Leuk Res.* 1999;23:499–505.

124. Tsuji T, Yasukawa M, Matsuzaki J, et al. Generation of tumor-specific, HLA class I-restricted human Th1 and Tc1 cells by cell engineering with tumor peptide-specific T-cell receptor genes. *Blood.* 2005;106:470–476.

125. Uhrberg M, Valiante NM, Shum BP, et al. Human diversity in killer cell inhibitory receptor genes. *Immunity.* 1997;7:753–763.

126. van Der Velden VH, te Marvelde JG, Hoogeveen PG, Bernstein ID, Houtsmuller AB, Berger MS, van Dongen JJ. Targeting of the CD33-calicheamicin immunoconjugate Mylotarg (CMA-676) in acute myeloid leukemia: in vivo and in vitro saturation and internalization by leukemic and normal myeloid cells. *Blood.* 2001;97:3197–3204.

127. Wu CJ, Yang XF, McLaughlin S, et al. Detection of a potent humoral response associated with immune-induced remission of chronic myelogenous leukemia. *J Clin Invest.* 2000;106:705–714.

128. Xue SA, Gao L, Hart D, et al. Elimination of human leukemia cells in NOD/SCID mice by WT1-TCR gene-transduced human T cells. *Blood.* 2005;106:3062–3067.

129. Yanada M, Matsuo K, Emi N, Naoe T. Efficacy of allogeneic hematopoietic stem cell transplantation depends on cytogenetic risk for acute myeloid leukemia in first disease remission: a metaanalysis. *Cancer.* 2005;103:1652–1658.

130. Zeis M, Siegel S, Wagner A, et al. Generation of cytotoxic responses in mice and human individuals against hematological malignancies using survivin-RNA-transfected dendritic cells. *J Immunol.* 2003;170:5391–5397.

Therapy of Acute Myelogenous Leukemia in Adults

Gautam Borthakur and Elihu E. Estey

Introduction

The clinical and biological heterogeneity of adult acute myelogenous leukemia (AML) is increasingly apparent. Incorporation of cytogenetic data into the WHO classification of AML [67] is a testament to the importance of disease biology in treatment outcomes. Therefore the approach to treatment of AML in adults need to be based on a risk-stratified approach, the risk being that of relapse or refractory disease not responding to induction chemotherapy. Another risk that needs to be considered in making treatment decisions particularly in elderly patients or those with poor-performance status and organ dysfunction is that of induction-related death. Co-morbidity indices can conceivably identify patients with higher possibility of induction-related death [16, 57] and prospective application of such indices will help to resolve the question of whether less than intensive chemotherapy can impact outcome in this group. This review will focus on drug therapy of AML and not on the hematopoietic stem cell transplant (HSCT).

Risk Categories

Cytogenetic abnormalities at diagnosis have emerged as important prognostic factors for adult AML [13, 33, 56]. Three risk categories are recognized based on cytogenetic features: favorable, intermediate, and adverse. Minor differences exist among reports from various collaborative groups but broad generalizations are possible (Table 1). The favorable group includes t(15;17) and core binding factor (CBF) leukemias, i.e., inv16 and t(8;21). Normal karyotype, -Y are included in the intermediate group. Adverse risk group includes complex

G. Borthakur (✉)
Department of Leukemia, The University of Texas MD Anderson Cancer Center, Houston, TX 77030, USA
e-mail: gborthak@mdanderson.org

L. Nagarajan (ed.), *Acute Myelogenous Leukemia,*
Cancer Treatment and Research 145, DOI 10.1007/978-0-387-69259-3_15,
© Springer Science+Business Media, LLC 2010

Table 1 Cytogenetic risk groups in acute myelogenous leukemia (agreed upon by major studies)

Risk group	Cytogenetics
Favorable	t(8;21), inv (16)/t(16;16), t (15;17)
Intermediate	Normal karyotype, -Y
Adverse	Complex karyotype, inv(3)/t(3;3), -7

Refs. [13, 33, 56].

(three or more chromosomal abnormalities) karyotype, -5, -7, abnormalities of 3q. According to the Medical Research Council (MRC) AML10 trial data, 5-year survival was 65, 41, and 14% in patients with favorable, intermediate, and poor-risk cytogenetics, respectively [33]. However, outcomes are quite variable within the intermediate-risk group, leading to the search for additional molecular genetic abnormalities.

More recent identification of mutations/internal tandem duplication (ITD) of *FLT3* receptor tyrosine kinase gene [46], mutations of *NPM 1* gene leading to its cytoplasmic retention [26], partial tandem duplication (PTD) of *MLL* gene [19], etc., in patients with AML and normal cytogenetics have added to the appreciation of disease heterogeneity among adult patients with AML. Impact of these molecular karyotypic abnormalities on AML outcome is discussed later. Influence of *KIT* mutations [15] on relapse free survival in patients with AML with favorable cytogenetics (core binding factor leukemias), a group that has done well historically with high-dose cytarabine, further underscores the need for refining risk-stratification of AML. The issue of "standard therapy" of adult AML needs to be considered against this backdrop.

Age is another well-recognized adverse prognostic factor for patients with AML. Patients age 55–60 years and above are considered "elderly" for AML therapy. Remission rates and relapse-free survival rates after achieving CR are lower in elderly patients with AML [36, 43, 59]. Older age is associated with adverse risk cytogenetics and secondary (history of antecedent hematologic disorder including MDS, prior chemotherapy) AML [36, 43]. Poorer-performance status and abnormal organ function also account for increased treatment-related mortality among the elderly [4]. Even after accounting for these known variables, older patients have worse outcomes than younger patients. This reflects influences of unknown prognostic markers.

Standard Induction and Post-remission Therapy for Adult AML

Standard dose cytarabine (100 mg/m^2 by continuous intravenous infusion [CI] over 7 days) with daunorubicin, an anthracycline (45–60 mg/m^2), i.v. for 3 days has been accepted as induction therapy for adult AML particularly for younger adults and results in remission rates between 65 and 75% [42, 62] in patients 18–60 years of age. Post-remission therapy usually comprises of few cycles of

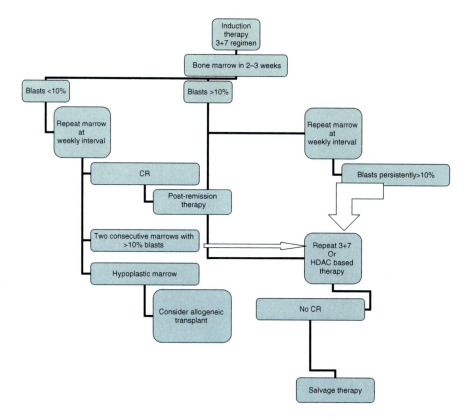

Fig. 1 Standard management of acute myelogenous leukemia in "younger" adults

high-dose cytarabine-based regimen in patients with good-risk and standard-risk cytogenetics adult AML (Fig. 1). The optimum number of cycles of post-remission therapy is not established. Patients with poor-risk cytogenetics are usually transplanted in first remission if a hematopoietic stem cell donor is available. Attempts to improve quality and rates of remissions by incorporating alternate anthracyclines like idarubicin, topoisomerase inhibitors (e.g., topotecan, etoposide), nucleoside analogs (e.g., fludarabine) [22, 23, 24, 51], modulators of multi-drug resistance [41, 66], and growth factors [6, 11, 65] did not yield desired results.

High-dose cytarabine (HDAC) in induction or post-remission therapy potentially adds to quality of remission. A randomized trial [44] reported by the Cancer and Leukemia Group B (CALGB) supported the idea of dose-dependent response to cytarabine in post-remission therapy (after standard dose cytarabine based induction) with best disease-free survival and probability of remaining free of relapse at 4 years in the group that received high-dose cytarabine as post-remission therapy (cumulative dose 18 gm/m^2). In this trial cytarabine 100 mg/m^2 was considered low dose, 400 mg/m^2 intermediate dose, and 3 gm/m^2 twice daily (all given for 5 days) was considered high dose. Participants in this trial also received further consolidation with low-dose cytarabine and daunorubicin. If HDAC is used for induction, additional HDAC in post-remission therapy does not offer further benefit [7]. Similarly in a randomized study sequential multi-agent chemotherapy was found to be not superior to HDAC alone as post-remission therapy [45].

Impact of Quality of Remission on Survival

Another question is the following: Does quality of remission matter? Outcomes definitely differ between responders to initial induction therapy versus non-responders [52, 69], while outcomes may be comparable between responders with ≤5% blasts and ≥5% blasts [69, 53]. Response in these studies were defined variably as ≥50% reduction of marrow blasts or ≤15–25% blasts in marrow after initial induction therapy not requiring recovery of cytopenias. However, data from our institution suggested that patients who achieve CR$_p$ (defined as CR without recovery of platelet counts) rather than CR tend to do worse with time (unpublished data). Thus quality of remission may impact outcome.

Risk Adapted Post-remission Treatment of Younger Adults with AML

Risk-stratification of AML is currently based on cytogenetics. As stated earlier, age and performance status also contribute significantly to therapy-related outcomes. Persistence of blasts 2 weeks after induction therapy may define a high-risk group [69]. After standard induction therapy as described above, a risk-stratification based on cytogenetics/persistence of blasts can be used to guide post-remission therapy in adults 18–60 years of age (Table 2). Variation of intensity of post-remission therapy (e.g., higher numbers of consolidation courses, incorporation of allogeneic, and autologous stem cell transplant) using a risk-adapted approach has been attempted with suggestions of improved outcome with such an approach [52, 8, 35]. Optimal risk-based approach is yet to be defined.

Good-Risk Cytogenetics Leukemias

Core binding factor (CBF) AMLs include AML with t(8;21) and inv 16 cytogenetic abnormalities. These cytogenetic abnormalities result in creations of dominant negative fusion products involving a group of heterodimeric transcriptional regulators known as core binding factors and thus the associated AMLs are

Table 2 Current "best options" for post-remission therapy of younger adults with AML

Risk group	Best option	Comments
Favorable	Repetitive cycles of HDAC +/− additional drug	Optimal dose, schedule, no. of cycles unknown
		3–4 cycles better than one cycle
		Autologous SCT not better than HDAC-based therapy
		Allogeneic SCT not better than HDAC-based therapy
Intermediate	Repetitive cycles of HDAC +/− additional drug	Optimal dose, schedule, no. of cycles unknown
		Multi-agent chemotherapy not better than HDAC alone
		Autologous SCT not better than HDAC-based therapy
		Allogeneic SCT not better than HDAC-based therapy
		Allogeneic SCT should be explored in biological
Adverse	Allogeneic SCT	Dismal outcome after intense chemotherapy
		Dismal outcome after autologous SCT allogeneic SCT may improve outcome

HDAC = high-dose cytarabine; SCT = stem cell transplant.

known as CBF AML. The pathogenesis of CBF AML is due to transcriptional dysregulation of normal myeloid differentiation by such dominant-negative fusion products. CBF AMLs share a characteristic of sensitivity to HDAC and thus HDAC is considered standard post-remission therapy for CBF AMLs [12]. Standard AML induction therapy and post-remission therapy with HDAC result in overall survival (OS) rates of between 60 and 75% [13, 56, 53] in CBF AMLs. The ideal number of post-remission therapies is not established but CALGB studies suggest that repetitive cycles of HDAC (three or more; cumulative dose: 54–72 gm/m^2) are superior to only one cycle (18 gm/m^2) [12, 14]. A meta-analysis of 392 adults with CBF AML enrolled in German AML trials did not identify a relation between the intended dose of post-remission cytarabine (ranged from 26.8 to 56.8 gm/m^2) and RFS [53]. Allogeneic HSCT produces inferior outcomes compared to that after HDAC-based consolidation.

Within this group of good-risk AML subgroups with relatively poorer outcomes can be identified. These include patients with high WBC count at presentation [53, 48, 47] (for t(8;21)), presence of trisomy 22 (for inv16) [53], age [18] (for inv16), c-kit tyrosine kinase domain (TKD) mutations in codon 816 [15]. Presence of mutations in the receptor tyrosine kinases like *FLT-3, KIT* is thought to promote proliferation of AML cells in CBF leukemias. Optimal treatment approaches for these subgroups are not defined because of small numbers in each series.

Intermediate-Risk AML

Intermediate-risk group cytogenetics include normal and -Y. Autologous and allogeneic HSCT have not been proven to improve outcome compared with conventional consolidation treatments that include HDAC [52, 5]. Thus HDAC consolidation remains the standard post-remission therapy in this group.

Presence of *FLT3* ITD identifies a subgroup within intermediate-risk group that has shorter remissions and thus at a higher probability of relapse [54]. Overall and relapse-free survival in patients with *FLT3* ITD may depend on the length of ITD [58].

PKC412 [68, 60] and CEP-701 [38, 39] (lestaurtinib) are *FLT3* inhibitors that are in clinical trial. Both have a single-agent clinical activity in patients with *FLT3* mutations. Combination treatment of PKC412 with daunorubicin and cytarabine has shown promising activity [61]. Sorafenib, a Raf kinase inhibitor, has been identified as *FLT3* inhibitor by investigators at our institution and clinical trials are in progress. Whether HSCT can improve outcome in this subgroup carrying *FLT3* ITD is not yet defined. Analysis of data from two MRC trials indicated that HSCT did not negate the higher relapse rates associated with presence of *FLT3* ITD and *FLT3* ITD status should not influence the decision to proceed to HSCT [29].

Mutations in exon-12 of *NPM* gene leading to abnormal cytoplasmic localization of NPM identify patients with AML and normal cytogenetics with higher response to induction therapy [26]. AML patients with *NPM1 + /FLT-ITD* are expected to have good response to standard induction therapy [55] and better overall survival and are thus unlikely candidates for allogeneic HSCT.

MLL-PTD is associated with short remission duration [19]. Since expression of wild-type (WT) *MLL* is silenced in AML cells with *MLL*-PTD, use of DNA methyl transferase (DNMT) inhibitors or histone deacetylase inhibitors is of potential therapeutic benefit [70].

High-Risk AML

Traditional definition of high-risk AML includes the group with adverse cytogenetics. However, non-response to first induction cycle (>10–15% blasts in marrow after week 2 of induction) also defines another group with poor outcomes [69, 34, 37]. Strategies to improve outcome in these groups need to focus on improving CR rates and maintaining remission with effective post-remission therapy. Though no significant strides have been made in improving CR rates, post-remission therapy with allogeneic HSCT is usually recommended in this group. Using donor to no donor comparison, allogeneic HSCT results in lower relapses but this has not translated into overall survival advantage [62, 63]. Newer conditioning regimens, HSCT in early aplasia rather than remission, etc., are in clinical trials to improve outcome. Since CR rate with first-line therapy is low, allogeneic HSCT from related or unrelated donors during aplasia following first or second induction therapy appears to be an effective and attractive option for patients with high-risk AML [50].

Treatment of Elderly Patient with AML

As mentioned earlier, patients with AML are considered elderly if age 55–60 years or above. The presence of adverse cytogenetics, secondary AML, etc., partly explains poor outcome in this group. The cytogenetic risk-stratification described before holds good for the elderly too [32] even though CR rates and OS are lower compared with younger patients in each cytogenetic risk category. Secondary AML is considered with antecedent hematologic disorder (probably indicating presence of preceding MDS), history of prior malignancies. Adverse cytogenetics and secondary AML are associated with resistance to induction therapy. On the other hand, co-morbidities including organ dysfunctions increase induction-related mortality in this group [36]. Even after accounting for all known predictors for poor outcome, elderly patients have worse outcomes than younger patients indicating unknown biologic factors.

Standard remission induction therapy in elderly patients with adequate organ function and performance status is similar to that used in younger patients (Table 3). For post-remission therapy low-to-intermediate dose cytarabine arabinoside for 2–3 cycles or 2–3 courses of the drugs used for remission

Table 3 Treatment recommendations for older patients

Relative risk of treatment-related mortality	Relative risk of resistance to anthracycline + cytarabine (AC) in absence treatment-related mortality	Therapeutic options (in order of preference)
Low (age 55/60–69, good performance status, no infection, normal organ function)	Low (intermediate cytogenetics and de novo AML)	(1) Investigational cytotoxic and (2) 3 + 7
Low	High (adverse cytogenetics or secondary AML)	(1) Investigational non-cytotoxic and (2) Low-dose ara-C (LDAC)
High (either ECOG performance status 2, mildly-moderately abnormal organ function, infection, or age 70–79)	High or low	(1) Investigational non-cytotoxic and (2) LDAC
Very high (either ECOG performance status 3–4, severely abnormal organ function, or age ≥ 80)	High or low	(1) Investigational non-cytotoxic and (2) supportive care only

induction at attenuated doses are administered. Across the board there is 15–20% mortality in elderly patients in the 4–6 weeks of remission induction therapy [25]. Attenuation of intensity of induction therapy to reduce induction-related mortality results in lower CR rates. Concomitant use of growth factors [65] or induction therapy with gemtuzamab ozogamycin [2] has not impacted survival in elderly patients with AML. The UK MRC trial 11 randomized 1314 older patients to three induction regimens containing lower dose cytarabine [30]. The combination of daunorubicin, cytarabine, and 6-thioguanine arm yielded better CR rates but survival rates were comparable among three arms because of continued relapses occurring over time. Moreover, addition of G-CSF resulted in shorter duration of neutropenia but did not improve CR rates or survival.

Elderly patients with good-risk cytogenetics and those in the 60–69 years of age with normal cytogenetics, performance status of 0–2, normal creatinine, and bilirubin are candidates for standard induction therapy (3 + 7 regimen) keeping in mind the fact that remission rate at 3 years is still far from satisfactory in these patients. Also, randomized comparison between low-dose cytarabine and standard chemotherapy (rubidazone [a daunorubicin-derived agent], 100 mg/m^2 for 4 days, Ara-C 200 mg/m^2 for 7 days) showed no survival benefit in the elderly AML patient with standard chemotherapy [64].

The proportion of elderly patients who fit a favorable profile of good-risk cytogenetics and preserved organ function (bilirubin and creatinine less than 1.5 mg/dL, no cardiac dysfunction, etc.) is less than 10%. Thus, in the vast majority of elderly patients standard induction therapy is not an option and thus participation in clinical trials should be the recommended option for these patients. Gemtuzumab ozogamycin has been evaluated as single-agent induction therapy in such patients but CR rates are low and induction mortality is significant [2]. Sequential use of gemtuzumab and chemotherapy is reported to produce CR in half the patients with 1-year survival of 34% [3]. Phase III studies are underway to evaluate this strategy.

For elderly patients not fit for induction chemotherapy use of low-dose cytarabine at 20 mg twice daily for 10 days every 4–6 weeks has been proposed as standard based on NCRI (formerly MRC) trial AML14 where low-dose cytarabine was compared with hydroxyurea and the study reported survival advantage for low-dose cytarabine (hazard ratio 0.61, 95% CI = 0.45–0.82, $p = 0.001$). However, CR rate of 17% with low-dose cytarabine leaves room for improvement. Even though patients enrolled in this study were deemed unfit for standard induction therapy by their treating physicians, only 12% of patients had a performance status of 3–4. This highlights physician's perception of appropriate therapy for elderly patients with AML. Targeted therapies with farnesyltransferase inhibitors, multi-kinase inhibitors, etc., are being pursued to expand therapeutic options for elderly patients with AML.

For elderly patients with poor performance status (Zubrod 3–4), organ dysfunction, or age over 80 years, supportive care should be offered as an option since treatment-related mortality is high in this group and there is dearth of data to support use of standard or low-dose cytarabine.

Post-remission Therapy for Elderly

Randomized comparison is not available for no post-remission therapy versus post-remission therapy in the elderly. However, survival was similar when 1 versus 4 cycles of consolidation therapy (MRC AML trial 11) was compared.

Allogeneic Stem Cell Transplant in the Elderly

Reduced intensity or non-myeloablative HSCT (RI-HSCT) has been explored to exploit the benefit of graft-versus-leukemia effect and lower upfront treatment-related mortality in elderly patients with AML. In comparison with myeloablative HSCT [1], acute graft-versus-host disease (grade II–IV) and treatment-related mortality are lower and relapse rates are higher with reduced intensity stem cell transplant (RI-HSCT). No leukemia-free survival benefit was seen with RI-HSCT.

More relevant is the question of general applicability of RI-HSCT to all older patients with AML. In a study at MDACC [20] only 14% of elderly AML patients in first CR underwent RI-HSCT despite 86% being considered transplant eligible, thus highlighting the potential limitation of general applicability. Thus RI-HSCT in elderly AML should be undertaken in the context of clinical trial.

Salvage Treatment for AML in Adults

Since most adult patients with AML will relapse or a smaller proportion will fail to enter CR after initial treatment (primary refractory), salvage therapy will be needed for most. Outcome after first relapse can be prognosticated based on duration of first remission, cytogenetics at diagnosis, age at first relapse, and prior SCT [9]. For patients with first-remission duration more than 1 year, standard induction regimens can be used for salvage since rates of second CR is 40–50% in this group. Analysis of data from MDACC showed that for patients with first CR duration of over 2 years and 1–2 years, the respective second CR rates were 73 and 46% [21]. A high-dose cytarabine-based salvage strategy is appropriate in this patient population. On the other hand, shorter first CR (particularly if less than 6 months) duration predicts second CR rates of 10–20% only. For patients who are beyond first salvage and have relapsed/refractory disease, the possibility of achieving CR is dismal. Thus investigational therapy should be offered to these patients. Once a second CR is achieved, patients should be considered for HSCT. For patients with CBF leukemia, longer first CR duration of over 10 months and donor availability at second remission predicted for better relapse-free survival [17]. Presence of FLT3-ITD predicted for lower rates of second CR in patients with normal cytogenetics and survival. The fact that HSCT at second CR holds promise for patients with CBF leukemia or normal cytogenetics has to be tempered with the reality that these two groups of patients together comprise approximately only half of the patients with relapsed AML.

At present HSCT is restricted to patients whose blast count in the marrow can be brought down to 10% or less with salvage chemotherapy. There is need for investigations on the role of HSCT for patients with persistent marrow blasts >10% or circulating blasts after salvage chemotherapy. Reports of HSCT for primary refractory AML suggest that approximately one-third of such patients can potentially be cured [28]. HSCT thus needs to be evaluated both in the setting of persistent marrow or circulating blasts after salvage and as initial salvage therapy at relapse.

Treatment of Minimal Residual Disease

With use of multi-parametric flow-cytometry it has been possible to determine minimal residual disease (MRD) after induction or consolidation in patients with AML [27]. Prognostic value of MRD detection is being validated by studies showing that patients with MRD negative status after consolidation therapy tend to have superior overall outcome compared with MRD positive patients [10]. Since CBF leukemias are associated with fusion transcripts, levels of these transcripts can be monitored for detection of minimal residual disease [49]. In CBF leukemias MRD predicts for higher chances of relapse [49]. In patients with normal cytogenetics, presence of Flt-3, NPM 1 [31], or PLL mutations can be used for MRD monitoring. Apart from mutations and fusion transcripts, high WT1 gene expression after induction therapy correlates with relapse in pediatric studies [40].

Effectiveness of therapy based on MRD monitoring is well proven for APL. Though there is increasing evidence linking MRD to disease relapse in other AMLs, the value of therapy based on MRD is not substantiated. This is possibly a reflection on the lack of effective therapy for relapses in AML other than APL.

References

1. Alyea EP, Kim HT, Ho V, et al. Comparative outcome of nonmyeloablative and myeloablative allogeneic hematopoietic cell transplantation for patients older than 50 years of age. *Blood.* 2005;105:1810–1814.

2. Amadori S, Suciu S, Stasi R, et al. Gemtuzumab ozogamicin (Mylotarg) as single-agent treatment for frail patients 61 years of age and older with acute myeloid leukemia: final results of AML-15B, a phase 2 study of the European Organisation for Research and Treatment of Cancer and Gruppo Italiano Malattie Ematologiche dell'Adulto Leukemia Groups. *Leukemia.* 2005;19:1768–1773.

3. Amadori S, Suciu S, Willemze R, et al. Sequential administration of gemtuzumab ozogamicin and conventional chemotherapy as first line therapy in elderly patients with acute myeloid leukemia: a phase II study (AML-15) of the EORTC and GIMEMA leukemia groups. *Haematologica.* 2004;89:950–956.

4. Appelbaum FR, Gundacker H, Head DR, et al. Age and acute myeloid leukemia. *Blood.* 2006;107:3481–3485.

5. Appelbaum FR, Pearce SF. Hematopoietic cell transplantation in first complete remission versus early relapse. *Best Pract Res Clin Haematol.* 2006;19:333–339.

6. Becker PS. Growth factor priming in therapy of acute myelogenous leukemia. *Curr Hematol Rep.* 2004;3:413–418.

7. Bishop JF, Matthews JP, Young GA, et al. A randomized study of high-dose cytarabine in induction in acute myeloid leukemia. *Blood.* 1996;87:1710–1717.

8. Bloomfield CD, Lawrence D, Byrd JC, et al. Frequency of prolonged remission duration after high-dose cytarabine intensification in acute myeloid leukemia varies by cytogenetic subtype. *Cancer Res.* 1998;58:4173–4179.

9. Breems DA, Van Putten WL, Huijgens PC, et al. Prognostic index for adult patients with acute myeloid leukemia in first relapse. *J Clin Oncol.* 2005;23:1969–1978.

10. Buccisano F, Maurillo L, Gattei V, et al. The kinetics of reduction of minimal residual disease impacts on duration of response and survival of patients with acute myeloid leukemia. *Leukemia.* 2006.

11. Buchner T, Hiddemann W, Wormann B, et al. Hematopoietic growth factors in acute myeloid leukemia: supportive and priming effects. *Semin Oncol.* 1997;24:124–131.

12. Byrd JC, Dodge RK, Carroll A, et al. Patients with t(8;21)(q22;q22) and acute myeloid leukemia have superior failure-free and overall survival when repetitive cycles of high-dose cytarabine are administered. *J Clin Oncol.* 1999;17:3767–3775.

13. Byrd JC, Mrozek K, Dodge RK, et al. Pretreatment cytogenetic abnormalities are predictive of induction success, cumulative incidence of relapse, and overall survival in adult patients with de novo acute myeloid leukemia: results from Cancer and Leukemia Group B (CALGB 8461). *Blood.* 2002;100:4325–4336.

14. Byrd JC, Ruppert AS, Mrozek K, et al. Repetitive cycles of high-dose cytarabine benefit patients with acute myeloid leukemia and inv(16)(p13q22) or t(16;16)(p13;q22): results from CALGB 8461. *J Clin Oncol.* 2004;22:1087–1094.

15. Cairoli R, Beghini A, Grillo G, et al. Prognostic impact of c-KIT mutations in core binding factor leukemias: an Italian retrospective study. *Blood.* 2006;107:3463–3468.

16. Charlson M, Szatrowski TP, Peterson J, Gold J. Validation of a combined comorbidity index. *J Clin Epidemiol.* 1994;47:1245–1251.

17. de Labarthe A, Pautas C, Thomas X, et al. Allogeneic stem cell transplantation in second rather than first complete remission in selected patients with good-risk acute myeloid leukemia. *Bone Marrow Transplant.* 2005;35:767–773.

18. Delaunay J, Vey N, Leblanc T, et al. Prognosis of inv(16)/t(16;16) acute myeloid leukemia (AML): a survey of 110 cases from the French AML Intergroup. *Blood.* 2003;102:462–469.

19. Dohner K, Tobis K, Ulrich R, et al. Prognostic significance of partial tandem duplications of the MLL gene in adult patients 16 to 60 years old with acute myeloid leukemia and normal cytogenetics: a study of the Acute Myeloid Leukemia Study Group Ulm. *J Clin Oncol.* 2002;20:3254–3261.

20. Estey E, de Lima M, Tibes R, et al. Prospective feasibility analysis of reduced intensity conditioning regimens (RIC) for hematopoietic stem cell transplantation (HSCT) in

elderly patients with acute myeloid leukemia (AML) and high-risk myelodysplastic syndrome (MDS). *Blood.* 2007;109:1395–1400.

21. Estey E, Kornblau S, Pierce S, et al. A stratification system for evaluating and selecting therapies in patients with relapsed or primary refractory acute myelogenous leukemia. *Blood.* 1996;88:756.

22. Estey E, Thall P, Andreeff M, et al. Use of granulocyte colony-stimulating factor before, during, and after fludarabine plus cytarabine induction therapy of newly diagnosed acute myelogenous leukemia or myelodysplastic syndromes: comparison with fludarabine plus cytarabine without granulocyte colony-stimulating factor. *J Clin Oncol.* 1994;12:671–678.

23. Estey EH, Kantarjian HM, O'Brien S, et al. High remission rate, short remission duration in patients with refractory anemia with excess blasts (RAEB) in transformation (RAEB-t) given acute myelogenous leukemia (AML)-type chemotherapy in combination with granulocyte-CSF (G-CSF). *Cytokines Mol Ther.* 1995;1:21–28.

24. Estey EH, Thall PF, Cortes JE, et al. Comparison of idarubicin + ara-C-, fludarabine + ara-C-, and topotecan + ara-C-based regimens in treatment of newly diagnosed acute myeloid leukemia, refractory anemia with excess blasts in transformation, or refractory anemia with excess blasts. *Blood.* 2001;98:3575–3583.

25. Estey EH. General approach to, and perspectives on clinical research in, older patients with newly diagnosed acute myeloid leukemia. *Semin Hematol.* 2006;43:89–95.

26. Falini B, Mecucci C, Tiacci E, et al. Cytoplasmic nucleophosmin in acute myelogenous leukemia with a normal karyotype. *N Engl J Med.* 2005;352:254–266.

27. Feller N, van der Pol MA, van Stijn A, et al. MRD parameters using immunophenotypic detection methods are highly reliable in predicting survival in acute myeloid leukaemia. *Leukemia.* 2004;18:1380–1390.

28. Fung HC, Stein A, Slovak M, et al. A long-term follow-up report on allogeneic stem cell transplantation for patients with primary refractory acute myelogenous leukemia: impact of cytogenetic characteristics on transplantation outcome. *Biol Blood Marrow Transplant.* 2003;9:766–771.

29. Gale RE, Hills R, Kottaridis PD, et al. No evidence that FLT3 status should be considered as an indicator for transplantation in acute myeloid leukemia (AML): an analysis of 1135 patients, excluding acute promyelocytic leukemia, from the UK MRC AML10 and 12 trials. *Blood.* 2005;106:3658–3665.

30. Goldstone AH, Burnett AK, Wheatley K, et al. Attempts to improve treatment outcomes in acute myeloid leukemia (AML) in older patients: the results of the United Kingdom Medical Research Council AML11 trial. *Blood.* 2001;98:1302–1311.

31. Gorello P, Cazzaniga G, Alberti F, et al. Quantitative assessment of minimal residual disease in acute myeloid leukemia carrying nucleophosmin (NPM1) gene mutations. *Leukemia.* 2006;20:1103–1108.

32. Grimwade D, Walker H, Harrison G, et al. The predictive value of hierarchical cytogenetic classification in older adults with acute myeloid leukemia (AML): analysis of 1065 patients entered into the United Kingdom Medical Research Council AML11 trial. *Blood.* 2001;98:1312–1320.

33. Grimwade D, Walker H, Oliver F, et al. The importance of diagnostic cytogenetics on outcome in AML: analysis of 1,612 patients entered into the MRC AML 10 trial. The Medical Research Council Adult and Children's Leukaemia Working Parties. *Blood.* 1998;92:2322–2333.

34. Haferlach T, Kern W, Schoch C, et al. A new prognostic score for patients with acute myeloid leukemia based on cytogenetics and early blast clearance in trials of the German AML Cooperative Group. *Haematologica.* 2004;89:408–418.

35. Heil G, Krauter J, Raghavachar, A et al. Risk-adapted induction and consolidation therapy in adults with de novo AML aged $\leq$ 60 years: results of a prospective multicenter trial. *Ann Hematol.* 2004;83:336–344.

36. Kantarjian H, O'Brien S, Cortes J, et al. Results of intensive chemotherapy in 998 patients age 65 years or older with acute myeloid leukemia or high-risk myelodysplastic syndrome: predictive prognostic models for outcome. *Cancer.* 2006;106:1090–1098.

37. Kern W, Haferlach T, Schoch C, et al. Early blast clearance by remission induction therapy is a major independent prognostic factor for both achievement of complete remission and long-term outcome in acute myeloid leukemia: data from the German AML Cooperative Group (AMLCG) 1992 Trial. *Blood.* 2003;101:64–70.

38. Kern W, Haferlach T, Schoch C, et al. Risk-adapted therapy of AML: the AMLCG experience. *Ann Hematol.* 2004;83(suppl 1):S49–S51.

39. Knapper S, Mills KI, Gilkes AF, et al. The effects of lestaurtinib (CEP701) and PKC412 on primary AML blasts: the induction of cytotoxicity varies with dependence on FLT3 signaling in both FLT3 mutated and wild type cases. *Blood.* 2006;108:3494–3503.

40. Lapillonne H, Renneville A, Auvrignon A, et al. High WT1 expression after induction therapy predicts high risk of relapse and death in pediatric acute myeloid leukemia. *J Clin Oncol.* 2006;24:1507–1515.

41. List AF, Kopecky KJ, Willman CL, et al. Benefit of cyclosporine modulation of drug resistance in patients with poor-risk acute myeloid leukemia: a Southwest Oncology Group study. *Blood.* 2001;98:3212–3220.

42. Lowenberg B, Griffin JD, Tallman MS. Acute myeloid leukemia and acute promyelocytic leukemia. *Hematology (Am Soc Hematol Educ Program).* 2003;82–101.

43. Lowenberg B, Suciu S, Archimbaud E, et al. Mitoxantrone versus daunorubicin in induction-consolidation chemotherapy—the value of low-dose cytarabine for maintenance of remission, and an assessment of prognostic factors in acute myeloid leukemia in the elderly: final report. European Organization for the Research and Treatment of Cancer and the Dutch-Belgian Hemato-Oncology Cooperative Hovon Group. *J Clin Oncol.* 1998;16:872–881.

44. Mayer RJ, Davis RB, Schiffer CA, et al. Intensive postremission chemotherapy in adults with acute myeloid leukemia. Cancer and Leukemia Group B. *N Engl J Med.* 1994;331:896–903.

45. Moore JO, George SL, Dodge RK, et al. Sequential multiagent chemotherapy is not superior to high-dose cytarabine alone as postremission intensification therapy for acute myeloid leukemia in adults under 60 years of age: Cancer and Leukemia Group B Study 9222. *Blood.* 2005;105:3420–3427.

46. Nakao M, Yokota S, Iwai T, et al. Internal tandem duplication of the flt3 gene found in acute myeloid leukemia. *Leukemia.* 1996;10:1911–1918.

47. Nguyen S, Leblanc T, Fenaux P, et al. A white blood cell index as the main prognostic factor in t(8;21) acute myeloid leukemia (AML): a survey of 161 cases from the French AML Intergroup. *Blood.* 2002;99:3517–3523.

48. O'Brien S, Kantarjian HM, Keating M, et al. Association of granulocytosis with poor prognosis in patients with acute myelogenous leukemia and translocation of chromosomes 8 and 21. *J Clin Oncol.* 1989;7:1081–1086.

49. Perea G, Lasa A, Aventin A, et al. Prognostic value of minimal residual disease (MRD) in acute myeloid leukemia (AML) with favorable cytogenetics [t(8;21) and inv(16)]. *Leukemia.* 2006;20:87–94.

50. Platzbecker U, Thiede C, Fussel M, et al. Reduced intensity conditioning allows for upfront allogeneic hematopoietic stem cell transplantation after cytoreductive induction therapy in newly-diagnosed high-risk acute myeloid leukemia. *Leukemia.* 2006;20:707–714.

51. Russo D, Malagola M, de Vivo A, et al. Multicentre phase III trial on fludarabine, cytarabine (Ara-C), and idarubicin versus idarubicin, Ara-C and etoposide for induction treatment of younger, newly diagnosed acute myeloid leukaemia patients. *Br J Haematol.* 2005;131:172–179.

52. Schlenk RF, Benner A, Hartmann F, et al. Risk-adapted postremission therapy in acute myeloid leukemia: results of the German multicenter AML HD93 treatment trial. *Leukemia.* 2003;17:1521–1528.

53. Schlenk RF, Benner A, Krauter J, et al. Individual patient data-based meta-analysis of patients aged 16 to 60 years with core binding factor acute myeloid leukemia: a survey of the German Acute Myeloid Leukemia Intergroup. *J Clin Oncol.* 2004;22:3741–3750.

54. Schnittger S, Schoch C, Dugas M, et al. Analysis of FLT3 length mutations in 1003 patients with acute myeloid leukemia: correlation to cytogenetics, FAB subtype, and prognosis in the AMLCG study and usefulness as a marker for the detection of minimal residual disease. *Blood.* 2002;100:59–66.

55. Schnittger S, Schoch C, Kern W, et al. Nucleophosmin gene mutations are predictors of favorable prognosis in acute myelogenous leukemia with a normal karyotype. *Blood.* 2005;106:3733–3739.

56. Slovak ML, Kopecky KJ, Cassileth PA, et al. Karyotypic analysis predicts outcome of preremission and postremission therapy in adult acute myeloid leukemia: a Southwest Oncology Group/Eastern Cooperative Oncology Group Study. *Blood.* 2000;96:4075–4083.

57. Sorror ML, Maris MB, Storb R, et al. Hematopoietic cell transplantation (HCT)-specific comorbidity index: a new tool for risk assessment before allogeneic HCT. *Blood.* 2005;106:2912–2919.

58. Stirewalt DL, Kopecky KJ, Meshinchi S, et al. Size of FLT3 internal tandem duplication has prognostic significance in patients with acute myeloid leukemia. *Blood.* 2006;107:3724–3726.

59. Stone RM, Berg DT, George SL, et al. Granulocyte-macrophage colony-stimulating factor after initial chemotherapy for elderly patients with primary acute myelogenous leukemia. Cancer and Leukemia Group B. *N Engl J Med.* 1995;332:1671–1677.

60. Stone RM, DeAngelo DJ, Klimek V, et al. Patients with acute myeloid leukemia and an activating mutation in FLT3 respond to a small-molecule FLT3 tyrosine kinase inhibitor, PKC412. *Blood.* 2005;105:54–60.

61. Stone RM, Fischer T, Paquette R, et al. Phase IB study of PKC412, an oral FLT3 kinase inhibitor, in sequential and simultaneous combinations with daunorubicin and cytarabine (DA) induction and high-dose cytarabine consolidation in newly diagnosed patients with AML. ASH Abstract. 2006;#404.

62. Stone RM, O'Donnell MR, Sekeres MA. Acute myeloid leukemia. *Hematology (Am Soc Hematol Educ Program).* 2004;98–117.

63. Suciu S, Mandelli F, de Witte T, et al. Allogeneic compared with autologous stem cell transplantation in the treatment of patients younger than 46 years with acute myeloid leukemia (AML) in first complete remission (CR1): an intention-to-treat analysis of the EORTC/GIMEMAAML-10 trial. *Blood.* 2003;102:1232–1240.

64. Tilly H, Castaigne S, Bordessoule D, et al. Low-dose cytarabine versus intensive chemotherapy in the treatment of acute nonlymphocytic leukemia in the elderly. *J Clin Oncol.* 1990;8:272–279.

65. Usuki K, Urabe A, Masaoka T, et al. Efficacy of granulocyte colony-stimulating factor in the treatment of acute myelogenous leukaemia: a multicentre randomized study. *Br J Haematol.* 2002;116:103–112.

66. van der HB, Lowenberg B, Burnett AK, et al. The value of the MDR1 reversal agent PSC-833 in addition to daunorubicin and cytarabine in the treatment of elderly patients with previously untreated acute myeloid leukemia (AML), in relation to MDR1 status at diagnosis. *Blood.* 2005;106:2646–2654.

67. Vardiman JW, Harris NL, Brunning RD. The World Health Organization (WHO) classification of the myeloid neoplasms. *Blood.* 2002;100:2292–2302.

68. Weisberg E, Boulton C, Kelly LM, et al. Inhibition of mutant FLT3 receptors in leukemia cells by the small molecule tyrosine kinase inhibitor PKC412. *Cancer Cell.* 2002;1:433–443.

69. Wheatley K, Burnett AK, Goldstone AH, et al. A simple, robust, validated and highly predictive index for the determination of risk-directed therapy in acute myeloid leukaemia derived from the MRC AML 10 trial. United Kingdom Medical Research Council's Adult and Childhood Leukaemia Working Parties. *Br J Haematol.* 1999;107:69–79.

70. Whitman SP, Liu S, Vukosavljevic T, et al. The MLL partial tandem duplication: evidence for recessive gain-of-function in acute myeloid leukemia identifies a novel patient subgroup for molecular-targeted therapy. *Blood.* 2005;106:345–352.

Subject Index

L. Nagarajan (ed.), *Acute Myelogenous Leukemia*,
Cancer Treatment and Research 145, DOI 10.1007/978-0-387-69259-3,
© Springer Science+Business Media, LLC 2010